THE PARTICLE EXPLOSION

THE PARTICLE EXPLOSION

FRANK CLOSE, MICHAEL MARTEN, & CHRISTINE SUTTON

New York Tokyo Melbourne
OXFORD UNIVERSITY PRESS

Oxford University Press, Walton Street, Oxford OX2 6DP

Oxford New York Toronto
Delhi Bombay Calcutta Madras Karachi
Kuala Lumpur Singapore Hong Kong Tokyo
Nairobi Dar es Salaam Cape Town
Melbourne Auckland Madrid
and associated companies in
Berlin Ibadan

Oxford is a trade mark of Oxford University Press

Published in the United States
by Oxford University Press Inc., New York

First published 1987
Reprinted 1987
First published as paperback 1994

British Library Cataloguing in Publication Data
Close, F. E.
The particle explosion.
1. Particles (Nuclear physics)
I. Title II. Marten, Michael III. Sutton, Christine, 1950-
539.7'21   QC793.2
ISBN 0 19 853999 1 (pbk)

Library of Congress Cataloging in Publication Data
Close, F. E.
The particle explosion
Bibliography: p.
Includes index.
1. Particles (Nuclear physics) I. Marten, Michael.
II. Sutton, Christine. III. Title.
ISBN 0 19 853999 1 (pbk)

Design: Richard Adams/AdCo Associates
Original photography: David Parker
Photographic colouring: David Parker
Diagrams: Neil Hyslop/David Parker
Additional photo research: Angela Murphy

Printed in Hong Kong

CONTENTS

CHAPTER 1
THE WORLD OF PARTICLE PHYSICS

Take a deep breath! You have just inhaled oxygen atoms that have already been breathed by every person who ever lived. At some time or other your body has contained atoms that were once part of Moses or Isaac Newton. The oxygen mixes with carbon atoms in your lungs and you exhale carbon dioxide molecules. Chemistry is at work. Plants will rearrange these atoms, converting carbon dioxide back to oxygen, and at some future date our descendants will breathe some in.

If atoms could speak, what a tale they would tell. Some of the carbon atoms in the ink on this page may have once been part of a dinosaur. Their atomic nuclei may have arrived in cosmic rays, having been fused from hydrogen and helium in distant, extinct stars. But whatever their various histories may be, one thing is certain. Most of their basic constituents, the fundamental particles—the electrons and quarks— have existed since the primordial Big Bang at the start of time. In recent years, physicists have learned to make these particles in the laboratory, and by studying them, they hope to learn about the origin of the Universe.

In this introductory chapter we present an overview of these founder members of the cosmos, and of the methods used to create and investigate them. Later chapters tell in detail the story of how they came to be discovered, and provide 'portraits' of all the important and better-known particles.

Atoms are the complex end-products of creation. Their basic constituents were created within the first seconds of the Big Bang. Several thousand years elapsed before these particles combined to make atoms. The cool conditions where atoms exist today are far removed from the intense heat of the Big Bang. So to learn about our origins we have to see within the atoms, and study the seeds of matter.

The inner labyrinths of an atom are as remote from daily experience as are the hearts of stars, but to watch the atomic constituents we have to reproduce the intense heat of stars in the laboratory. This is the world of high-energy particle accelerators, which create feeble imitations of the Big Bang in small volumes of a few atomic dimensions. The science that studies the basic particles and attempts to understand the nature of matter and energy is known as particle physics.

Particle physics developed out of a number of discoveries made in the last quarter of the 19th century. The most important were the discoveries of X-rays, of radioactivity, and of the electron. These revelations have led to our detailed modern understanding of the structure of the atom and to the finding of hundreds of new subatomic particles. Today, physicists face a new challenge: to develop a theory that unifies the elementary constituents and forces of nature in a single coherent explanation.

At the turn of the century, the existence of atoms was little more than hypothesis. The popular belief was that atoms were small solid spheres. Today we know that the atoms of all elements are combinations of electrons, protons, and neutrons, (except for atoms of the lightest element, hydrogen, which consist of a single electron and a proton, but no neutron). Moreover atoms contain an enormous amount of empty space in which the lightweight negatively-charged electrons gyrate. By contrast, the more bulky neutral neutrons and positively-charged protons are tightly bunched in a dense central nucleus. The number of protons identifies the element. For example, hydrogen, the lightest element, has one proton; uranium, the heaviest naturally occurring element, has 92.

This understanding of the basic structure of the atom has transformed the 20th century. The exploration of the atomic nucleus led to the development of nuclear

Fig. 1.1 *The tracks of electrically charged particles spill out across a table, bearing a message about the workings of the subatomic world they inhabit. The job of particle physicists is to interpret such messages, and to build theories that describe the behaviour of the particles and the forces that control them. Here bubble chamber scanner Renee Jones of the Fermi National Accelerator Laboratory (Fermilab), near Chicago, is working on a photograph taken at the laboratory's 15 foot (4.6 m) bubble chamber. The bubble chamber is a large tank filled with liquid hydrogen, which in this case was subjected to a beam of particles called neutrinos. Because they are electrically neutral, the neutrinos do not produce any tracks themselves, but one of them interacted violently with a proton in the bubble chamber's liquid, producing the spray of charged particles that shoots out towards the bottom of the picture. The photograph of the interaction, or 'event', is projected onto the table, enabling Jones to measure accurately the positions, lengths, angles, and curvature of the tracks. From these measurements physicists can identify the various types of particles involved in the event.*

bombs and nuclear power stations. The detailed understanding of the behaviour of electrons around the nucleus revolutionized the chemical industry and created electronics.

Today particle physicists have a much richer picture of the subatomic world. They have discovered hundreds of new varieties of particle—a veritable explosion, which gives this book its title. There are pions and kaons, omegas and psis, 'strange' particles and 'charmed' ones. The members of this subatomic 'zoo' have been named with apparent disregard for logic. Many particles are called after letters of the Greek alphabet, and physicists habitually refer to them simply by their Greek letter. The pion, for instance, is π. Most of the new particles exist not as single entities, but come in a number of varieties, according to their varying electric charge. There are in fact three kinds of pion—positive, negative, and neutral. Thus there is a pi-plus (π^+), a pi-minus (π^-), and a pi-zero (π^0).

An added complication is that every particle has its antimatter equivalent—a particle that is the same in almost every respect except that its electric charge is reversed. The pi-minus, for example, is the antimatter equivalent of the pi-plus. The first antimatter particle to be discovered, in 1932, was the positron, the antiparticle of the electron. Despite their name, antimatter particles are just as real as matter particles, and physicists can create them at will at particle accelerators.

This host of particles at first seemed an embarrassment of riches, but in the past 20 years an underlying pattern has emerged. Now particle physicists believe that their subatomic zoo can ultimately be reduced to only two types of particles: quarks and leptons. All that is needed in addition are a few particles called 'gauge bosons' to transmit nature's fundamental forces, such as gravity and electromagnetism.

All forms of matter are built from these basic constituents. What is more, the quarks, leptons, and gauge bosons are believed to be elementary—in other words they do not consist of anything smaller. They are points of congealed energy, and they are what everything else in the Universe, from cottage cheese to quasars, is made of.

Quarks are what protons and neutrons and many other particles are made of. As of 1986, it seems probable that there are six varieties—or 'flavours'—of quark. They are called up, down, charm, strange, top (or truth), and bottom (or beauty). In forming particles, quarks combine either in threesomes or in the company of an antiquark; they apparently never exist alone.

Leptons are a second distinct type of fundamental particle. The most familiar lepton is the electron. As electric current, electrons bring the subatomic world into our daily experience. There are also two heavier versions of the electron. These are the muon (which is a major by-product of cosmic rays) and the tau. No one is sure why they exist. In addition, there are three neutrinos, or 'little neutral ones': the electron-neutrino, the muon-neutrino, and the tau-neutrino. Neutrinos are so light that no one has yet detected their masses, if indeed they have any at all. They pervade the Universe. If they do have mass, their collective bulk could outweigh the visible galaxies and eventually cause the Universe to collapse under its own weight.

So we have six leptons accompanying six quarks. Particle physicists believe that these 12 particles occur in three separate 'generations'—the up and down quarks with the electron and its neutrino; the charm and strange quarks with the muon and its neutrino; and the top and bottom quarks with the tau and its neutrino. What distinguishes the three generations is that they exist at very different levels of energy. At the energy level of the Universe today, only the first generation can survive. This is why our atoms consist of protons and neutrons (made of up and down quarks), and electrons. But at extremely high energies, such as existed in the first instants after the birth of the Universe in the Big Bang, exotic forms of matter built from the second or third generations of particles would have been almost as plentiful as first generation matter is today.

Identifying the particles is only half the task. An important aim is to understand the nature of the forces that cluster the individual particles into bulk matter and fuel the cycle of life. These forces are carried by the action of a third major family of particles, the 'gauge bosons'.

Fig. 1.2 *Richard Feynman (b.1918) gives a lecture at CERN, the European centre for research in nuclear physics near Geneva. Theorists like Feynman, who works at the California Institute of Technology in Pasadena, play an important role in interpreting the results of particle physics experiments. In 1965, the year this photograph was taken, Feynman shared the Nobel prize for physics with Sin-itiro Tomonaga and Julian Schwinger, for work on quantum electrodynamics, or QED, the theory that describes the electromagnetic interactions of subatomic particles.*

The most familiar of these are the photons, which impart the electromagnetic force. Photons are bundles of electromagnetic radiation, ranging from gamma rays and X-rays at one end of the electromagnetic spectrum, through ultraviolet, visible, and infrared light, to heat, microwaves, and radio waves at the other end of the spectrum. Though they appear very different to us, the only physical difference between these radiations is their wavelength—gamma rays have very short wavelengths and radio waves very long ones. They are all 'carried' by the same particle, the photon. A gamma ray photon has a high energy, a radio wave photon has a low energy.

The electromagnetic force is one of three forces that govern the behaviour of atoms. It keeps the negatively-charged electrons in their 'orbits' around the positively-charged nucleus, making atoms and bulk matter seem solid. The theory of electromagnetism was published by James Clerk Maxwell in 1865 and has matured into the modern theory of quantum electrodynamics (QED). This theory is so elegant and powerful in its description of many subatomic phenomena that it has become *the* paradigm for other theories of particle behaviour.

The second of the subatomic forces, the weak force, is familiar to us because it is the force that gives rise to radioactivity. It transmutes the nuclei of radioactive atoms by converting neutrons within the nuclei into protons, electrons, and neutrinos. The carriers of the weak force—the W and Z particles—are similar to the photon but with one notable difference: whereas the photon has zero mass, the W and Z are enormously heavy even by comparison with the proton.

The third subatomic force is the most powerful of all. It is mediated by gauge bosons called gluons, grips the quarks within neutrons and protons, and holds the atomic nucleus together. We do not notice this 'strong' force directly because it acts only within the confines of the nucleus, where it is over a hundred times more powerful than the electromagnetic force. But in concert with the weak force the strong force is responsible for the basic interactions between protons at the heart of the Sun, which initiate nuclear fusion and cause the Sun to shine.

There is a delicate balance here. The Sun burns slowly because the weak force is indeed feeble. This has kept the Sun burning long enough for the Earth and life to form. The strong force fuses the protons and neutrons together, converting hydrogen

Fig. 1.3 *This computer display of a so-called 'three-jet event' is evidence for the existence of gluons—the carriers of the strong force which acts between quarks and is ultimately responsible for binding protons and neutrons within the atomic nucleus. The image shows a cross-section through the High Resolution Spectrometer detector at the Stanford Linear Accelerator Center (SLAC) in California. An electron and its antimatter equivalent, a positron, have come in at right angles to the image, from in front of and behind the page, and collided at the centre of the detector. In the resulting annihilation, their mass is converted to energy which rematerializes in the form of a quark and an antiquark, one of which almost instantaneously radiates a gluon. These three particles are not visible themselves, but they leave a distinctive signature by immediately materializing further quarks and antiquarks which then group together to form composite particles. It is these particles which spill out into the detector in three separate jets—one each for the original quark, antiquark, and gluon.*

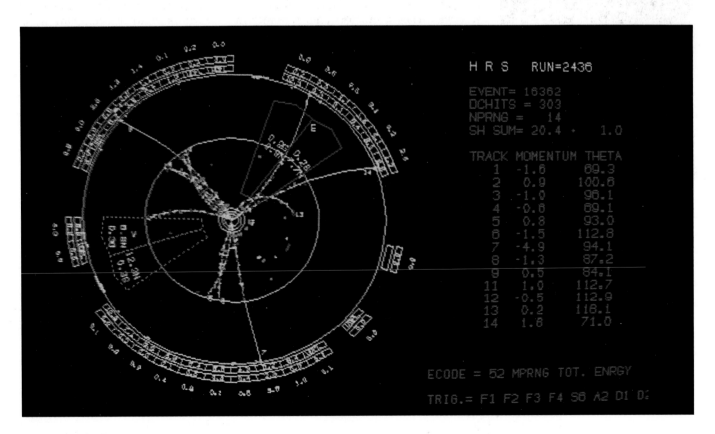

into helium and releasing small amounts of mass as radiant energy. The electromagnetic radiation streams across space to where, 150 million kilometres away, we bask in the glow of this nuclear furnace.

We are held on Earth in orbit around the Sun courtesy of nature's fourth great force: gravity. Without gravity there would be no Universe. The gravitational force has contradictory characteristics. It is so feeble that it has no measurable actions within individual atoms. But it acts over enormous distances, each little atom making its trifling contribution to the great collective whole. Gravity holds the planets in their orbits, and clusters the stars together to make the galaxies. It also holds us on the ground and caused apocryphal apples to fall on Newton's head. It is ironic that although the action of gravity was explained by Newton in the 17th century, it is today the least understood of the fundamental forces. We understand what it does, but not how it works. There are experiments seeking to detect gravity waves, but none have yet been found. Its gauge particle, the graviton, remains hypothetical.

PARTICLE PHYSICS TODAY

THE IDEA that the Universe consists of three types of elementary particles, which are governed by four fundamental forces, is backed up by a considerable body of experimental evidence. But it would be quite wrong to suggest that this is the end of the story. On the contrary, there remain major mysteries. Why do the particular quarks, leptons, and gauge bosons exist? Why do they have the masses that they have rather than other masses? And why are there four forces? Are they possibly different manifestations of a single underlying force? Are there other forces at work, as yet unperceived?

It is in search of answers to these questions that particle theorists have for several years been engaged in the search for 'unification'. The classic example of unification is Maxwell's theory of electromagnetism, which explained the apparently diverse phenomena of electricity and magnetism as due to a single force. In the late 1960s, electromagnetism and the weak force were unified in a similar way in the 'electroweak' theory. Among its many consequences, electroweak theory predicted the existence of the Z particle as one of the gauge bosons that mediate the weak force. The discovery of the Z in 1983 helped to confirm that the electroweak unification was correct, and it encouraged theorists in their efforts to develop a grand unified theory that would unify the electromagnetic and weak forces with the strong force. Such grand unified theories, or GUTs, predict that the strong force, which binds the proton together, and the weak force, which causes radioactive decays, spring from the same basic force. This implies that protons will decay occasionally. The search for this major proof of GUTs is currently underway.

Theorists conjecture that in the initial heat of the Big Bang all four forces, including gravity, might have been as one. But the forces split apart as the Universe cooled, so that their unity is now obscured. Thus it is that the search for unification is a search for the physics of the Big Bang. It carries a significance beyond particle physics itself. One of the unexpected developments in the past 10 years is that particle theories and experiments have become increasingly intertwined with astrophysics and cosmology. For example, at Fermilab, the major particle physics laboratory near Chicago, astrophysicists studying the large-scale implications of current ideas in particle physics inhabit offices sandwiched between the particle theorists and the computer tape library. Their work concerns some of the major questions posed by the very existence of the Universe. How did it begin? Why does it have the form it does? Will it continue expanding forever or will it eventually begin to contract?

These theoretical constructs are not mere flights of fancy; this is not a modern analogue of ancient theological debates on the number of angels on the head of a pin. Theories survive or fall by experimental tests. There is a symbiosis between two breeds of particle physicist: the experimenter and the theorist.

The theorist organizes what has been discovered into a theory which may predict the existence of new particles. The experimenter tries to discover the predicted

Fig. 1.4 *Much of the early history of particle physics, in the 1920s and 30s, centred around the study of cosmic rays—high-energy particles that reach Earth from outer space. Detecting their interactions as they pass through the atmosphere involved taking equipment to the highest possible altitudes. Here one of the pioneers of cosmic ray research, Robert Millikan (1868–1953), is trekking with his team up Mt Whitney, which at an altitude of about 4350 m is the highest point in California. The picture shows, from left to right, Glenn Millikan (Robert's son), Otto Oldenberg, Robert Millikan, Ted Cooke (the guide), and C. H. Prescott jr.*

Fig. 1.5 *By the 1970s, particle physicists were building huge accelerators to raise particles to high energies artificially. This 1975 photograph shows the machine used to excavate the circular tunnel, beneath French and Swiss farmland, which eventually housed the Super Proton Synchrotron (SPS) in a ring 7 km in circumference.*

particles, but may well end up discovering something quite different, quite unpredicted, which the theorist must then explain in a modified or entirely new theory. It is a measure of the growth of the science that the time is long gone when individual physicists could lay claim to have both experiment and theory at their fingertips. Now specialization is the order of the day, though theorists and experimenters still need to appreciate the subtleties of the other's craft as they feed off each other's work.

Another characteristic of modern particle physics is its internationalism. A typical experiment today involves dozens, sometimes over a hundred, people. It is not something that one institution can develop, build, and operate. A recent experiment at Fermilab involved ten institutions, six from America, one from Britain, another from Italy, another from Japan, and one from the Soviet Union. CERN, the European particle physics laboratory on the outskirts of Geneva, is itself a multinational effort funded by 14 European countries. Enter the canteen there and you are immersed in a multilingual babble. Furthermore, CERN has forged links with its counterpart in Eastern Europe—JINR, at Dubna 100 km north of Moscow—which brings together physicists from the Soviet Union and its satellites.

Fig. 1.6 *Modern particle accelerators are monuments to high technology. The 2 km diameter circular tunnel of the main accelerator at Fermilab in Illinois is buried beneath the ground, but its location is marked by a service road on the surface. The main laboratory buildings lie to the left of the ring, dominated by the high-rise administration building designed by Robert Wilson, who was the laboratory's director from 1967–1978. The city of Chicago lies in the distance to the top of the picture.*

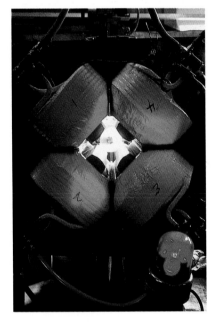

Fig. 1.7 *Many types of particle accelerator contain electromagnets to bend and focus a circulating beam of particles, which travel inside a pipe held at extremely high vacuum. This is a view of the type of magnet that is used to focus the beam, rather as a lens focuses a beam of light. It is a quadrupole magnet, with two north poles and two south poles arranged so that poles of the same kind are diametrically opposite each other. The coils of wire that generate the magnetic field are clearly visible, wrapped in a cream-coloured covering. The beam pipe passes between the four faces of the magnet.*

Because modern particle physics laboratories are so large and expensive, there are only a handful of major ones around the world. In Europe there is CERN, which straddles the French–Swiss border, and DESY (which stands for Deutsches Elektronen Synchrotron) in a suburb of Hamburg. In the US, there are the Brookhaven National Laboratory on Long Island, the Fermi National Accelerator Laboratory ('Fermilab') on the plains outside Chicago, the Stanford Linear Accelerator Center (SLAC) at Stanford, California, and the Cornell Electron Storage Ring (CESR) at Cornell University, New York. In the Soviet Union there are major laboratories at Dubna, at Novosibirsk in Siberia, and at Serpukhov, south of Moscow. The Japanese have a laboratory called KEK, in the 'science city' of Tsukuba. And in China there is the Institute for High Energy Physics in Beijing.

Each of these laboratories is a giant enterprise, employing hundreds of physicists, technicians, administrators, and support personnel. They consist of many buildings spread over a large area. They operate with annual budgets of hundreds of millions of pounds. The centrepiece of each—its *raison d'être*—is a large particle accelerator.

Many homes contain a small particle accelerator—the television tube. At the rear of the tube electrons boil off from a heated metal filament. An electric field accelerates the electrons towards the screen, and magnets steer them to strike different points so that their separate glows make a sensible coherent image.

A modern particle accelerator is a giant machine for firing particles at a target at extremely high energies. Typically it is built in an underground tunnel, several kilometres in length. In cross section the tunnel is reminiscent of a metropolitan underground railway tunnel. On one side it is empty, allowing access for people and equipment. On the other side a small oval pipe, perhaps 20 centimetres wide by 10 centimetres high, threads through a succession of electromagnets. It is through this 'beam pipe', which is kept at a very high vacuum, that the particles travel. The particles most commonly used are electrons and protons. Electric fields kick the particles millions of times each second, propelling them forward to high energy.

'Driving' a particle accelerator is like flying a spacecraft. The 'bridge' is the accelerator control room, consisting of rows of computer monitors. While the particles whirl around a few kilometres of beam pipe at almost the speed of light, nothing much seems to be happening. Two or three people may be drinking coffee, consulting a computer display, or telephoning someone in the experimental areas that the machine is feeding.

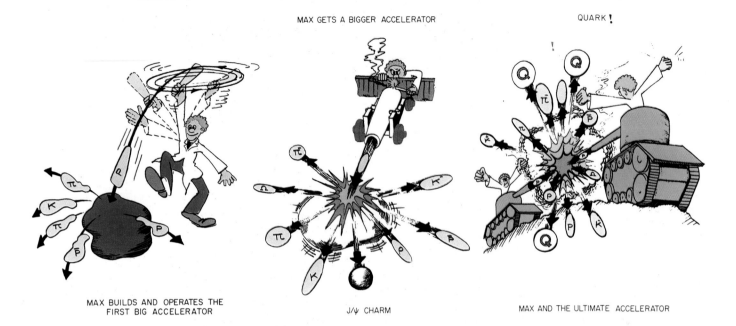

MAX BUILDS AND OPERATES THE
FIRST BIG ACCELERATOR

J/ψ CHARM

MAX AND THE ULTIMATE ACCELERATOR

The automatic pilot is in control. The path of the particles is preprogrammed. The constant adjustments of accelerating units and magnets, of coolants and vacuum pumps and electricity supply, are all controlled by the computers, which teams of experts have spent hours programming. The people in the control room have little to do, except to make periodic checks. But there are moments of high stress, as when the pilot prepares to land the spacecraft. For example, the machine physicists at CERN can prepare beams of antimatter, which survive only so long as they keep out of the way of the matter that is all around them. It takes a whole day to prepare the beam, accumulating enough antimatter particles to be of use for the experimenters who await. Then the controllers must pilot the beam correctly so that it eventually arrives at the experimental apparatus. One push of the wrong button at the wrong moment and all will be lost. A whole day will be needed to put it all right again.

Why do particle physicists need to accelerate electrons and protons to high energies? The energetic beams are used to probe the structure of matter in one of two basic ways. First, a beam can be directed at a stationary 'fixed' target, which can be a structurally simple element, like hydrogen, or something physically more complex but easier to handle, such as copper. Or, as is increasingly the case, the accelerator may handle two beams of particles travelling in opposite directions, which are kept apart until they reach maximum energy, when they are brought to meet head on. In either case, the aim is to encourage some of the particles involved to interact with each other, for it is the results of such interactions that reveal what is happening at the sub-microscopic level of matter.

The interaction of two particles can be a violent collision, in which the particles that meet are destroyed, only to materialize again in some new guise. Or it may be a glancing collision, in which two particles brush against each other like billiard balls. Or it may take a variety of forms between these two extremes. Whatever happens, it turns out that we can learn more if we use higher energies to produce the interactions.

In some instances, the energy can assist in materializing additional particles, in accordance with Albert Einstein's famous equivalence of mass and energy: $E = mc^2$. An extreme example is when matter and antimatter mutually annihilate into pure energy which can rematerialize as new, different particles. In this way, particle physicists have been able to create particles and forms of matter that do not occur naturally on Earth, but which may be commonplace in more violent parts of the Universe.

Fig. 1.8 *Max, the friendly physicist from the Brookhaven National Laboratory on Long Island, New York, illustrates how particle physics has developed through experiments at bigger and bigger accelerators. From studies of protons (p), pions (π), kaons (K), and antiprotons (p̄) at accelerators in the 1950s (left), Max progressed to the bigger machines of the 1970s (centre), which revealed an ever-growing 'zoo' of particles including the omega (Ω) and the J/Ψ. These two particles in particular played an important role in showing that most of the particles we observe are built from quarks, including the 'charm' quark of the J/Ψ. In the 1980s, Max works at the 'ultimate accelerator' (right)—the type of machine that collides beams of particles head on, and which gives the best glimpses yet of the world of quarks (q).*

TRACKING THE PARTICLES

ACCELERATING PARTICLES to high energies and firing them at fixed or moving targets is the only way we know of exploring the interior of the atom. Nature has a fundamental rule: the finer the detail you want to resolve, the more energetic your probe must be. The particles created in today's high-energy collisions can be as small as 10^{-16} cm across, and live for only a few hundredths of millionths of a second or less. Recording these tiny and ephemeral pieces of matter is the job of the detectors.

Detectors come in a variety of types and sizes, but the majority today are huge, multi-layered pieces of apparatus. Despite their differences, they all rely on the same basic principles. They never 'see' the particles directly; what they see are the effects of the particles on their surroundings—much as an animal leaves tracks in the snow, or a jet plane forms trails of condensation across the sky.

Electrically charged particles leave trails as they gradually lose energy when they travel through a material, be it a gas, a liquid, or a solid. The art of particle detection is to sense this deposited energy in some manner that can be recorded. Then, in the way that measurements of the footprints of our ancestors reveal something about their size and the way they walked, the information recorded can reveal details of a particle's nature, such as its mass and its electric charge. All the techniques described in later chapters rely on this same principle, from the simple photographic emulsions of the 1930s and 40s to the metre-long gas-filled chambers, criss-crossed by thousands of wires, of the 1980s.

The detector is a kind of ultimate microscope, which records what happens when a particle strikes another particle, either in a fixed target such as a lump of metal or a chamber filled with a gas or a liquid, or in an on-coming beam in a collider. The 1950s and 60s were the age of the bubble chamber, so called because electrically charged particles moving through it produce trails of tiny bubbles in the liquid filling the chamber. Photographs taken in bubble chambers dominated the imagery of particle physics for many years, but today only a few of the devices are still in use and most experiments are based on electronic detectors.

Electronic detectors tend to be built on a very large scale. It seems to be one of the ironies of science that the smaller the subject of your investigation, the larger your detector needs to be. The UA1 and UA2 detectors at CERN, which discovered the carriers of the weak force, the W and Z particles, sit in underground pits. You can look down on UA1 from ground level if you do not suffer from vertigo. From above, it does not seem so large until you notice the human ants working on it. When you get down to its own level, you realize it is as big as a two-storey house.

Detectors such as UA1 and UA2 are hybrid devices, which consist of many different sub-detectors—scintillation counters, wire chambers, drift chambers, Čerenkov counters—whose job is to measure the paths, angles, curvatures, velocities, or energies of the particles created in an event. The many sub-detectors are sandwiched together, sometimes in a series one behind the other (in a fixed-target experiment), sometimes in a kind of Swiss roll wrapped around the beam pipe (in a collider experiment). And every part of the detector has hundreds of cables running from it, each of which goes to a particular place in the control system.

A typical detector at a modern particle physics laboratory is a major undertaking. It will take 3–5 years to design, operate for 5–10 years, and its results will continue to be analysed for a further 2–4 years. Someone involved in the project from beginning to end may spend almost 20 years on this one detector. It is not something that a handful of individuals can set up on a laboratory bench. It requires computer experts, draughtsmen, engineers, and technicians, as well as some 50 or 60 physicists from a number of institutions.

Detectors rarely record *all* the particle collisions that occur in a particular experiment. Usually collisions occur thousands of times a second and no equipment can respond quickly enough to record all the associated data. Moreover, many of the collisions may reveal mundane 'events' that are relatively well understood. So the experimenters often define beforehand the types of event that may reveal the particles they are trying to find, and program the detector accordingly. This is what a major

part of the electronics in a detector is all about. The electronics form a filter system, which decides within a split second whether a collision has produced the kind of event that the experimenters have defined as 'interesting' and which should therefore be recorded by the computer. Of the thousands of collisions per second, only one may actually be recorded. One of the advantages of this approach is its flexibility: the filter system can always be reprogrammed to select different types of event.

The results of all this detection are a large number of computer tapes recording the events that have been selected. Often, computer graphics enable the events to be displayed on computer monitors as images, which help the physicists to discover whether their detector is functioning in the correct way and to interpret complex or novel events.

Imaging has always played an important role in particle physics. In earlier days, much of the data was actually recorded in photographic form—in pictures of tracks through cloud chambers and bubble chambers, or even directly in the emulsion of special photographic film. Many of these images have a peculiar aesthetic appeal, resembling abstract art. Even at the subatomic level nature presents images of itself that reflect our own imaginings.

The essential clue to understanding the images of particle physics is that they show the *tracks* of the particles, not the particles themselves. What a pion, for instance, really looks like remains a mystery, but its passage through a substance—solid, liquid, or gas—can be recorded. Particle physicists have become as adept at interpreting the types of track left by different particles as the American Indians were at interpreting the tracks of an enemy.

Fig. 1.9 *If the accelerators are huge, then so too are the particle detectors. This is the UA2 detector at CERN, photographed in 1981 during its construction. It is typical of the electronic detectors used at modern colliding beam machines. When installed, the two sections to the left and right move in to fit against the central section. The whole assembly moves into the tunnel of the Super Proton Synchrotron, and the accelerator's vacuum pipe, through which the counter-rotating beams of protons and antiprotons circulate, passes horizontally through the middle of the detector.*

Detectors like UA2 are highly complex, and consist of many different components all designed to perform complementary tasks in the identification of particles. The central section of UA2, for example, bristles with hundreds of phototubes—devices that convert into electronic signals the flashes of light produced by charged particles as they traverse plastic 'scintillating' material. UA2's moment of glory came in 1983, when it shared in the discovery of the W and Z particles—the carriers of the weak force—with its sister experiment at CERN, UA1.

Fig. 1.10 *Not the kind of bubbles that physicists working on bubble chambers want to see! This foam bath occurred at the Brookhaven National Laboratory in 1974, when the fire protection system for the 7 foot (2.1 m) bubble chamber came on automatically because of an electrical design fault. Clean-up operations took one day.*

Fig. 1.11 *BEBC—the Big European Bubble Chamber—typifies the large bubble chambers used in the 1970s to study the interactions of the high-energy particles produced at the new big accelerators. The chamber is a cylinder 3.7 m in diameter which, when operating, contains liquid hydrogen at −173°C. Here we see the chamber surrounded by its vacuum tank which helps to keep the hydrogen cool. In normal operations this tank is hidden from view, being surrounded by the coils of a superconducting magnet and various detectors to assist in identifying particles that escape the liquid. BEBC recorded its first tracks in 1973 and ceased operation 11 years later, after taking some 6.3 million photographs of particle interactions.*

A number of simple clues immediately narrows down the possibilities. For instance, many detectors are based around a magnet. This is because the tracks of electrically-charged particles are bent in a magnetic field. A curving track is the signature of a charged particle. And if you know the direction of the magnetic field, then the way that the track curves—to left or right, say—tells you whether the particle is positively or negatively charged. The radius of curvature is also important, and depends on the particle's velocity and mass. Electrons, for instance, which are very lightweight particles, can curve so much in a magnetic field that their tracks form tight little spirals—a common feature in many of the pictures in this book.

Most of the subatomic zoo of particles have brief lives, less than a billionth of a second. But this is often long enough for the particle to leave a measurable track. Relatively long-lived particles leave long tracks, which can pass right through a detector. Shorter-lived particles, on the other hand, usually decay visibly, giving birth to two or more new particles. These decays are often easily identified in images: a single track turns into several tracks.

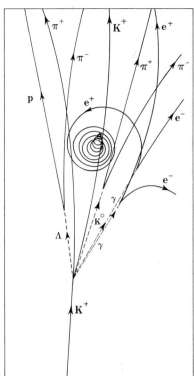

Fig. 1.12 *The tracks of particles through a bubble chamber have a beauty of their own. In this image from the Lawrence Berkeley Laboratory, where bubble chambers were pioneered, irrelevant tracks have been removed to leave only those involved in the event. The tracks curve in the magnetic field of the chamber, and the direction of curvature reveals their charge. The event shows several features typical of bubble chamber photographs. A positive kaon (K^+) comes in from the bottom of the picture and collides with a proton in the chamber liquid, producing a positive pion (π^+), a positive kaon, and three neutral particles—a pi-zero, a neutral kaon (K^0), and a lambda (Λ). The neutral particles leave no tracks, because they do not ionize the chamber's liquid, but we can infer their presence from the 'V's created when each decays into charged particles that do leave tracks. The lambda decays into a proton (p) and a negative pion (π^-), the neutral kaon into a positive and a negative pion. The pi-zero is more complicated; it lives so briefly that it is not shown even by a dotted track; it decays almost immediately to two gamma rays (γ), which are also neutral and therefore invisible; but the gammas betray their presence when each converts into an electron–positron pair (e^-, e^+). It is by identifying and analysing these charged decay products that physicists can deduce the nature of the neutral particles, and draw in their tracks as in the dotted lines in the diagram. Note also how one of the positrons forms a tight spiral in the magnetic field; this is a characteristic signature of low-energy electrons and positrons and is to be seen in bubble chamber photographs throughout this book.*

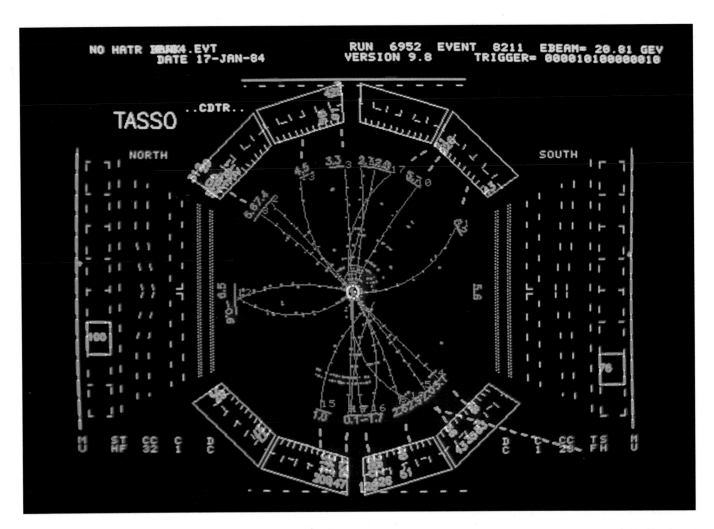

Fig. 1.13 *The TASSO detector at the DESY laboratory in Hamburg records the tracks of particles produced in an electron–positron annihilation. The computer display shows a cross-section through the detector; the electron and positron have come in at right angles to the image, from in front of and behind the page, and annihilated at the centre. The resulting energy has materialized as a pair of extremely short-lived particles, which probably include a bottom particle (containing a bottom quark). These two particles immediately decay into other particles, which shoot out through the detector, scoring 'hits' in its various instruments. The blue dots mark wires hit in the 2.6 m diameter 'drift chamber' that surrounds the beam pipe (yellow circle). The red lines show the tracks of particles drawn in by the computer to fit the pattern of hits. The tracks bend in the field of an electromagnet (not shown), whose coil surrounds the drift chamber. The purple blocks represent 'shower counters' which detect neutral particles from the energy they deposit. The blue bar at the top of the image indicates a detector that has fired beyond the iron of the magnet and the lead of the shower counters. Muons are the only charged particles that could penetrate this far, so a muon must have passed this way.*

Neutral particles present more of a headache to experimenters. Particles without an electric charge leave no tracks in a detector, so their presence can be deduced only from their interactions or their decay products. If you see two tracks starting at a common point, apparently arising from nowhere, you can be almost certain that this is where a neutral particle has decayed into two charged particles.

Our perception of nature has deepened not only because the accelerators have increased in power, but also because the detection techniques have grown more sophisticated. The quality of particle imagery and the range of information it provides have both improved over the years. Whereas bubble chamber photographs show the tracks of all charged particles directly, often in messy confusion, computers can highlight or select the important tracks, and even draw in where neutral particles have passed.

Modern high-quality computer graphics produce full-colour displays of considerable beauty. In the most sophisticated systems, the view of the event can be rotated so that it can be seen from any angle. The computer can also be instructed to eliminate irrelevant tracks, so that the essential features of an event stand out clearly. By refining the structure of a complicated event, and by enabling physicists to see it three-dimensionally, from any viewpoint, today's computers can give a richer picture of particle interactions.

The accelerators that produce high-energy particles and the detectors that record the results of their collisions are the main tools of the experimenters. But their work would not be possible without effects that are understood only with the aid of the two pillars of modern physics: *special relativity* and *quantum theory*. Each of these enables theoretical physicists to make sense of the behaviour of matter at miniscule distances and high energies. They have also allowed experimenters to develop the techniques for tracking down particles.

For example, the particles in the accelerator are moving extremely close to the speed of light. They obey the rules of Einstein's theory of special relativity, in which moving bodies gain mass as they gain energy, and in which times and distances depend critically upon the reference frame in which they are measured. Indeed, particle physicists are so conversant with the notions of special relativity that they find it more convenient to talk of the masses of particles in terms of equivalent amounts of energy.

A proton, for example, weighs about 1.67×10^{-27} kg—a minute amount. But, according to Einstein, energy E is the product of mass m and the square of the velocity of light c: $E = mc^2$. In the units of energy used by particle physicists, the proton's mass becomes 938 MeV$/c^2$. (An electron volt is the amount of energy that an electron gains when it accelerates through an electric potential of one volt. Particle physicists deal in keVs [thousands of electron volts], MeVs [millions of electron volts], and GeVs [thousands of millions of electron volts].) To make life easier, the particle physicists also choose to work in units where the velocity of light c is equal to 1, so the proton's mass becomes simply 938 MeV. The electron's mass is significantly smaller than the proton's at 0.51 MeV. On this scale the average human being would 'weigh' 4×10^{31} MeV!

The mass of a proton at rest is 938 MeV, but when it is accelerated close to the speed of light, its mass increases as it gains energy. Fermilab's big accelerator, the Tevatron, can accelerate a proton to a mass of 1000 GeV (or 1 teraelectronvolt, 1 TeV). Its mass has effectively been increased 1000 times. Accelerated to the same extent, the average human would weigh as much as a juggernaut.

Relativity plays another vital role in particle physics. An energetic particle which has a lifetime of only one hundred millionth of a second—10^{-8} seconds—before it breaks up into other particles, can in fact travel several metres before it does so, thanks to an effect in special relativity called *time dilation*. This says that a time measured in a stationary reference frame—the laboratory—is lengthened with respect to a moving reference frame—the particle—to an extent that depends on the speed of the motion. The faster the particle is travelling, the slower time elapses for it. It is like the twin who ages less in a high-speed rocket than the sibling who stays at home. In this way, short-lived particles, such as pions and kaons, can be produced in beams that survive long enough to be useful in experiments.

Quantum theory plays a more important role in the work of theorists than experimenters, though it underlies the physical processes that operate in much of the microelectronics used in experiments today. For the theorists, it provides the basic ground rules to explain the behaviour of particles when they interact. Most theorists are concerned with the interpretation of the measurements the experimenters make. An ideal theory has no loose ends—no unknown parameters that have to be measured—and it makes predictions that can be tested in further experiments.

The application of quantum theory to the electromagnetic interactions of particles has led to what is perhaps the most powerful of all physical theories—that of quantum electrodynamics. This theory, which was fully developed in the 1940s, enables calculations of atomic behaviour to be made with extreme accuracy. It has since become the model for theories of other kinds of interaction between particles.

This book is the story of how we have come to our modern understanding of the subatomic particles. It is also a showcase of particle imagery, from early cloud chamber and emulsion photographs to the latest computer graphic displays. These pictures not only show that the subatomic world is real and accessible; they also have their own peculiar abstract beauty.

The following chapters are both a journey to the heart of matter and a voyage through time. The even-numbered chapters describe the history of particle physics, and the techniques that have been developed over the past 100 years to generate and to study the particles. Chapters 3, 5, 7, and 9 introduce the particles discovered by this succession of techniques. Finally, the last chapter explains how some of the particles have been put to work in the service of medicine, industry, and even the detection of art forgeries!

Fig. 1.14 *Albert Einstein (1879-1955). He aptly summed up the problems experimental particle physicists face when he described detecting particles as 'shooting sparrows in the dark'.*

CHAPTER 2
EXPLORING THE ATOM

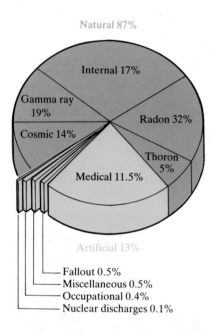

Fig. 2.1 *Natural sources provide 87 per cent of the total radiation that an average resident of the United Kingdom receives. This natural radiation comes from outer space (cosmic rays), from radioactive nuclei in the food we eat (internal), but mainly from rocks, soil, and building materials, which release gamma rays and radioactive gases (radon and thoron). The remaining 13 per cent of the total radiation dose comes from artificial sources, mainly in medical diagnosis and therapy.*

Fig. 2.2 *Radiation is everywhere, though in normal circumstances this sign in Utah, giving directions to a radiation shelter, can go unheeded.*

RADIOACTIVITY: the very word conjures up a multitude of images, from the frightening fall-out from atomic bombs and the hazards associated with nuclear power stations on the one hand to treatments for cancer on the other. It has become a byword of the 20th century, and one that arouses suspicion in many people. Yet although radioactivity was discovered only in the closing years of the 19th century, it is a natural process, happening constantly all around us and even within us. And not only is it natural and universal, but it is also essential. Without radioactivity stars would not shine and the ingredients from which we are built would never have been formed. Moreover, it provides scientists with a window on the nucleus at the heart of the atom—a window that allows us to look in at the fundamental nature of all matter.

Today we know that radioactivity is the result of atoms changing from one variety to another. An atom of uranium, for example, can change spontaneously to an atom of thorium, a slightly lighter element. The process is random and infrequent; in 4 500 000 000 years only half the atoms in a lump of uranium will have changed like this. But in transmuting to thorium, the uranium atom spits out a form of *radiation*. The change occurs as the inner parts of the atom rearrange themselves into a more stable form. The atom is moving towards equilibrium—towards a more 'comfortable' form—just as when water finds its own level.

The natural transmutations of uranium, and of other elements, make the rocks and soil about us radioactive. The materials from which we commonly build our houses emit radiation. Even our bodies are radioactive! To give an idea of quantity, suppose that a person living in the United Kingdom receives on average 1000 units of radiation in a year from natural sources: 160 of these will be due to radiation from outer space (cosmic rays), the remainder from rocks, soil, building materials and food. But an inhabitant of certain parts of south-west England, where the underlying granite rock contains pockets of higher concentrations of uranium, can receive three times as much radiation from the local surroundings. And because the atmosphere shields us from cosmic rays, people living at high altitudes receive more radiation; at 1000 m in the Alps, for instance, a person is exposed to about 35 units more than someone living at sea level in Holland. For most people, other sources of radiation are small by comparison. A two-hour flight high up in an aeroplane adds an extra two units, while a typical X-ray examination at the dentist is equivalent to about 10 units.

The best advertised property of radiation is probably its ability to destroy human tissue and ultimately to kill. Natural radiation accounts for some 1000 deaths a year in the UK, less than 1 per cent of the number of deaths from cancer (150 000 deaths) or due to smoking cigarettes (100 000 deaths).

So what is radioactivity? All atoms, from hydrogen, the lightest, to uranium, the heaviest naturally-occurring element, consist of lightweight particles called electrons, carrying negative electric charge, which encircle a bulky, dense, positively charged nucleus. The balanced negative and positive charges give rise to an atom that is neutral overall, but which contains intense electrical forces within it. Chemical reactions occur when the motions of the peripheral electrons are disturbed by neighbouring atoms. Radioactivity, by contrast, arises from within the nucleus at the atom's heart. Indeed, it is the attempts of the *nucleus* to come closer to equilibrium that give rise to radioactivity.

There are three common types of radioactivity, which are characterized by their emissions, known as alpha, beta, and gamma 'rays' or 'radiation'. Though these

emissions are all quite different, historically they all became referred to as rays or radiation—the words are interchangeable. The gamma radiation is perhaps most like the popular concept of 'rays', for it consists of an invisible, but very energetic form of electromagnetic radiation, roughly a million times more energetic than visible light.

The other two kinds of radioactivity—alpha and beta—do not involve the emission of electromagnetic radiation, but of electrically-charged subatomic particles. The beta rays, the most penetrating of the two radiations, consist of electrons. These electrons are indistinguishable from the electrons found in the periphery of atoms, but their origin is different. They are ejected from the nucleus, and they move so fast that they can penetrate sheets of lead a millimetre thick. For some time physicists thought that the electrons of beta radiation were actually present within the nucleus, but this is not so. Electrons in beta rays are created by the nucleus and immediately ejected; they are no more part of the nucleus than a dog's bark is part of the animal.

The alpha rays—the least penetrating kind of radiation—also consist of particles, but in this case positively-charged particles, now known to be the nuclei of atoms of helium. When one element emits an alpha particle, it turns into a different element— this is what happens when uranium transmutes to thorium. This discovery, made in 1902, created great consternation. Here, for the first time, was evidence that atoms of one element can turn spontaneously into other elements. The implication was that atoms of different elements must be composites of more elementary particles and not truly fundamental themselves.

Today we know that this is correct. Atoms consist of electrons and a nucleus, which itself is built from positively-charged particles called protons and neutral particles called neutrons. This chapter tells the story of how scientists in the early part of the 20th century discovered these facts; how the discovery of X-rays in 1895 led to the accidental discovery of radioactivity, and how that in turn led to a new view of the nature of the atom and the birth of nuclear physics.

ELECTRONS AND X-RAYS

DURING THE latter quarter of the 19th century the industrial revolution in Europe and America was beginning to bring a new standard of living to the working classes. Machines performed tasks that had previously involved dirty, even dangerous, manual labour. Nature was being tamed and exploited with the aid of science.

The steam engine was the epitome of applied thermodynamics, electricity was beginning to light the home, and electromagnetic motors were being developed. The existence of atoms as the smallest indivisible particles of matter was becoming generally accepted, but they seemed of little practical importance. There was a feeling that there was nothing new to be discovered in physics; the task was only to work out all the benefits that might accrue from the new mastery of nature. A sense of omniscience beckoned.

The development of new technologies in this period provided opportunities for extending the domains of scientific investigations, in particular into the nature of electricity. Electricity was one of the great scientific adventures of the 19th century, but its origins and properties were still largely a mystery.

One way of investigating electricity was to examine what happens when it passes through all manner of substances, including gases. Early in the previous century, scientists had observed the beautiful glow that appears when an electric current flows through a gas at low pressure—a phenomenon common today in mercury and sodium streetlights. By the 1880s, new improved vacuum pumps, one of the many inventions of the 19th-century boom, enabled other scientists to follow up these investigations more thoroughly. The basic equipment was a thin glass tube into which metal electrodes had been fitted at each end, and a pump to remove the air. When an electric current passed through the rarefied air an eerie cold glow appeared in the gas and around the surface of the tube.

These tubes became known as Crookes tubes after the British physicist William Crookes, who began a systematic study of the electrical glows in 1879. The lights

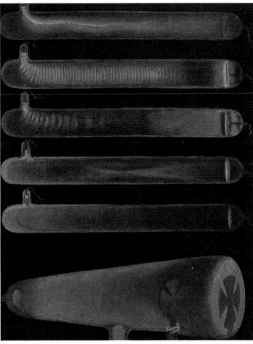

Fig. 2.3 *William Crookes (1832–1919). This cartoon by Spy, which appeared in the magazine* Vanity Fair *in 1903, shows Crookes holding one of the glass tubes named after him.*

Fig. 2.4 *As gas is evacuated from a Crookes tube, a changing pattern of colourful lights emerges when an electric current passes through the tube. Eventually (fifth tube down) the inner surface of the glass shines with a greenish glow. Evidence that rays of some kind are traversing the tube from the negative electrode (cathode) to the positive one (anode) appears in the shadow cast by an object, such as the cross within the bottom tube.*

that appeared in his tubes were indeed strange and beautiful and may have played no small part in Crookes's interest in spiritualism and psychic manifestations. He became a founder member of the Society for Psychical Research; his tube turned out to be a forerunner of modern television.

But what makes the eerie glow in a Crookes tube, and how does an electric current pass through the rarefied gas in the tube? Investigators in the 1870s found that as they lowered the pressure in the tube further and further, the gas in fact ceased to glow, though a current still flowed, and a luminous spot appeared on the wall of the tube opposite the cathode—the negative electrode. Objects placed in the tube would cast shadows in this glow, showing that a stream of rays must emanate from the cathode, causing the glass to glow where they struck it. The emanations were called *cathode rays*.

By the mid-1890s there were two schools of thought as to what the cathode rays might be—wave-like vibrations or energetic charged particles. In 1895 in Paris, Jean Perrin showed that the rays carry negative charge and that he could deflect the fluorescent spot, and hence the rays themselves, with a magnet. But there was a general reluctance to believe in a new kind of particle. Then, in 1897, J. J. Thomson, professor of physics at Cambridge University, performed a series of experiments that were to prove conclusively that the cathode rays are indeed streams of particles.

Thomson found that he could deflect the rays by electric fields as well as by magnetic fields. His discovery was possible because he could produce a better vacuum than other investigators; the residual gas in a poor vacuum is sufficient to conduct electricity, so a static electric field cannot be maintained. It was in measuring the motion of the rays in both magnetic and electric fields that Thomson came to a remarkable conclusion: the negatively-charged particles have a mass approximately two thousand times less than the mass of a hydrogen atom, the lightest known atom in the Universe. But since atoms were at the time considered indivisible, nothing lighter than a hydrogen atom was expected to exist.

Thomson obtained the same results irrespective of the material of the cathode or the gas in the tube. He was led to conclude that the cathode rays had to be 'matter in a new state, a state in which the subdivision of matter is carried very much further ... this matter being the substance from which all the chemical elements are built up'. The new particles became known as electrons, and in 1906 Thomson was rewarded with the Nobel prize for his work.

Electrons, we now know, carry the electric current across a Crookes tube, and give

rise to the eerie glow. Electrons from the cathode gain energy in the electric field along the tube. They can pass this energy on to atoms in collisions in the rarefied gas in the tube, and these 'excited' atoms then divest themselves of the extra energy by emitting light: the gas glows. Once the gas pressure is low enough, however, the electrons can travel along the tube without colliding at all, so the main glow disappears and the cathode rays leave only a fluorescent spot where they strike the opposite end of the tube.

Today, nearly a century later, Thomson's insight remains correct. Electrons still appear to be fundamental, and even the most powerful 'microscopes' have failed to find evidence for substructure within them. The discovery of the electron was the fundamental work that made possible the modern revolution in electronics and computing. It is perhaps the prime example of pure research in basic physics leading, a few decades later, to fundamental changes in human society. In 1897, however, Thomson's revelations provided the first evidence that atoms are not like featureless billiard balls but have a complicated inner structure.

If atoms contain negatively-charged electrons, then there must also exist positive charges to render the atoms neutral overall. Where are these positive charges in the atom? How can we ever hope to look inside minute atoms and see them? The tool to answer all these questions was radioactivity, discovered the year before Thomson established the existence of the electron. Yet radioactivity was found only as a result of investigations into another new phenomenon of the mid-1890s—the mysterious X-rays.

One of the many people to investigate the strange lights in Crookes tubes was a respected German professor, Wilhelm Röntgen, who worked in Würzburg. In 1895, he inadvertently left some unexposed photographic plates, tightly wrapped, near his tube. Later, upon taking the plates out for use, he found that they were fogged. Moreover, when he repeated the sequence of events, he found the same results: the wrapped, unexposed plates always became fogged when left near the Crookes tube.

One night as he was leaving his laboratory Röntgen remembered that he had forgotten to switch off the Crookes tube. Returning to the room in the dark he noticed a glow coming from a sheet of paper on a nearby table. The paper was coated with barium platinocyanide, a substance known to give off a cold glow in a strong light. But there was no light. And the Crookes tube was covered by thin black cardboard!

Röntgen realized that the cause of the glow must be the same as that of the fogged photographs: invisible rays of an unknown type were coming from the Crookes tube.

Fig. 2.6 *Wilhelm Röntgen (1845–1923).*

He called them *X-rays*. He soon discovered their most startling property, for which they became best known—their ability to penetrate as easily through many objects as ordinary light passes through glass. We now know that X-rays are light of very short wavelength, so materials that are opaque to the longer wavelengths of visible light can easily transmit the shorter wavelength X-rays. The rays can pass through skin and tissue, casting a shadow only when they meet more solid bone.

One other important property of X-rays, which was soon discovered, is their ability to 'electrify' air—to make it a good conductor of electricity at normal pressures. J. J. Thomson in Cambridge was particularly interested in this phenomenon, known as *ionization*, and he investigated the effect thoroughly in 1896, with the assistance of the young Ernest Rutherford, of whom we shall hear more later. Together they found evidence that X-rays split the air into equal numbers of positively and negatively-charged atomic particles, or *ions*.

Later, once Thomson had established the existence of the electron, he could identify the negative particles with electrons, and he began to think of the positive ions as atoms with one or more of their electrons missing. This is indeed the case. A single 'parcel' of X-ray energy—an X-ray photon—can knock an electron out of an atom in air, or another medium, leaving behind a positive ion. The photon loses all its energy in the interaction and therefore ceases to exist—we say it is 'absorbed'; the electron and ion can later recombine. This process of ionization turns out to be of vital importance in particle physics, for it is a property not only of X-rays, but of energetic charged particles, such as electrons themselves. With particles, the details of the process are different; rather than surrendering all its energy in one interaction like a photon generally does, a high-energy particle loses a little of its energy at a time, creating a trail of ion-electron pairs in its wake. It is through making this ionization visible that we are able to 'see' the effects of the smallest constituents of matter.

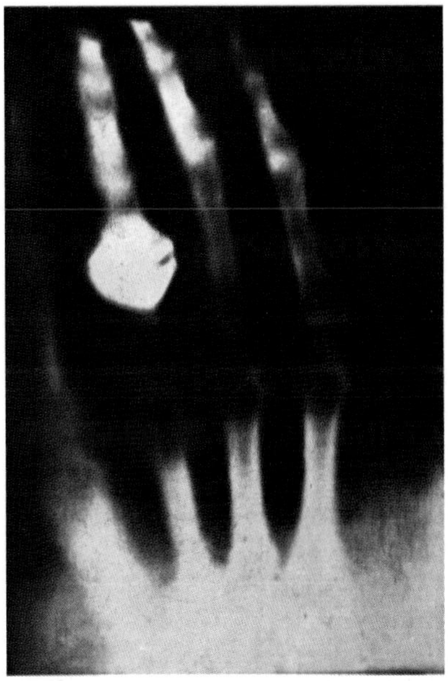

Fig. 2.7 *Röntgen's first X-ray photograph of a human shows the hand of his wife with the ring she was wearing.*

THE DISCOVERY OF RADIOACTIVITY

RÖNTGEN RECEIVED the first Nobel prize for physics, for his discovery of X-rays, in 1901. By that time, popular magazines had seized on the bizarre photographs showing the insides of things, revealing a world previously unseen. The public were as fascinated as the scientists; the prudish Victorians even worried about whether ladies could be seen naked beneath their several layers of petticoats! For the scientists, however, the question remained: where did the X-rays come from?

On 20 January 1896, Henri Poincaré put forward his ideas on the origins of X-rays at the French Academy of Sciences. The X-rays appeared to come from the fluorescent spot where the cathode rays hit the wall of the Crookes tube. This caused Poincaré to speculate that X-rays were not unique to Crookes tubes but were emitted by all fluorescent bodies—materials that glow on exposure to a strong source of light, such as the Sun. Within days many people were performing experiments with fluorescent materials—it was so easy. Wrap a photographic plate in black paper, put a piece of the chosen substance on top, lay it in some sunlight for a while and develop the plate.

Fig. 2.8 *The discovery of X-rays captured the public imagination, as this cartoon from about 1900 shows. It is titled 'Beach party à la Röntgen'.*

Fig. 2.9 *Henri Becquerel (1852–1908).*

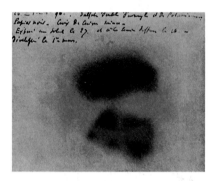

Fig. 2.10 *Becquerel's first evidence for radioactivity. These blurred images were formed on a photographic plate left for a few days under some uranium salts in a drawer, in February 1896.*

Fig. 2.11 *Pierre Curie (1859–1906) and Marie Curie (1867–1934). This magazine cover from 1904 portrays them at work in their laboratory.*

But Henri Becquerel had already had the same idea. Becquerel was then 44 years old, with a flowing beard tinged with white. He was a professor at the École Polytechnique and the Museum of Natural History in Paris, and came from a family of distinguished scientists. Several years before, he had been helping his father do an experiment involving a uranium salt, and had noticed that the crystals would glow for some time after they had been removed from sunlight. He decided to use the same salt in his X-ray experiments, hoping that it might be a powerful source of X-rays and leave a clear imprint on photographic plates.

On 26 February 1896, Becquerel prepared for his experiments with the uranium salt, but clouds shut out the Sun. So he put everything away in a drawer just as it was: plates wrapped in black paper; a metal cut-out pattern on top between the paper and the uranium to produce a shadow and leave no doubt as to the cause of the photographic image; and finally the all-important uranium salt—impotent without sunlight to stimulate it into action. Or so Becquerel thought.

The Sun did not come out for three days and on 1 March Becquerel decided to develop the plate anyway—presumably to prove that without the Sun there was very little effect. He was astonished to find instead a very clear image. The uranium salt evidently gave out invisible rays even in pitch darkness.

Becquerel soon found that the rays emanate from the uranium in the salts, and he formed images on photographic plates from samples of pure uranium metal. He found the rays to be similar to X-rays in some respects—they were invisible, would act on photographic plates and would ionize air, splitting it into positive and negative components that would conduct electricity. But the rays from uranium were different in two crucial ways. First, the uranium rays did not penetrate materials. Second, uranium and its compounds emitted the rays spontaneously. Day and night, for weeks on end—and as we now know for millennia—uranium gives out its invisible rays. X-rays, on the other hand, are produced only when cathode rays—high-energy electrons—strike a material, such as the glass at the end of a Crookes tube.

It is worth realizing that the quality of photographic materials in the 1890s was very variable. Several experiments that reported fogging from non-radioactive salts probably had bad materials. To detect the radiation from uranium one needed unblemished plates, great care, and skill. Becquerel succeeded by skill where others had failed through poor technique. Often Becquerel's discovery is described dismissively as 'accidental', as if he became rich by the arbitrary fortune of winning a lottery. This does him discredit; like Röntgen, he noticed the unexpected and pursued it. And like Röntgen and Thomson, Becquerel was rewarded with the Nobel prize, which he received in 1903. He shared it with two people whose names had by then become synonymous with radioactivity—Pierre and Marie Curie.

Five years before the discovery of radioactivity, Manya (Marie) Sklodowska had come to Paris from her native Poland, intent on becoming a scientist. Her life was a hard one, and she earned her living by cleaning apparatus at the Sorbonne, giving lessons and accumulating just enough to rent a small room in an attic. She was top student at the university and soon after graduating in 1895 she married Pierre Curie, a professor of physics, and started work in her husband's laboratory. Once the Curies heard of Becquerel's discoveries, Marie decided to investigate the new kind of radiation for her doctoral thesis. In particular, she wanted to know if uranium was the only element that emitted the rays and to quantify the amount of radiation emitted by different substances.

Photography could show, by the density of dark spots forming on a plate, whether the radioactivity in a sample was weaker or stronger. But the technique was not very accurate, and Marie decided that it would be better to measure the amount of radioactivity by another method. She did this by making a small capacitor—two metal plates separated by air. The idea was to charge up one plate and connect the other to ground. Dry air is a poor conductor of electricity and so the capacitor should normally pass very little current; but when uranium was sprinkled on one plate, its rays would ionize the air between and a current could flow. The more intense the rays, the larger the current, which Marie could measure by a sensitive technique invented by her husband.

Seizième année. — N° 779.   Huit pages : CINQ centimes   Dimanche 10 Janvier 1904.

Le Petit Parisien

SUPPLÉMENT LITTÉRAIRE ILLUSTRÉ

TOUS LES JOURS
Le Petit Parisien
(six pages)
5 centimes

CHAQUE SEMAINE
LE SUPPLÉMENT LITTÉRAIRE
5 centimes

DIRECTION: 18, rue d'Enghien (10e), PARIS

ABONNEMENTS

PARIS ET DÉPARTEMENTS:
12 mois, 4 fr. 50. 6 mois, 2 fr. 25

UNION POSTALE:
12 mois, 5 fr. 50. 6 mois, 3 fr

UNE NOUVELLE DÉCOUVERTE. — LE RADIUM

M. ET Mme CURIE DANS LEUR LABORATOIRE

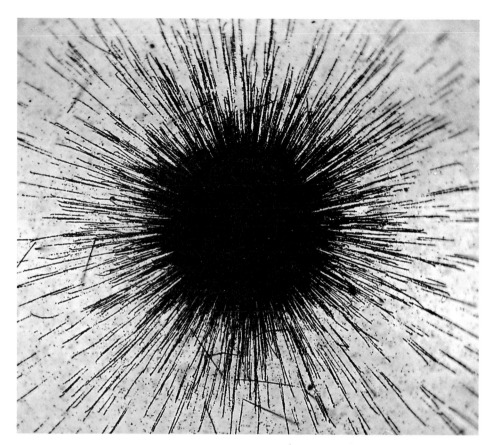

Fig. 2.12 *Alpha particles shoot out from a speck of radium salt on the surface of a photographic plate covered with a special emulsion. The electrically charged alphas leave tracks in the emulsion which appear as dark lines on the negative image formed on the developed plate. (The central blob is about a tenth of a millimetre across.)*

Fig. 2.13 *The words accompanying this advert for 'Radon Water' extol the water's revitalizing effect: 'Radon Water keeps the uric acid compounds in their easily soluble forms (carbonic acid and ammonia) which, being gases, are promptly expelled through the system by natural process.' The idea was to use a bulb containing 'a fixed quantity of the essential radioactive element which is the source of Radon' with a standard syphon.*

'Larger' is perhaps a misleading word here: the currents were only billionths of an ampere! But even so they were detectable. Marie tested a vast number of materials and found effects from only one element apart from uranium—thorium. She also measured the amount of radioactivity for several chemically pure compounds of uranium. The current was always directly in proportion with the percentage of uranium in the sample; the radioactivity from a substance containing 10 per cent uranium was only one tenth that of pure uranium. But Marie also found to her surprise that raw, impure uranium ores showed more radioactivity than she could explain in terms of the uranium they contained. She suspected that the raw minerals must contain something over and above uranium, a more powerful emitter yet.

From one tonne of the uranium ore known as pitchblende, the Curies extracted a few grams of the culprits during 1898. Two new radioactive elements emerged: polonium, named after Marie's native Poland, and radium, the most powerful radioactive substance known.

Radium emits a million times more intensely than uranium—as the impact of a bullet is to the brush of a feather. Every sample of radium pours out energy in torrents. Crystals containing radium glow so brightly that they allow you to read in an otherwise dark room. Radium can burn the skin as Pierre Curie found to his cost. Uranium leaves an imprint in photographs after hours; radium gives one instantly. When better quality photographic emulsions were developed in the early 1900s, it became possible to photograph directly the radiations from radium.

Luminous watch dials, chorus girls wearing radium-painted costumes on a darkened stage, and gambling with luminous chips were but some of the many exploitations of the substance. The power in radioactivity was staggering. The press reported that the energy contained within one gram of radium, released by radioactivity, would drive a 50-horsepower ship all the way around the world at 50 km/hour; and so it would, were it possible to convert the energy to power with 100 per cent efficiency. The radiation could also kill cancer cells, and the media announced that a cure for cancer was at hand.

Marie Curie welcomed such euphoria and started up the famous Radium Institute in Paris. Only later did people begin to notice that in addition to destroying isolated cancer cells, radium and radioactivity could cause cancerous mutations of healthy tissues. Several workers died. Pierre was tragically killed in a road accident in 1906, but he already had severe symptoms almost certainly due to his prolonged unprotected contact with the radiations.

RUTHERFORD AND RADIOACTIVITY

ALTHOUGH BECQUEREL discovered radioactivity and the Curies isolated radium, it was Ernest Rutherford who put their findings to use as a scientific tool. He used the new radiations as a form of artillery with which to bombard atoms and probe their inner secrets. From these investigations, first as a research student in Cambridge in the 1890s, then as professor at McGill University in Canada, and later in Manchester, he cracked the secrets of the atom and showed it to be similar in some respects to a miniature solar system. In 1919, he finally returned to Cambridge, succeeding his mentor, Sir J. J. Thomson, as Cavendish Professor and the greatest experimental physicist of the day.

Rutherford's first researches into radioactivity were carried out while he was still a young research student at Cambridge, where he had arrived in 1895 from his native New Zealand. It was here in 1897-8 that he discovered two components of radio-activity—the alpha and beta rays.

Rutherford had been working with Thomson on the ionizing effects of X-rays in gases; in particular they had studied the conduction of electric currents through ionized gas. With the discovery of radioactivity, Rutherford turned to studying the ionization produced by the radiation from uranium; but not for long, for he soon became far more interested in the radiation itself, and began to use the ionization of gases as a means of studying radioactivity, rather than the other way about.

The instrument that Rutherford generally used for these investigations was an electrometer. The details of operation vary from one design to another, but the basic principle is to measure the deflection of a charged metallic strip in an electric field. If the air around the strip becomes ionized, the charge leaks away—a current flows— and the strip moves. Rutherford could measure the rate of leakage, and hence the amount of ionization, by timing the movement: the faster the leakage, the more the ionization and the stronger the radiation.

Fig. 2.14 *Ernest Rutherford (1871–1937), when he was a student in New Zealand in 1892.*

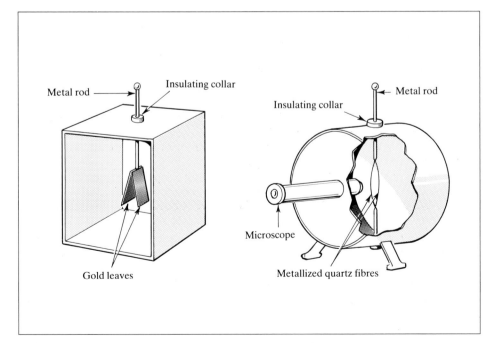

Fig. 2.15 *The gold-leaf electroscope (left) was one of the earliest instruments used in studying electrical phenomena. It consists basically of a box with a window. A metal rod, which passes through an insulating collar in the top of the box, has two thin sheets of gold foil attached. When the rod is electrically charged, the two gold leaves acquire the same charge and repel each other. If the air in the box is ionized (for example by radiation), the charge on the leaves becomes gradually neutralized and the leaves collapse together. More advanced instruments, which allow amounts of electricity to be measured accurately, are called electrometers. The device shown here (right) is of the kind designed by Theodor Wulf (p.68). It contains two metallized quartz fibres held under tension, which repel each other when charged. The degree of separation between the wires can be measured by the microscope attachment.*

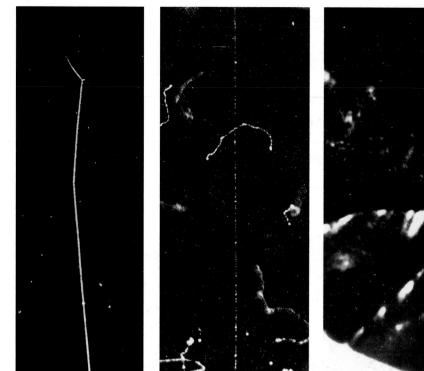
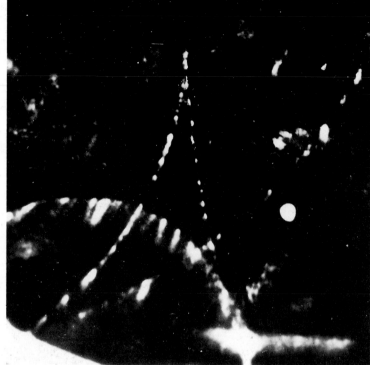

Figs. 2.16–2.18 *Rutherford discovered that radioactive materials emit three distinct types of radiation—alpha, beta, and gamma—which produce different characteristic tracks in a cloud chamber.*

Fig. 2.16 *(left) The final portion of the track of an alpha particle. The track changes direction where the alpha collides with atoms in the air inside the chamber. Finally, close to the end, the track becomes fainter as the positively-charged alpha particle captures electrons, losing its charge and hence its ability to ionize.*

Fig. 2.17 *(centre) Electrons—beta rays—have a much smaller mass than alpha particles and so have far higher velocities for the same energy. This means that fast electrons do not lose energy so readily in ionizing the atoms they pass. Here we see the intermittent track of a fast beta-ray electron. (The short thick tracks are not caused by the beta ray; they are due to other electrons knocked from atoms in the gas filling the chamber by invisible X-rays. Their tracks are thicker because they are moving more slowly than the beta ray and are therefore more ionizing; and they wiggle about because they are frequently knocked aside in elastic collisions with electrons in the gas atoms.)*

Fig. 2.18 *(right) Gamma rays are non-ionizing and therefore leave no tracks in a cloud chamber. However, they can convert into equal amounts of matter and antimatter if they have high enough energy. Here an invisible gamma ray from a radioactive source materializes as an electron and an antielectron (a positron); being oppositely charged, they curl away in opposite directions in the cloud chamber's magnetic field.*

In the course of his studies, Rutherford covered a sample of uranium with sheets of aluminium foil, which absorbed the radiation. As he increased the thickness of foil to around a hundredth of a millimetre, he found that less and less radiation penetrated. This much Rutherford expected—the radiation is progressively absorbed. But as he increased the thickness still more, he was surprised to find that the radiation appeared to maintain its intensity. Only when he added much more covering did he find that the radiation was indeed diminishing, though very slowly; it took several millimetres of aluminium to absorb the remaining radiation completely. Rutherford concluded that there were in fact two types of radiation. One of these was absorbed very quickly; the other was a low-intensity but more penetrating radiation. He named the easily-absorbed radiation 'alpha' and the penetrating component 'beta'.

Rutherford later looked at the radiation from thorium, and found alpha and beta rays like those from uranium. But in addition he discovered an extremely penetrating radiation, more penetrating than even the beta variety. This became known as gamma radiation.

In 1898, Rutherford moved from Cambridge to McGill University in Montreal. His first work there was to measure the energy output in radioactivity from uranium. He was astonished to discover that it was far greater than in any known chemical process. Here was the first glimpse of the latent power that lies deep within the heart of the atom. And in 1900, Rutherford suggested that the energy might be released in the regrouping of constituents within the atom. This was a remarkable assertion to make at a time when the structure of atoms had hardly begun to be explained.

At about the same time, Rutherford became intrigued by some odd behaviour in the radioactivity of thorium compounds: the amount of radiation seemed to vary and to be very sensitive to currents of air. After a detailed series of experiments, Rutherford came to the conclusion that the thorium was emitting something that was also radioactive. He referred to the unknown substance as 'an emanation', and found that it remained radioactive for only a short time, quite unlike thorium or uranium. Rutherford was convinced that the emanation was a gas, but he needed the assistance of a chemist to analyse it properly. To this end he enlisted the talents of the young Frederick Soddy, newly arrived at McGill from Oxford.

Rutherford the physicist and Soddy the chemist together made a formidable team. In a series of detailed investigations which involved the careful chemical separation

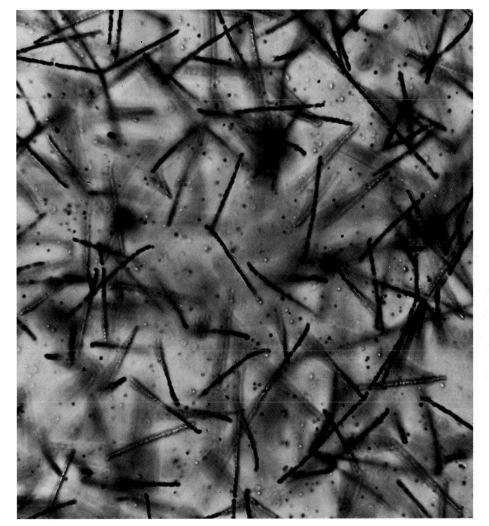

Fig 2.19 *A photographic emulsion impregnated with a radioactive thorium salt reveals the tracks of alpha particles when it is developed. As Rutherford and Soddy found, the decay of thorium initiates a chain of decays. First one radioactive element is formed, then that decays to another, and so on. This is why several alphas can be seen emerging from the same spot: they are the result of successive stages in the decay of a single nucleus. The length of the tracks in this negative print is about 0.03 mm.*

Fig. 2.20 *Rutherford at McGill University in 1901, with some of his students.*

of different radioactive materials, they found conclusive proof that the emanation was not only a gas, but a new element, chemically quite different from thorium and more akin to the so-called inert gases, such as argon. This gas is now known as radon; with its discovery, Rutherford and Soddy had found the first amazing evidence of one element transmuting into another. This was alchemy at work, but naturally.

Still more surprises were in store. Soddy showed that he could separate out from thorium compounds a substance which appeared to contain much of the original thorium's radioactivity and which produced all the emanation. This material, which they called thorium-X, was again chemically quite different from thorium; it proved, eventually, to be a form of radium. The sequence of events that Rutherford and Soddy discovered is that thorium transmutes first to this form of radium, which in turn changes into radon, at each stage spitting out radiation. (The full sequence of events is in fact still more complicated!)

This last step in the chain of reasoning—that radiation is a direct result of transmutation—was possible once Rutherford had shown in 1902 that alpha rays, the main form of radiation from the thorium chain, are particles of matter. Other researchers had discovered that the beta rays consist of negatively-charged particles, which were eventually identified as electrons. But alpha rays appeared at first to be undeflected by electric and magnetic fields. Rutherford, however, succeeded where others had failed and managed to deviate alpha rays in strong fields. These experiments showed that the rays consist of positively-charged particles, which must be thousands of times heavier than electrons. Here was conclusive proof that heavy atoms change from one type to a slightly lighter type by ejecting tiny atomic fragments.

Fig. 2.21 *Frederick Soddy (1877–1956).*

RUTHERFORD AND THE ATOM

THE DISCOVERY of radioactive transmutations was a crucial step towards dispelling the notion of atoms as indivisible building bricks. Thomson's work at Cambridge had already shown the cathode rays to be tiny negatively-charged fragments of matter—electrons. And in 1903, the year that Rutherford and Soddy published their dramatic conclusions, Thomson put forward a model of the atom in which he thought of the neutral atom as composed of electrons embedded in a positively-charged sphere, like plums in a pudding. Rutherford was later—in 1914—to describe Thomson's model as 'not worth a damn'. By that time he had performed a series of brilliantly simple experiments that showed him just what an atom looks like.

In 1907, Rutherford moved from Montreal to become professor of physics at Manchester University. There he attracted all sorts of people to work with him. His origins in New Zealand freed him from the class-consciousness of Edwardian England. The brilliant and the best came from the north of England, and others joined from overseas. In Rutherford's team we find the makings of the first international research group—the norm today but a very new idea in the first decade of this century.

Among those in Rutherford's team at Manchester was a young German, Hans Geiger, who is famed today for the 'Geiger counter', which he was to develop in the 1920s. Using a prototype of that counter, he and Rutherford were able to go still further towards discovering the nature of the alpha rays. Rutherford suspected that the positive particles he had detected at McGill were charged atoms—ions—of helium, the second lightest element after hydrogen. This idea was strengthened by the discovery of helium gas in association with radium. To settle the question, Rutherford needed to be able to detect alpha particles one at a time, so that he could measure their exact charge and mass.

The apparatus Rutherford and Geiger used to tackle this problem was located in a gloomy cellar in the physics department at Manchester. Their detector exploited the fact that radiation passing through a gas at low pressure and in a strong electric field produces a greatly enhanced amount of ionization. They used a brass tube some 60 cm long, with a thin wire passing along the centre, and pumped it out to a low pressure. The wire and tube were insulated from each other and 1000 volts applied between them. This voltage set up an electric field, which became much stronger nearer the wire. Ions created by the passage of alpha particles would be attracted towards the wire; where the field became stronger nearer the wire, the ions would move faster and in turn ionize more of the gas, amplifying the initial effect. One ion could produce thousands of ions, which would all end up at the central wire. There they would produce a pulse of electric charge large enough to be detected by a sensitive electrometer connected to the wire.

Geiger and Rutherford used their device—which nowadays we would call a 'proportional counter'—to count individual alpha particles coming down a narrow tube from a thin film of radium. From this, they could calculate how many alpha particles were radiated by the whole sample of radium, and compare the answer with the total electric charge emitted, which could be measured with the electrometer. The calculations showed that the charge of the alpha particle is twice that of the hydrogen ion, and this in turn indicated almost certainly that the alphas are helium ions.

Rutherford realized that he could not prove that single alpha particles entered the wire detector, so with characteristic thoroughness he looked for an alternative technique. The vital ingredient came in a letter from Otto Hahn in Berlin, who had worked with Rutherford in Canada. One of Hahn's colleagues, Eric Regener, had been detecting alpha particles by letting them hit a screen coated in zinc sulphide. When a particle struck, the screen emitted a flash of light—a phenomenon known as *scintillation*. Inspired by these results, Rutherford and Geiger built improved zinc-sulphide screens and were astonished to find the technique every bit as good as the electrical methods they were using. By combining the two techniques, they were able to prove that they were detecting individual alpha particles. A characteristic flash on a scintillation screen announced the arrival of an alpha particle at the wire detector and deflected the electrometer.

Fig. 2.22 *Rutherford (right) and Hans Geiger (1882–1945) counting alpha particles in their laboratory at Manchester University.*

Later in the same year, 1908, Rutherford confirmed that alpha particles are helium ions by collecting some in a tube, and allowing them to neutralize themselves by picking up electrons from their surroundings. In this way he collected a gas, which he could stimulate into emitting light, in the same manner as in a sodium lamp. The spectrum of this light provided a fingerprint that identified the gas as helium without a doubt. Rutherford announced this result in his speech when he was awarded the Nobel prize in 1908—not for physics, but for chemistry—for his work with Soddy on transmutations.

It is but a small step from Rutherford's findings to our modern view of alpha particles as the *nuclei* of helium atoms. There was however no way that Rutherford could make that last step in 1908 as the idea of the nucleus was still unknown. That was to be his next dramatic contribution: deducing what the atom looks like inside, and creating a picture that has survived the tests of the past 70 years.

The scintillation screen had proved crucial in Rutherford's work on alpha particles, and he was so impressed that he made it his primary research tool, and dropped the electrical methods he had used since his early days at Cambridge. The technique was soon put to work to help him solve a problem that had puzzled him for some time. At McGill he had noticed that when a beam of alpha particles passed through thin sheets of mica, they produced a fuzzy image on a photographic plate. The alphas were being scattered by the mica and deflected from their line of flight.

This result was a surprise because the alphas were moving at 15 000 kilometres per second, or one-twentieth the speed of light, and had an enormous energy for their size. Strong electric or magnetic forces could deviate the alphas' flight paths a little, but nothing like as much as when they passed through a few micrometres (millionths of a metre) of matter. This suggested that there must be unimaginably powerful forces at work within the atoms of the mica sheet.

In 1909, Rutherford assigned to Ernest Marsden, a young student of Geiger's, the task of discovering if any alphas were deflected through very large angles. Marsden used gold leaf rather than mica, and a scintillating screen to detect the scattered alphas. He could move the screen not only behind the gold foil, but to the sides, and round next to the radioactive source itself. This way he could detect alphas scattered through all angles.

The effort in counting the scintillations was enormous. Individual flashes were observed through a low-power microscope focused on the screen, one small patch at a time, to reveal the number of alphas scattered through a particular angle. The flashes were faint and sparse and could be counted by eye only in a suitably darkened room. This placed a great strain on the eye muscles and also on the observer's powers of concentration. Work could continue for only a few minutes at a stretch, so Rutherford, the master, assisted Marsden, the student. One watched and one recorded for a few minutes only and then changed places with the other.

To the amazement of Rutherford and Geiger, Marsden discovered that 1 in 20 000 alphas bounced right back from whence they had come, to strike the screen when it was next to the source. This was an incredible result. Alpha particles, which were hardly deflected at all by the strongest electrical forces then known, could be turned right around by a thin gold sheet a few hundreds of atoms thick! No wonder that Rutherford exclaimed, 'It was as though you had fired a 15-inch shell at a piece of tissue paper and it had bounced back and hit you'.

At first neither Marsden, Geiger, nor even Rutherford could understand these results at all. Geiger took up some earlier work again, and Marsden left the team for a while to do some research on the atmosphere at a meteorological station; but Rutherford kept on puzzling, his normal output of scientific papers falling dramatically. Then, late in 1910, with the aid of a very simple calculation, Rutherford at last saw the meaning of the results. The key was that he knew the energy of the incoming alphas, which were all emitted at the same speed. He also knew that each alpha particle carries a double dose of positive charge. The positive charge within the atom repels the alphas, slowing them to a halt as they approach and eventually deflecting them; the closer the alphas approach the more they are deflected, until in extreme cases they are turned round in their tracks.

Rutherford could calculate just how close to the positive charge the alphas should get, and the result astonished him—in his notebooks we can see his writing change

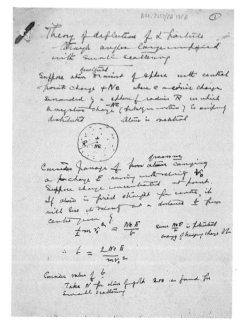 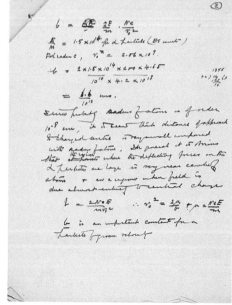

Fig. 2.23 These pages from Rutherford's notebook show where he calculates how close an alpha particle must approach the positive charge within an atom if it is to be turned back completely in its tracks. The answer, 6.6×10^{-12} cm, astonished Rutherford—you can see his writing change—for it showed that the positive charge is concentrated in a tiny core deep within the atom. This is the moment the atomic nucleus was 'discovered'.

as the significance of his calculation hits him. The alpha particles can on rare occasions come to within 10^{-12} cm of the atom's centre, one ten thousandth of the atom's radius, before they are turned back. This means that the positive charge is found *only* at the very centre of the whole atom, not distributed throughout the atom as Thomson had proposed in his plum-pudding model.

Rutherford had discovered that atoms consist of a compact positively-charged nucleus, around which circulate the negative electrons at a relatively large distance. The nucleus occupies less than one thousand million millionth (10^{-15}) of the atomic volume, but it contains almost all the atom's mass. If an atom were the size of the Earth, then the nucleus would be the size of a football stadium. The atom's volume is mostly empty space.

These experiments brought Rutherford to the inescapable conclusion that the atom resembles a solar system in miniature. But it was the Danish physicist Niels Bohr who developed the theoretical aspects of the 'planetary atom' in detail. Bohr went to Cambridge in 1911 to work under J. J. Thomson, after finishing his PhD on the theory of electrons in metals, but he soon encountered Rutherford during one of his visits from Manchester. Bohr was so impressed that in March 1912 he moved to Manchester, where he learned all about Rutherford's concept of the atom. In particular, he began to consider the solar system atom from the point of view of the electrons.

The idea of electrons whirling round a nucleus seemed at first to suffer one fatal flaw. In their circular motion, the electrons should, according to the laws of electromagnetism, continuously radiate electromagnetic energy. This is manifestly not the case, and to overcome this problem Bohr turned to the new ideas of quantum physics. He proposed that electrons in atoms possess only fixed amounts of energy, related to distinct orbits about the nucleus. This means that an atomic electron cannot gradually lose energy through continuous radiation, but must emit it only in lumps—or 'quanta'—of fixed size. With this simple model of the atom, Bohr could account for the lines observed in the spectra from elements such as hydrogen and sodium; in so doing, he paved the way for our modern understanding of the atom.

Rutherford's discovery that the positive charge of an atom is concentrated in a central nucleus made him wonder what precisely carries the charge. The negative charge of the atom is carried by the tiny electrons; are similar objects responsible for the positive charge of the nucleus?

The experiments with alpha particles and gold foil had shown that the nucleus is small; the positive charge in the atom is concentrated at its centre. But atoms of gold contain 79 electrons, and their total negative charge must be balanced by an equally large positive charge in the nucleus. This charge is many times greater than that of an alpha particle, and so an encroaching alpha is repelled long before it reaches the nucleus at the heart of a gold atom. The experiments at Manchester had penetrated the atom to reveal the nucleus, but had not probed the nucleus itself. Rutherford realized that for light atoms, with only a few electrons, the situation should be different. There the nuclear charge must be smaller, the repulsive force less powerful, so the alpha particles should make a closer approach. Hydrogen is the lightest element of all, with only one electron, so its nucleus should be the smallest and least repulsive electrically to alpha particles.

Rutherford and Marsden turned to firing alpha particles through hydrogen. After losing energy in successive collisions with hydrogen molecules, the alphas would all come to a halt at more or less the same distance from the radioactive source, as they were all emitted with the same energy. So, beyond a certain distance—the 'range' of the alphas in hydrogen—the particles could no longer penetrate the gas to strike a zinc sulphide screen.

But Marsden and Rutherford found that a few scintillations *did* occur beyond the range of the alphas, and that whatever produced them could travel up to four times farther through the gas. The scintillations had to be caused by charged particles, and deflection by a magnetic field showed that they were positive. Rutherford argued that the new particles could be nothing other than the nuclei of hydrogen, knocked out from their place in atoms in the gas by the energetic alphas. The hydrogen nuclei, each carrying only a single positive charge compared with the double charge of the

Fig. 2.24 *Niels Bohr (1885–1962).*

Fig. 2.25 *Alpha particles of the same energy have the same range, and radioactive materials emit alphas at one or more specific energies. Here the majority of alphas from a source of thorium C′ (polonium-212) travel 8.6 cm in a cloud chamber filled with air before stopping, while a single higher-energy alpha travels 11.5 cm.*

Fig. 2.26 *Rutherford and research students at the Cavendish Laboratory in 1920. On Rutherford's right is his predecessor and mentor, J. J. Thomson.*

The full complement (left to right) are:

Back row: A. L. McAulay, C. J. Power, G. Shearer, Miss Slater, Miss Craies, P. J. Nolan, F. P. Slater, G. H. Henderson, C. D. Ellis.

Middle row: J. Chadwick, G. P. Thomson, G. Stead, J. J. Thomson, E. Rutherford, J. A. Crowther, A. H. Compton, E. V. Appleton.

Front row: A. Muller, Y. Ishida, A. R. McLeod, P. Burbidge, T. Shimizu, B. F. I. Schonland.

Chadwick, Compton, Appleton, and G. P. Thomson (J. J.'s son) were all destined to win the Nobel prize.

alpha particles (helium nuclei) could travel four times as far through the gas. Rutherford at first called the hydrogen nuclei 'H-particles'. Today we call them protons—the particles that carry the positive charge in nuclei. But it took a few more years for Rutherford to establish that the nucleus of hydrogen is a fundamental building block that occurs in the nuclei of all atoms.

The first evidence came in some experiments in which Marsden was measuring the distances that alpha particles travel in air, between a source and a zinc sulphide screen. He noticed some particles with particularly long ranges, and wondered if they too could be H-particles. This was in 1914, and shortly afterwards the First World War began. Geiger had gone back to Germany, Marsden departed to become a professor in New Zealand, and many of the students went off to the war. Rutherford, for his part, became involved in work on submarine detection for the Board of Inventions and Research, under the auspices of the Admiralty. But he was able to continue with a little research, in his spare time, more or less on his own.

It was not until 1917 that Rutherford finally decided that the long-range particles that Marsden had observed were indeed H-particles, chipped off nitrogen atoms in the air in the detector. He came to this conclusion only after three years of patient study, in which he eliminated all the other possible identities for the penetrating particles. Using alpha particles as bullets, Rutherford managed to knock H-particles—hydrogen nuclei—out of the atoms of six elements: boron, fluorine, sodium, aluminium, phosphorus, and nitrogen. In 1919, he announced his discoveries and named the particles *protons*, from the Greek for 'first', for they were the first identified building blocks of the nuclei of all elements.

WILSON AND THE CLOUD CHAMBER

IN 1918, J. J. Thomson retired as Cavendish Professor at Cambridge, and Rutherford left Manchester to take the place of his old master. By now Rutherford was nearing his fifties, and at the Cavendish he began increasingly to direct the researches of the new generation of young scientists around him, rather than to do experiments himself. One area to be researched was the disintegration of elements, which he and Marsden had uncovered at Manchester. This task he assigned to a newly-graduated student of physics, Patrick Blackett, and he suggested using a remarkble new tool that had been invented at Cambridge—the cloud chamber.

We are all familiar with the vapour trails of a jet aircraft, which can remain in the sky for several minutes, providing a record of the plane's movements. The trails consist of fine water droplets condensed on the jet engine's exhaust fumes, creating a long thin cloud. This same principle underlies the cloud chamber, the first device to produce visual images of particle tracks.

The story goes back to September 1894, when Charles Wilson, a young research physicist at Cambridge, was working in the meteorological observatory on the summit of Ben Nevis. While there he was struck by the beauty of coronas (coloured rays around the Sun) and glories, where the Sun glows around shadows in the mist. So startling was the experience that Wilson decided to investigate these natural phenomena back at the Cavendish Laboratory in Cambridge. To do so he needed a ready-made mist, so he built a glass chamber fitted with a piston and filled with water vapour. When he withdrew the piston quickly, the sudden expansion cooled the gas so that a mist formed in the cold damp atmosphere.

In the course of his investigations, Wilson found that if he made repeated small expansions of the chamber—without allowing in fresh air—the mist would disappear. He could explain this because he knew the droplets in the mist formed on specks of dust; on repeated expansions the droplets would slowly move by gravity towards the bottom of the chamber and so gradually remove all the dust, and there would be nothing left on which the mist droplets could form. The surprise came when Wilson, having already cleared the chamber of dust in this way, then made very large expansions. He found that these large expansions always produced a thin mist, no matter how many times he expanded the chamber. But what could the droplets be

Fig. 2.27 *Charles (C. T. R.) Wilson (1869–1959).*

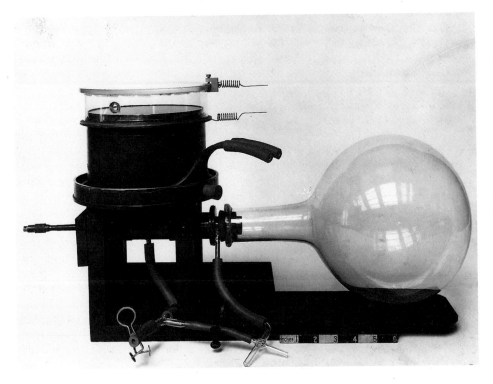

Fig. 2.28 *Wilson's first cloud chamber. The chamber itself is the squat glass cylinder at the top left of the picture; the coils are where an electric field was applied to clear away stray ions between expansions. Below the chamber is a cylinder containing the piston. The glass bulb to the right was pumped out to a low pressure. When a valve between the bulb and the chamber beneath the piston was opened, air would rush into the bulb, causing the piston to fall and the air in the glass cylinder to expand suddenly. Water vapour in the air would then condense out on any ions present, so making the ionized tracks of particles visible.*

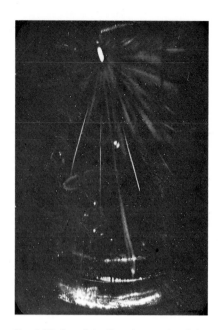

Fig. 2.29 *One of the first photographs of the tracks of ionizing particles in a cloud chamber, obtained by Wilson early in 1911. The tracks are due to alpha particles emitted by a small amount of radium on the tip of a metal tongue inserted into the chamber.*

forming around? Wilson surmised—correctly as we now know—that the mysterious drops were condensing on the electrically-charged particles, or ions, known to cause conductivity in the atmosphere.

Wilson was soon able to test his theory with the newly discovered X-rays. X-rays ionize air, so if Wilson was correct they should generate water drops upon passing through a cloud chamber. Early in 1896, Wilson exposed his primitive cloud chamber to X-rays and saw an immense increase of rainlike condensation. This left no doubt that he was making ions in the air 'visible'. But at this point Wilson abandoned his cloud chamber experiments, side-tracked by another experience on a mountain-top—a thunderstorm in which his hair literally rose on end—and he spent the next few years developing a fundamental understanding of atmospheric electricity.

It was not until 1910 that Wilson returned to his work on the cloud chamber. Then he passed alpha and beta radiation through the device and saw for the first time the tracks of the individual particles, which he described as 'little wisps and threads of clouds'. Cloud drops formed instantly around the ions produced by the radiation and when illuminated the tracks stood out like dust motes in a sunbeam.

Wilson's cloud chamber provided the first visible records of the motion of particles smaller than an atom, and he was rewarded with the Nobel prize in 1927. Meanwhile at the Cavendish Laboratory, Wilson's technique had fallen into the hands of a man with a passion for gadgets—Patrick Blackett.

The scintillation technique with which Rutherford and Marsden had first observed nuclear disintegrations gave only a limited amount of information about the proton knocked out of the target nucleus; it could not reveal the recoil of the bombarding alpha particle or the residual nuclear fragment. The cloud chamber, on the other hand, offered the possibility of recording the tracks of all the players in the game. Blackett adapted Wilson's basic technique and devised a chamber that expanded automatically every 10 to 15 seconds and took a picture on ordinary cine film.

From 1921 to 1924, he obtained more than 23 000 photographs of alpha particles bombarding nitrogen in a cloud chamber, with 20 or so tracks per image. In most of these the alpha particles shot through the diffuse nitrogen gas without interruption; and in several pictures, the nitrogen nuclei had deflected the motion of the alpha particles, the nucleus and the alpha particle bouncing off each other like billiard balls in flight.

But most exciting of all were eight precious examples where the forked tracks had quite a different appearence from those in a normal elastic 'billiard ball' type of collision. In each one, the track of a proton was clearly visible, straight and fine, with less ionization than the alpha tracks; and a short, stubby track, similar to that of the nitrogen nucleus, was also apparent. But there was no sign of the recoiling alpha particle. The conclusion was that this had become bound to the chipped nucleus, to make a nucleus of a form of oxygen. The alpha particle had modified the nitrogen nucleus; nuclear transmutation had been captured on film.

Fig. 2.30 *Patrick Blackett (1897–1974).*

Fig. 2.31 *One of Blackett's eight examples of a nuclear transmutation induced by an alpha particle, which he found among some 23 000 cloud chamber photographs taken in 1925. Alpha particles from a radioactive source stream out from the bottom of the picture. Most travel the full length of their range, but the one on the far left interacts with a nitrogen atom in the air filling the chamber. The alpha is in fact captured, and a nucleus of a heavy isotope of oxygen—*^{17}O—*is formed, accompanied by a lone proton. The proton shoots off to the right, leaving a faint track; the recoiling oxygen nucleus leaves the thick track to the left, and collides again before coming to a halt not far from the scene of its formation.*

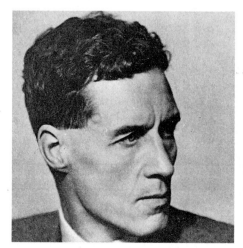

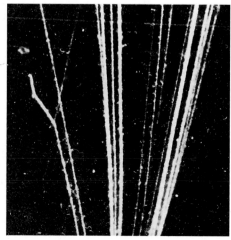

THE DISCOVERY OF THE NEUTRON

BY THE 1920s, the role of protons in carrying the positive charge of the nucleus had become firmly established. But the mystery of the nature of the atomic nucleus was not completely solved. The nucleus contains most of the atom's mass, as well as its positive charge, so the protons should presumably account for all this mass: a nucleus with twice the charge of another should have twice the number of protons and twice the mass. But this is not so. Nuclei have at least double the mass expected from the number of protons suggested by the total charge.

To account for this discrepancy, Rutherford speculated in 1920 that there are electrically neutral partners of the protons within nuclei—'neutrons'. But he was alone in this idea. The picture that physicists generally accepted was of a nucleus containing protons *and* electrons. The theory was that the nucleus contains twice as many protons as there are orbiting electrons: half the protons are neutralized by the orbiting electrons, while the other half are neutralized by electrons *inside* the nucleus. The phenomenon of beta decay, in which electrons are emitted from the nucleus, gave strong support to this notion.

The first indications that Rutherford might be right came in 1930 in experiments by Walther Bothe and Hans Becker in Germany, though they did not realize the significance of their work. They bombarded beryllium using alpha particles from polonium—an alpha source free from gammas—and they observed the emission of an extremely penetrating neutral radiation, which they assumed was gamma rays.

This work was followed up by Irène Curie, the daughter of Marie and Pierre Curie, and her husband Frédéric Joliot. They found the same neutral radiation, and observed that it had the power to knock protons out of paraffin wax—a substance rich in hydrogen—which they were using in an attempt to absorb the novel radiation. This should have been the clue that the radiation consisted of Rutherford's proposed neutrons, but the Joliot-Curies failed to recognize the signs. They believed the radiation to be gamma rays, although they were surprised by how readily it could scatter heavy

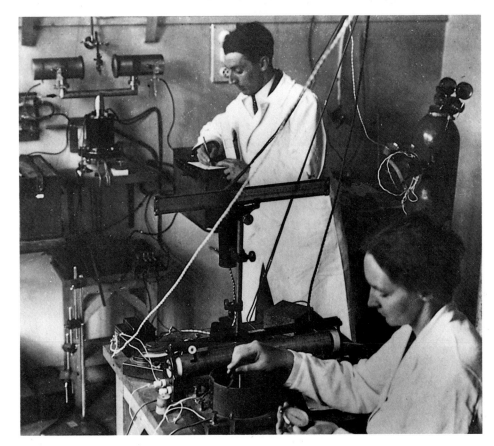

Fig. 2.32 *Irène Curie (1897–1956) and her husband Frédéric Joliot (1900–1958). They narrowly missed discovering the neutron, but in 1933 showed for the first time that by bombarding materials with radiation they could create new radioactive elements that do not occur naturally on Earth.*

Fig. 2.33 *Rutherford (with cigarette) and John Ratcliffe in 1932. Another member of the Cavendish Laboratory, Vivian Bowden, built the sign because one of the leads (coming in from the right of the picture) from the ionization chamber to the amplifier trolley (left) was sensitive to sound, including Rutherford's renowned loud voice. The open door leads to Chadwick's laboratory in which he discovered the neutron.*

protons. It should have been very much harder than knocking lightweight electrons out of atoms; instead it seemed to be just as easy.

The Joliot-Curies published their results in January 1932, and their paper had an immediate impact at Rutherford's Cavendish Laboratory. James Chadwick had worked with Rutherford at Manchester and had then gone to work with Geiger in Berlin, but had been interned during the First World War. In 1919, on his return to England, he rejoined Rutherford at Cambridge and worked on the transmutation of elements with alpha particles. Unlike the Joliot-Curies, Chadwick was prepared mentally for the existence of the neutron—indeed he had already made several unsuccessful attempts to search for neutrons. He realized at once that the neutral radiation from the Joliot-Curies' beryllium was not gamma radiation at all, but neutrons.

To prove this, Chadwick immediately set about similar experiments, but in addition he collided the neutral rays with a variety of gases—hydrogen, helium, and nitrogen. In this way, he could observe the differing amounts by which the atoms in the various gases recoiled. The heavier the atoms, the less they recoiled, and from this Chadwick could calculate the mass of the neutral projectiles themselves. They had more or less exactly the same mass as the proton; gamma rays, by contrast, have no mass. Thus it became clear that nuclei consist not only of positively-charged protons, but also of neutral neutrons.

Being neutral, neutrons freely penetrate the electric fields of atoms and nuclei. This is in complete contrast to protons, which feel an intense repulsive force as they approach positively-charged nuclei. But in the same year as the neutron's discovery another major revolution was taking place. For the first time scientists discovered how to speed up the subatomic bullets and smash nuclei apart artificially.

Fig. 2.34 *James Chadwick (1891–1974), discoverer of the neutron, seen here on the right with a fellow physicist from the Cavendish Laboratory, Peter Kapitza, who did important work on low-temperature physics both in Cambridge and back in his native USSR.*

THE COCKCROFT–WALTON MACHINE

THE HIGHEST-ENERGY alpha particles from radioactive sources are only just powerful enough to penetrate the nuclei of the lighter atoms. Rutherford noticed, however, that the faster alphas are more penetrating than the slower, less energetic ones. He therefore became interested in the possibility of somehow making alphas travel even faster, at greater energies than nature provides.

At first, Rutherford's goal seemed out of reach. Charged particles are accelerated by an electric field—this is what happens to the electrons in a Crookes tube. But to achieve energies similar to those from radioactive materials such as radium and polonium would require accelerating fields of several million volts, which was beyond the technology of the 1920s. But in 1928, a paper by the Russian theorist George Gamow arrived at the Cavendish Laboratory, and it set off an important chain of events.

Gamow showed that alpha particles could on rare occasions 'tunnel' their way through the repulsive barrier presented by the positive charge of the nucleus. In other words, the alphas did not always need the high energy required to leap over the barrier; at lower energies there was a small but real chance that they could penetrate the nucleus, although the lower the energy, the lower the probability of penetration.

One person who saw immediately the importance of Gamow's calculations was John Cockcroft. He had a degree in electrical engineering from Manchester and one in mathematics from Cambridge, and had been working under Rutherford at the Cavendish Laboratory since 1924. Cockcroft realized that Gamow's paper meant that millions of volts were not necessary to produce nuclear disintegrations artificially; lower voltages would do provided you had an intense enough beam of alphas or, indeed, other charged particles. He calculated that protons accelerated by 300 000 volts (300 kilovolts, or 300 kV) could penetrate the nuclei of boron at least 1 in every 1000 times. With Rutherford's backing and a grant of £1000 from the university, Cockcroft set about designing and building a 300 kilovolt proton accelerator. In this he was joined by Ernest Walton, a young Irishman who had recently come to the Cavendish Laboratory from Dublin, and who had already tried various schemes for accelerating particles.

Fig. 2.36 *A pair of alpha particles shoot out in opposite directions (towards top right and bottom left) following the disintegration of a lithium nucleus struck by an energetic proton. The proton was accelerated in the apparatus developed by Cockcroft and Walton. The lower end of the accelerating tube, which contained the lithium target, can be seen projecting into the cloud chamber. Single tracks are due to particles whose partners have not been in the correct direction to escape into the chamber through one of the mica windows in the tube wall.*

Fig. 2.37 *A view of the Cockcroft–Walton apparatus in late 1931. The two-section accelerating tube is the glass column in the centre; the column containing the rectifiers used in generating the high voltage is just visible on the left of the picture. A microscope focused on a piece of scintillator set up in the small hut below— seen here with Ernest Walton (b.1903) inside—revealed the first evidence on artificial disintegrations in April 1932.*

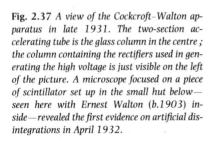

What does a voltage have to do with accelerating protons? Protons are electrically-charged particles, and in the presence of an electric field they feel a force. Protons, with positive charge, are accelerated towards a negative voltage; alternatively, a positive voltage presents a repulsive force that decelerates them.

Cockcroft and Walton generated protons using an electric discharge which stripped electrons off atoms of hydrogen gas in a small compartment. The protons produced in this way were attracted towards a negative plate and drawn off through a small hole. They were then accelerated towards another negative electrode at the far end of an evacuated glass tube.

With a system working at 280 kilovolts, Cockcroft and Walton could find no evidence for nuclear disintegrations when the accelerated protons struck the chosen targets. So they tried instead for higher voltages using a voltage multiplying system that Cockcroft had designed. The 'Cockcroft ladder', as it has become known, is a circuit that takes an *alternating* voltage and converts it to a *direct* voltage of double the magnitude. Why start with an alternating voltage—switching, as mains voltage does, for instance, between positive and negative voltages, 50 or 60 times a second? It happens that not only is electricity supplied with an alternating voltage, but also that low alternating voltages can be stepped up to high values, of several tens of kilovolts, using transformers.

The new voltage multiplying system allowed Cockcroft and Walton to apply up to 800 kV to their accelerating tube. And this time, late in April 1932, they were in luck. When they trained their high-energy beam of protons on a target of lithium they at once saw bright scintillations on a screen set up to detect charged particles emitted in nuclear transformations. They had detected the absorption of a high-velocity proton by a lithium nucleus and the subsequent splitting of the new compound nucleus into two alpha particles. The three protons and four neutrons of the lithium nucleus and the additional accelerated proton had rearranged themselves into two helium nuclei—alpha particles—each with two protons and two neutrons. Rutherford described the sight as 'the most beautiful in the world'.

NUCLEAR REALITY

IF, AS I have reason to believe, I have disintegrated the nucleus of the atom, this is of greater significance than the war.
—Ernest Rutherford, in an apology of absence for missing a meeting on anti-submarine warfare in 1917/18.

Since the discovery of the reality of the atom at the beginning of the 20th century, science and technology have developed along two closely related courses. One is 'applied atoms'—the application of the knowledge that atoms contain electrons—which has led for instance to television, computers, and microelectronics. The other direction has been the continued probing of atomic structure and the resulting realization that there is a yet deeper layer of reality, namely subatomic particles. In analogous fashion, once the nucleus had been shown to consist of a complex mix of neutrons and protons, science and technology had again a new layer of matter available for exploration and exploitation. Vast reserves of energy are contained within the nucleus, as evidenced by its release in radioactivity. But this occurs at random and at first it seemed a hopeless task to extract the energy in a way that could be turned to use. Rutherford saw the problem clearly. In 1933, in his presidential address at the British Association meeting in Leicester, he said, 'Anyone who expects a source of power from the transformation of these atoms is talking moonshine.'

But in 1933, Irène and Frédéric Joliot-Curie discovered that radioactivity was not only a spontaneous phenomenon but that under certain circumstances it can be made to happen. They found that by firing alpha particles at materials they could induce radioactivity, because the alphas created new, unstable nuclear species. Then came another surprise, late in 1938, when Otto Hahn and Fritz Strassman in Germany discovered that if they fired neutrons at uranium, the particles split the nuclei in two. This process of *fission* releases energy, but it also frees further neutrons, which can in turn trigger the fission of neighbouring nuclei, and so on. The result is a runaway chain reaction, which leads to an explosion unless properly controlled. When the process is controlled, we have nuclear power stations; when uncontrolled, the chain reaction will multiply catastrophically, and we have one of the most destructive weapons the human race has invented—the atomic bomb.

The technology of 'applied nuclei'—the application of our knowledge about the atomic nucleus—has provided us with nuclear power and weaponry. But applied nuclear physics has also led, for instance, to the development of artificial radioactive species, which are used as medical tracers, and to cancer therapy using beams of neutrons. Meanwhile, the science of nuclei has also progressed from Rutherford's time, though it has broadened into two complementary directions. One branch—nuclear physics—deals with the structure of the nucleus, and the complex behaviour of the conglomeration of protons and neutrons that form it.

The other branch—elementary particle physics—is the search for yet deeper structure, the world beyond the nucleus. Hints of this deeper structure emerged soon after the nuclear discoveries, when particles similar to, but quite distinct from, neutrons and protons were found. These particles first turned up, not in the nuclear experiments of the Joliot-Curies, Hahn and Strassman, nor at Rutherford's Cavendish Laboratory, but instead in studies of the cosmic radiation—showers of particles that stream down from outer space—as Chapter 4 describes.

Fig. 2.38 *Frédéric Joliot (right) and his research assistants Lew Kowarski (left) and Hans von Halban (centre) were the first to observe the release of neutrons in the fission of uranium, early in 1939.*

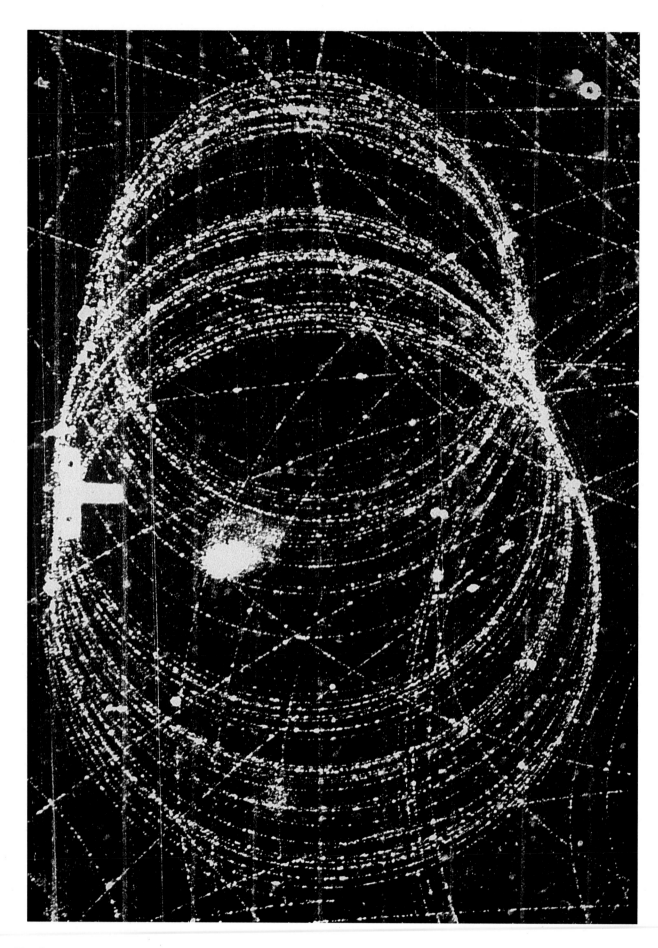

CHAPTER 3
THE STRUCTURE OF THE ATOM

THE IDEA that matter is made of invisibly small particles was first put forward in the 5th century BC by the Greek philosopher Leucippus. He called them *atomos*, meaning 'indivisible', to indicate that they were the elementary constituents of matter. However, the ideas of Leucippus and his pupil Democritus were comprehensively rejected by other philosophers in Ancient Greece. The view of matter that instead emerged triumphant, and which continued to be held in Europe for almost 2000 years, was put forward by Empedocles in about 440 BC. He proposed that everything consists of four substances, later termed 'elements': earth, air, fire, and water.

Though rejected by the physicists of classical times, the atomic theory was popularized by the Roman poet Lucretius in his *De Rerum Naturae* in the 1st century BC. It is primarily through this work that atomic theory was rediscovered in the early Renaissance, and then became widely known in Western Europe during the 16th century. By the late 17th century, physicists such as Robert Boyle, Robert Hooke, and Isaac Newton were suggesting that the properties of matter could be understood in terms of forces acting between some form of ultimate constituents. As Newton wrote in his *Opticks*, '. . . it seems probable to me that God in the Beginning formed Matter in solid, massy, hard, impenetrable, moveable Particles . . .'.

Two hundred years elapsed before Newton's successors began to discover that atoms are neither solid, nor impenetrable, nor even indivisible. By 1932, physicists knew of three basic constituents of atoms—the electron, the proton, and the neutron. They had also come to appreciate that the electromagnetic force, which binds the electrically-charged constituents together, is carried by the photon—the particle of familiar, visible light. The photon plays a crucial role in atoms; it is in a sense the 'mortar' that holds together the 'bricks' in building the matter of the everyday world.

The early experiments that initially revealed these particles often used detectors that produced no lasting visual image. However, these detectors did provide accurate information about the quantities, energies, directions, and so forth of the invisible particles, thus allowing experimenters to build up a picture of the unseen world. We have seen how Rutherford did much of his early research on radioactivity with the aid of electrometers, before turning to scintillation techniques which allowed particles to be counted as they hit a special screen. With these devices, Rutherford discovered the nature of alpha rays and the dense nucleus at the heart of the atom.

Notice the importance of being able to count the numbers of particles; in a sense, Rutherford had no need to 'see' the particles he studied. Yet in the 1920s, he did become a champion of the cloud chamber, the first device to show clearly the ionized trails of individual particles. The cloud chamber recorded not just the instant of a particle's passage past a certain point, but a whole series of points along a trail. Moreover, it provided a time-lapse picture, capturing a sequence of events in the patterns of trails on a single image, rather like tyre marks left on a sandy beach.

Later, in the 1940s, physicists brought to perfection another technique for recording the tracks of particles. In this case, the detector and the photographic film were one and the same. An electrically-charged particle travelling through layers of special photographic emulsion darkened the film along its track (see pp. 77–79). Once developed, the layers of emulsion provided a record of successive segments of the particle's track, and when pieced together these segments could reveal a complete trail.

Visual techniques like these do not show us the particles themselves, but they do create 'footprints' that allow us to recognize different particles. In this chapter we meet the inhabitants of the atom, and discover how varied their footprints are.

Fig. 3.1 *Under the influence of a magnetic field, an electron in a cloud chamber spirals around some 36 times, producing a track about 10 m long. The electron starts its life at the very bottom of the picture. It has been created together with an antielectron, or positron, by an invisible gamma ray entering from the right. (The chamber was exposed to gamma rays produced at the electron synchrotron at the Lawrence Berkeley Laboratory.) The positron travels towards the bottom left corner, while the electron's spiral moves slowly up the page due to a slight variation in the magnetic field. Notice how the spiral becomes significantly tighter about half way up the picture; this is because the electron has lost energy by radiating a photon. Notice also the small irregularities in the spacing between loops; this is due to 'scattering' when the electron's path is deviated slightly by collisions with atoms in the chamber's gas. The other tracks crossing the picture are also due to electrons and positrons, created by gamma rays beyond the right-hand edge of the picture; the electrons (negatively charged) all curl clockwise in the magnetic field, while the positrons (with positive charge) curl anticlockwise.*

THE ATOM

THE EXISTENCE of the tiny bundles of matter that we call atoms is accepted today as indisputable. We are familiar with the many different types of atoms, from hydrogen to uranium, that occur naturally on Earth, and we have created in the laboratory at least 15 types of atoms heavier than uranium. We know that atoms consist of negatively-charged electrons, positively-charged protons, and electrically-neutral neutrons, and we understand the structure and behaviour of these atomic constituents in great detail. We know how atoms combine together to form molecules and complex organic and inorganic chemicals.

This familiarity with atoms makes it difficult to appreciate how recently their existence became generally accepted. As late as 1900, even after J. J. Thomson's discovery of the electron, there were leading physicists and chemists who did not believe in atoms at all, or who considered them to be no more than convenient 'models'. The majority held that atoms—as the ultimate, irreducible particles of matter—really existed, but this was more a question of faith than of knowledge. Positivists like the influential German physicist Ernst Mach emphasized that the evidence was circumstantial. No one had ever *seen* an atom.

In the first ten years of the 20th century, however, the 'atomic hypothesis' became universally accepted. Ironically, this was not because atoms had finally been found, but because the work on cathode rays and radioactivity by Thomson, Rutherford, and others conclusively showed that atoms consist of smaller constituents: negative electrons and a positive nucleus. In a sense, no one 'discovered' the atom at all; instead, physicists went straight to the discovery of what atoms are made of.

More than half a century elapsed before the development of specialized, high-resolution electron microscopes made it possible to 'see' atoms directly. And as Fig. 3.2 shows, even today's best instruments reveal atoms as no more than fuzzy, featureless spots, although the magnification of the printed image is 90 million times. Atoms are at the limit of 'direct' vision. All the other particles that we meet in this book are 'seen' only by their effects, such as the tracks they leave in cloud chambers and bubble chambers or the signals they induce in electronic instruments.

The concept of the atom may owe its origins to the Ancient Greeks, but the story of the reality of atoms begins in the early years of the 19th century, and in particular with John Dalton, a weaver's son from the north of England. Dalton is generally acknowledged as a poor experimenter yet his insight brought a quantitative edge to studies of the chemical elements, and set in motion ideas that are still valid today. He found that certain recently-discovered 'laws' governing the combination of elements to form compounds could be explained if each element consists of its own unique brand of indivisible atom. Atoms of one element would all be alike, but they would differ from one element to another. In particular, Dalton assigned *atomic weights* to the atoms of the 20 elements he knew of in 1808. These he tabulated, beginning with hydrogen, whose atoms are the lightest, and ending with mercury, to which he mistakenly assigned a larger atomic weight than for lead (Fig. 3.3).

Dalton established one important characteristic of an atom—its weight. But it soon became apparent, as still more elements were discovered, that elements of quite different atomic weight often resembled each other. Potassium is a reactive metal like sodium, for example, but has almost twice the atomic weight. In 1869, Dmitry Mendeleyev, professor of chemistry at the University of St Petersburg, put forward the classification of the elements that we still use today—the Periodic Table (Fig. 3.4). He grouped the 62 elements he knew of into a number of families, leaving gaps corresponding to elements that should exist but which had not yet been discovered. The discovery of these 'new' elements, such as gallium and germanium, was a triumph for Mendeleyev's classification. But though the scheme recognized the validity of the atomic hypothesis, it still provided no insight into the precise nature of an atom, and that worried the purists.

Both Dalton and Mendeleyev were chemists; the physicists of the day were more involved in coming to grips with the nature of electricity and magnetism, heat and light, than with atoms. But it was discoveries made by physicists that were to bring

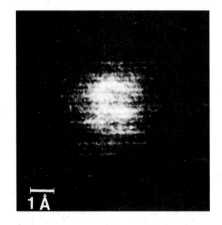

Fig. 3.2 *A scanning transmission electron micrograph of a single atom of gold. The scale bar represents 1 ångström unit, or 0.1 nanometres. Even at this magnification of some 90 million, the atom appears only as a featureless blur.*

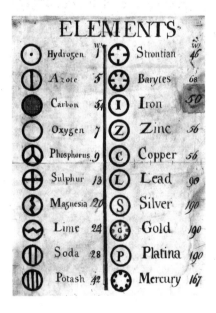

Fig. 3.3 *Dalton's list of atomic weights and the symbols he used for a number of 'elements' (we now know that some of the materials included here are compounds, not pure elements). Dalton worked out the weight of each substance relative to hydrogen, the lightest. His answers, and therefore the detailed ordering in the table, turned out to be wrong because he did not know the correct proportions in which the elements combined to form compounds such as water. However, what is important is that Dalton interpreted his measurements in terms of atoms, with the atoms of different 'elements' having different weights.*

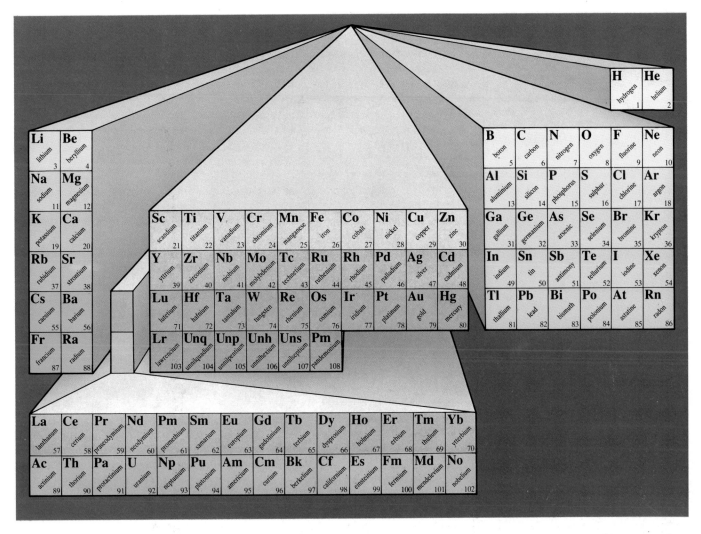

an understanding of the atom, and also of the meaning of atomic weight and the ranking of elements in the Periodic Table. Chapter 2 has recounted how the discovery of X-rays and radioactivity led to a series of findings that culminated in 1911 with Rutherford's model of the atom. He had found that atoms are mostly empty space, and he pictured the atom as a miniature 'solar system', with electrons whirling round a central nucleus. The electrons are spread throughout the volume, surrounding a tiny compact nucleus which is smaller relative to the atom than the hole is relative to a 500 metre fairway of a golf course.

The Danish theorist Niels Bohr seized on Rutherford's ideas and by 1913 he had developed a detailed theory of electrons in atoms which explained the spectra of light emitted by certain atoms, notably hydrogen, with great success. Although incomplete and in some respects wrong, Bohr's theory still underlies our understanding of how electrons behave in atoms.

The electrons in atoms occur with discrete energies in 'shells' at certain distances from the nucleus. An electron can 'jump' from one shell to another. If it gains energy, it jumps to a shell further from the nucleus; if it radiates energy, it falls to a shell closer in. Bohr's really innovative step was to propose that only certain energy levels are available in an atom of any particular element. And these are determined by the positive charge of the central nucleus. Bohr's theory for the first time related the energy of radiation absorbed and emitted by atoms to the properties of the nucleus that Rutherford had discovered.

Rutherford's physics department at Manchester included a young man, Harry Moseley, who at the time was interested in the X-rays emitted by atoms. He realized that if Bohr's theory was correct, the energy of X-rays emitted by the various elements

Fig. 3.4 *In 1869, Mendeleyev proposed arranging the elements in a table where elements with the same chemical and physical properties were grouped together in the same column; each row in the table corresponded to increasing atomic weight. However, it was not until earlier this century, in the wake of the discovery of the electron and the details of atomic structure, that chemists began to appreciate the reasons for the table's structure. To this day the 'periodic table', as this modern version shows, has the same basic form that Mendeleyev proposed, although we now know of many more elements.*

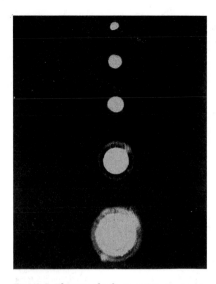

Fig. 3.5 *Electron clouds in argon atoms are magnified 126 million times in these images acquired by a 'holographic microscope' invented by L. S. Bartell at the University of Michigan. The technique is a two-stage process. First, electrons scatter from atoms to form a multitude of holograms which are superimposed to give a single pattern. Then a laser is used to reconstruct an image from the accumulated hologram. In these images the exposure time increases from top to bottom, and together they show that electrons occur more often closer to the nucleus than further out. Only electrons close in contribute to the shortest exposure; longer exposures reveal the more diffuse electron cloud in the outer reaches, in complete accord with the predictions of quantum theory.*

Fig. 3.6 *Illustrations of a variety of atoms, showing the electron clouds—or orbitals—that surround the nucleus. As the number of electrons increases, so additional orbitals are filled according to a regular pattern of spherical 's' orbitals (white to blue), dumbbell-shaped 'p' orbitals (orange), 'd' orbitals (yellow-green) which are usually shaped like crossed dumbbells, and 'f' orbitals (purple), which usually consist of three pairs of dumbbells. S orbitals can contain up to 2 electrons, p orbitals up to 6, d orbitals up to 10, and f orbitals up to 14.*

The single electron of hydrogen occupies, but does not fill, a single s orbital. Carbon, with 6 electrons, fills two s orbitals and partially fills a p orbital. Silicon has 14 electrons filling three s orbitals and one p orbital, and partially filling a second p orbital. Iron's 26 electrons fill three s orbitals and two p orbitals, with 1 electron in a fourth s orbital and the remaining 7 electrons occupying a d orbital. Silver, with 47 electrons, completely occupies four s orbitals, three p orbitals, and two d orbitals, with 1 remaining electron half filling a fifth s orbital. And finally, europium's 63 electrons take the pattern further and fill six s orbitals, four p orbitals, and two d orbitals, with 7 electrons over to occupy an f orbital.

would be a direct measure of the amount of charge in their atomic nuclei. So Moseley carefully measured the X-ray spectra and discovered that the positive charge of each nucleus is a whole multiple of the negative charge of the electron. He also found that this amount of positive charge increases one unit at a time as you move from one element to the next up the Periodic Table. Hydrogen has a positive charge of 1, equal in magnitude to the negative charge of one electron; helium has charge 2, lithium charge 3, and so on through to uranium with a total charge of 92 units.

Moseley had uncovered an important atomic parameter, now known as the *atomic number*, which gives the number of positive charges on the nucleus—in other words, as Rutherford later discovered, the number of protons it contains. The Periodic Table lists elements in order of the number of protons (and electrons) in the atom.

A further important contribution to the theory of atoms came in 1921, when the Austrian physicist Wolfgang Pauli put forward his so-called Exclusion Principle. This has the effect of restricting the number of electrons that can inhabit any one shell. For instance, the shell closest to the nucleus can accomodate only two electrons, and so can the next shell; the third shell can accomodate six electrons, the fourth two electrons, the fifth six electrons, the sixth ten electrons, and so on according to a complex but regular formula.

This pattern of electrons in shells underlies the relationships in the Periodic Table in the following way. In normal circumstances, the orbiting electrons fill the shells closest to the nucleus; when this occurs, the atom is said to be in its ground state. (Electrons can, however, be 'excited' to more distant shells, if the atom absorbs energy.) In the ground state of a hydrogen atom, its single electron only half fills the first shell, whereas in the helium atom—with two electrons—the first shell is completely filled. The 18 electrons of argon completely fill 5 shells, while the 92 electrons of uranium occupy, but do not completely fill 18 shells.

Whether an atom's electrons fill shells completely or not turns out to determine the chemical and physical properties of the element, and to a large extent the Periodic Table groups together atoms with the same number of electrons in the *outer* shell. The so-called 'noble gases'—helium, neon, argon, krypton, xenon, and radon—are all elements that react hardly at all with other elements. This is because they have just the right number of electrons to fill completely a number of shells, and filled shells are particularly stable configurations. These elements occupy the last column of Mendeleyev's table. The alkali metals—lithium, sodium, potassium, rubidium, caesium, and francium—all have similar chemical properties because they all have complete shells plus one extra electron; they are grouped in the first column of the Periodic Table.

Modern physicists still use the 'solar system' model of the atom as a kind of verbal and visual shorthand. Electrons are said to 'orbit' the nucleus and, when energized, to jump from one orbit to another, as if instantly relocating between Jupiter and Saturn. But while this is a useful way of visualizing the atom, it does not express what physicists now know about the behaviour of electrons in atoms. For one thing, the 'orbits' of electrons about the nucleus are not necessarily the majestic ellipses described by planets orbiting a star; on the contrary, an electron may move in a highly fluctuating and idiosyncratic manner. More fundamentally, the precise path around the nucleus of any individual electron can never be known. And the more we try to pin it down, the more it eludes us—a subatomic will-o'-the-wisp.

This inescapable uncertainty, enshrined in Werner Heisenberg's Uncertainty Principle, is one of the fundamental tenets of quantum mechanics, the theory developed in the 1920s and 1930s that has proved triumphantly successful in describing and predicting subatomic events and processes. Quantum theory replaces certainty with probability. While the path of any individual electron around an atomic nucleus is unknowable, the average paths of a million electrons in a million atoms can be statistically described with great accuracy. Physicists sometimes speak of electrons forming a 'cloud' around the nucleus, but it would be more accurate to describe them as producing a blur, like the spokes of a rapidly-revolving bicycle wheel. We cannot distinguish their individual motions, only the generalized effect of repeated motions (Fig. 3.6).

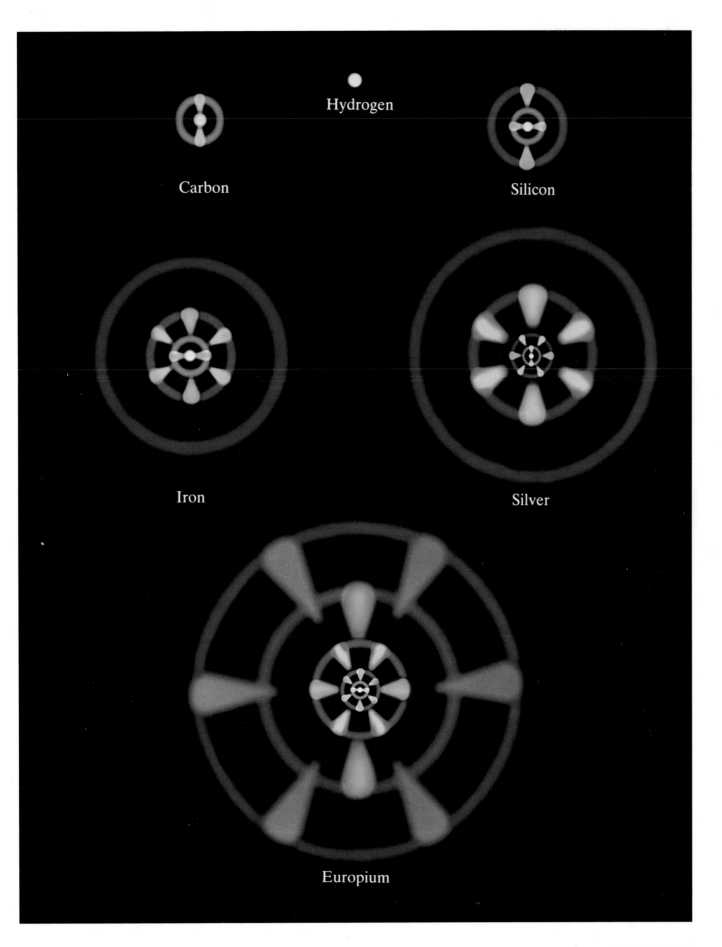

Hydrogen

Carbon

Silicon

Iron

Silver

Europium

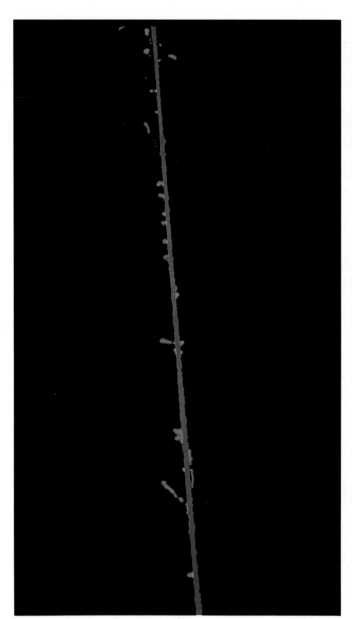

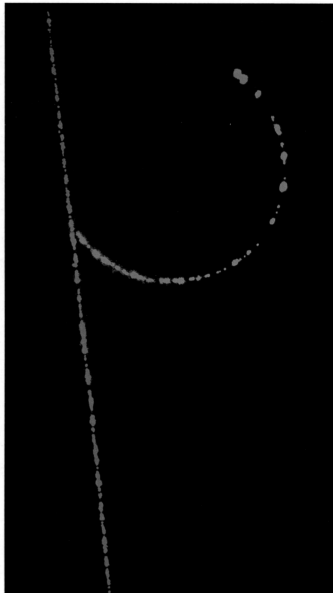

Figs. 3.7–3.8 *Images of electrons knocked from atoms at different speeds. Like many pictures that follow, these black and white cloud chamber photographs have been coloured to identify the tracks of particular particles.*

Fig. 3.7 (*left*) *The track of an alpha particle* (*blue*) *in a hydrogen-filled cloud chamber shows wispy side shoots* (*red*) *due to electrons. Only when the alpha comes very close to an atomic electron does the electron acquire a sufficient velocity to form a short trail. Such energetic electrons, which in this image travel as far as 2.9 mm, are known as 'delta rays'.*

Fig. 3.8 (*right*) *A high-energy cosmic ray particle* (*green*) *shoots into a cloud chamber and knocks an electron* (*red*) *out of the gas, giving it enough energy for the electron to leave a long track as it curls under the influence of a magnetic field. The cosmic ray track is about 10 cm long.*

THE ELECTRON

IN HUMAN terms, electrons are the most important of all subatomic particles. Electrons in the outer reaches of atoms give structure to the Universe: they are the means by which atoms are bonded together to form molecules and, eventually, large aggregates of matter such as ourselves and everything around us. The achievements of modern chemistry and biochemistry, from the invention of plastics to the synthesis of new drugs and the manipulations of genetic engineering, are ultimately based on our detailed understanding of the behaviour of electrons in atoms.

Remove electrons from atoms, set them in motion, and you have an electric current: a stream of electrons flowing through a substance capable of conducting them, such as the copper of electric wiring. Electrons not only provide us with electricity, they are the basis of electronics—from televisions to microchips. Electronics, in fact, is defined as the applied science of the controlled motion of electrons.

The electron was discovered by J. J. Thomson in his experiments on cathode rays in 1897, as described in Chapter 2. But 40 years elapsed before physicists came to understand the behaviour of electrons in detail. We now know that electrons are stable, lightweight, negatively-charged elementary particles. They appear to be stable

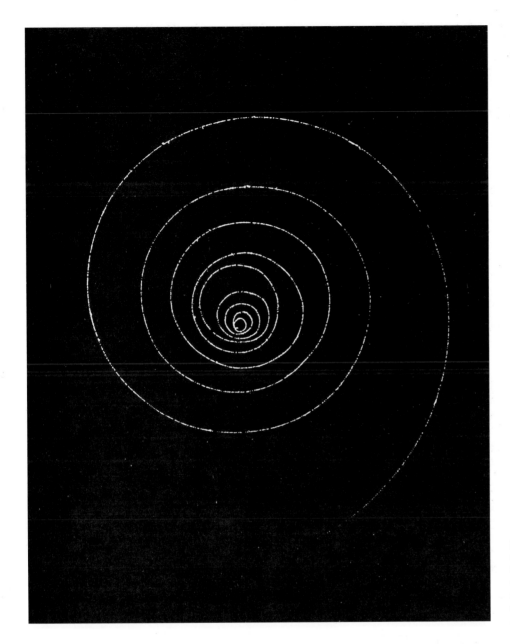

Fig. 3.9 *An electron spirals in the magnetic field of the 10 inch (25 cm) bubble chamber at the Lawrence Berkeley Laboratory. As it loses energy in ionizing the liquid hydrogen in the bubble chamber, the electron becomes less and less resistant to the force of the magnetic field and curls inwards on successively tighter curls.*

in the sense that when unperturbed they can live forever; unlike most of the particles described in this book, they do not transmute into other particles. They also seem to be elementary, in that they do not consist of smaller entities, in contrast to the other constituents of atoms—protons and neutrons—which are now known to be built from smaller particles called quarks. All the evidence suggests that electrons are truly indivisible, and are at least as small as physicists can measure—less than 10^{-18} m across.

An important characteristic of the electron is its lightness. It weighs in at 9.1×10^{-31} kilograms. To understand how light this is, consider the difference between the weight of a small bag of sugar (1 kg or 2.2 lb) and 15 million planets the size of Earth. The same ratio exists between the weight of an electron and the bag of sugar. For physicists, however, the vital statistic is that an electron weighs 1836 times less than a proton.

Partly because they are so light, electrons can easily be deflected from their paths or knocked out of their positions in atoms. Their paths also curve strongly in magnetic fields. These effects are illustrated in the two cloud chamber pictures (Figs. 3.7 and 3.8) and in the bubble chamber photograph (Fig. 3.9).

In Fig. 3.7, an alpha particle emitted in a radioactive decay is seen knocking electrons from the atoms of the gas that fills the cloud chamber. Figure 3.8 shows a single electron being knocked out of an atom by the passage of a cosmic ray. The electron's track is longer in this case because it has been given more energy in the collision; it is curved because the chamber is subject to a magnetic field. Electrons are also knocked from atoms by electromagnetic radiation, especially the higher-energy X-rays and gamma rays.

Figure 3.9 shows a single electron, which has been knocked out of an atom of hydrogen in a bubble chamber—a device containing superheated liquid in which streams of tiny bubbles form along ionized trails (pp. 113-14). A powerful magnetic field causes the track of the electron to curve; and because the electron loses energy as it travels, its track curves more and more tightly to form a perfect spiral. Such spiral tracks are a typical signature of electrons, and we shall see them recurring in pictures throughout this book.

A single electron carries the smallest quantity of electric charge ever detected in isolation—1.6×10^{-19} coulombs. This is a trifling amount, but its impact is tremendous. The strength of the electrical force between electrons and protons in an atom is comparable to the gravitational force between bulk matter: the electrical forces within a microgram speck of dust are equal to the Earth's gravitational pull on an object weighing several hundred tonnes. A sufficient excess of electrons on the Earth and the Sun would repel the Earth and send it spiralling out of the Solar System.

Electrons exist both on their own, as free particles, and as constituents of atoms, and they can change from one role to the other and back. An electron forming part of a carbon atom in the skin of your wrist could be knocked out of position by a passing cosmic ray and become part of the tiny electric current in your digital wristwatch, and then in turn become part of an oxygen atom in the air you breathe as you raise your arm to look at the time.

In hot and violent parts of the Universe—inside stars or in the super-hot shells of gas expanding after a supernova explosion—the particles of matter tend to disintegrate into their constituent parts and to be in constant flux. Stable atoms are rarely or never formed. Nuclei stripped bare of electrons mingle with clouds of free electrons, in a state of matter known as *plasma*. In comparison, conditions on Earth are cold and electrons and nuclei are 'frozen' into the relative stability of atoms.

Where did electrons come from in the first place? Physicists believe that most of them were created out of intense radiation in the first instants after the Big Bang when the Universe was still very hot. They were not created alone, but together with their antimatter equivalents—positively-charged electrons or 'positrons' (pp. 84-5). Such 'electron-positron pairs' are still produced all the time out of intense radiation, both naturally and in physicists' laboratories.

Electrons are also created in the beta-decay form of radioactivity. Indeed, the so-called 'beta rays' emitted when atomic nuclei decay in this way are nothing other than electrons, no different from the electrons in an electric current or in the atoms that make up our bodies.

Although an isolated electron can live forever, most electrons are constantly interacting with other particles. They disappear from the Universe, either when they turn back into energy after encountering and mutually annihilating with a positron or when they are occasionally absorbed by an atomic nucleus (converting a proton into a neutron—a process known for obvious reasons as 'electron capture').

For over a century, electrons have proved to be extremely useful tools in exploring the nature of matter, because they are charged and lightweight and therefore easy to accelerate. An electron in an electric field experiences a force and is accelerated; this is precisely the effect that impelled the cathode rays across the glass tubes that Crookes, Thomson, and others used. Nowadays we see a similar effect every time we watch television. Electric fields inside the television tube accelerate the electrons towards the screen, where they stimulate atoms to emit flashes of light. Millions of electrons in coordinated bursts produce the visual image on the screen. A similar process is at work in an electron microscope, which records the patterns of energetic electrons scattered from the specimen in its field of view.

With bigger 'tubes' and stronger electric fields, physicists can use electrons to probe

still further into matter, firing electrons right at the heart of the atom—the nucleus. Figure 3.10 shows an 'electron tree', produced by an intense beam of electrons, hundreds of times more energetic than those in a television tube, which has been fired into a block of Perspex (Lucite). The mutual repulsion of the negative electrons sends them out in a branched pattern, like frozen lightning. The image was created by a relatively small electron accelerator at the Stanford Linear Accelerator Center (SLAC) in California. SLAC also houses the world's longest electron machine—a 'tube' 3 km long. When the electrons leave this machine their energies are more than a million times greater than in the television tube, and they can resolve details not only within the atomic nucleus, but within the protons and neutrons themselves. Around 1970, SLAC's electrons provided the first detailed glimpses of the world within protons and neutrons.

Fig. 3.10 *An 'electron tree' produced in a block of plastic (15 cm square by 2.5 cm thick) by a beam of electrons. The electrons initially penetrate about 0.5 cm into the block and stop. As the number of electrons builds up, however, their mutual electric repulsion begins to force them apart. If the beam is switched off before this happens, a subsequent small tap to the block with a metal punch releases the electrons suddenly and they shoot out, rather like lightning, leaving a pattern of tracks that becomes 'frozen' in the plastic. (The colour has been added.)*

THE NUCLEUS

STRIP AN atom of its electrons and you are left with its nucleus, a compact bundle that occupies only a thousand-million-millionth of the atom's volume but which provides 99.95 per cent of its weight. The nucleus also provides the positive electric charge that balances the negative charge of the electrons to make the atom neutral overall. The electrons in the outer reaches of an atom determine its external relations—how it bonds with other atoms—but it is the nucleus that determines the atom's nature.

The nucleus is more than just a core. It is an entirely new level of reality, where the forces of electromagnetism and gravity that govern atoms, molecules, and larger conglomerations of matter play only a minor role. In the nucleus, different forces are at work, which are unfamiliar in the wider world: the 'weak' force governs beta-radioactivity, and the 'strong' force holds the constituents of the nucleus together.

All nuclei except for the hydrogen nucleus, which consists of a single proton, are

Fig. 3.11 *A cosmic ray magnesium nucleus, entering the picture from the top left, collides with a bromine nucleus in photographic emulsion and the two nuclei shatter into a multitude of fragments. The electrically charged fragments, many of which are protons that were contained in the two nuclei, leave visible tracks in the emulsion. (The track of the incoming magnesium nucleus is about 0.18 mm long.)*

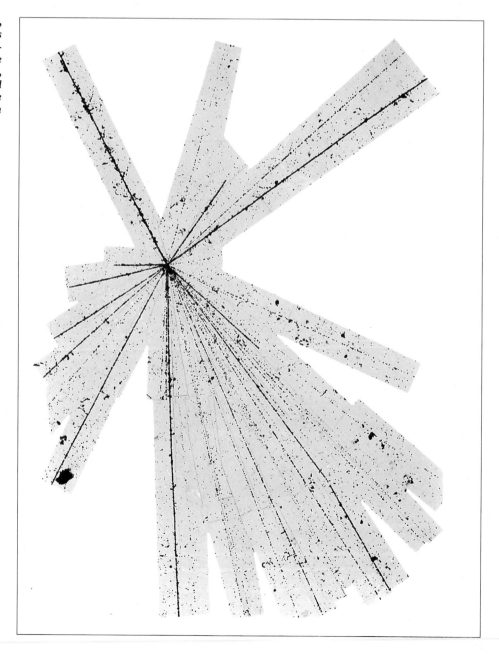

compound entities built from both protons and neutrons. This complex inner structure is illustrated in Fig. 3.11, which shows what happens when the nucleus of a magnesium atom (12 protons and 12 neutrons) collides head-on with the nucleus of a bromine atom (35 protons and 44 neutrons) in photographic emulsion. The two nuclei shatter into a multitude of fragments, although only those that are electrically charged produce tracks in the picture.

Positive protons contribute both electric charge and mass to the atomic nucleus, whereas the neutral neutrons contribute mass alone. Thus the number of protons determines the charge of an atom, while its mass depends on the sum of its protons and neutrons. Each proton carries one unit of electric charge. When we look at the Periodic Table of the elements, we find that they are ranked according to the number of protons in their nuclei, from hydrogen and helium, with one and two protons respectively, to the 'transuranic' elements, which go at least as far as 107.

The nuclei of heavy elements, with a large amount of electrical charge, exert a correspondingly large electrical influence on matter. When such nuclei pass through a photographic emulsion, for example, these powerful electrical forces ionize atoms and leave behind a trail of sensitized emulsion, which then appears on the developed image. Thus the greater the charge of the nucleus, the thicker is the track that it leaves in the emulsion.

This effect is seen in Fig. 3.12, which shows the tracks of 15 nuclei, from hydrogen to iron. With its single proton, hydrogen is all but invisible; the 26 protons of the iron nucleus, on the other hand, leave a thick track liberally bedecked by the curlicues of electrons knocked out of atoms in the emulsion as the iron nucleus ploughs through them.

The atom, as we have seen, is held together by the electromagnetic force: the positive nucleus attracts the negative electrons. The more protons there are in the nucleus, the greater its positive charge, and the more electrons it is able to attract. But what holds the nucleus itself together? Why do the protons, all with the same electric charge, not repel each other and cause the nucleus to disintegrate? The answer lies with the strong force, which holds the nucleus together despite the mutual electromagnetic repulsion of its protons. Within the nucleus the strong attraction is over 100 times more powerful than the electromagnetic disruption.

The strong force does not distinguish between protons and neutrons. In other words, neutrons and protons attract one another with the same strength as either one attracts its own kind. The neutrons, having no electrical charge, are not subject to electromagnetic repulsion, so extra neutrons help to stabilize nuclei. They provide sources of strong attraction for the protons and help them to fight the disruptive electrical force.

This is why nuclei tend to have more neutrons than protons. Too great an excess of neutrons, however, will destabilize a nucleus. If such a nucleus is created—in

Fig. 3.12 *The more protons a nucleus contains, the higher its positive charge and the greater its ability to ionize; hence the denser its trail in a detector. Here are the tracks of fifteen different nuclei, captured in emulsions exposed to the primary cosmic radiation high in the atmosphere. Hydrogen (H), with a single proton and therefore only one unit of positive charge, leaves a barely visible track. But because the ionization depends on the square of the charge, the track of lithium (Li), with only three protons, is nine times stronger and stands out clearly. Iron (Fe) has 26 protons, enough to produce a track 676 times stronger than hydrogen. The wispy side-strands are due to delta rays—electrons knocked from atoms in the emulsion, which have enough energy to produce their own short tracks.*

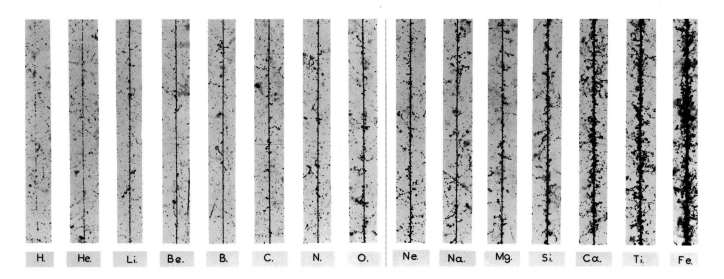

H.   He.   Li.   Be.   B.   C.   N.   O.   Ne.   Na.   Mg.   Si.   Ca.   Ti.   Fe.

high-energy collisions, for instance—then it becomes more stable by the process of beta decay, in which a neutron converts into a proton, at the same time emitting an electron and a particle known as a neutrino.

For a particular element, characterized by a specific number of protons, there may be several relatively stable nuclei with different numbers of neutrons. These are called *isotopes*. One example is uranium, which occurs naturally mainly as uranium-238; its 92 protons are accompanied by 146 neutrons, making 238 'nucleons' in total. Less than one per cent of naturally-occurring uranium is uranium-235. This isotope, with its 143 neutrons, is the one that is responsible for the chain reactions in nuclear bombs and reactors.

Just as the electrons in atoms arrange themselves in shells, so do protons and neutrons in nuclei, following very similar rules. In atoms, the shell with the lowest energy is full when it contains two electrons. In nuclei, it is full when it contains two neutrons and two protons. This makes an exceptionally stable configuration; in fact, it is the nucleus of helium—the alpha particle. The stability of alpha particles accounts for their appearance in radioactive decays, particularly of heavy elements such as uranium and thorium. These elements become lighter, or more stable, by shedding protons and neutrons in alpha particle clusters.

Alpha particles are also common among the fragments of high-energy nuclear collisions. Figure 3.13 shows a carbon nucleus in the cosmic radiation, moving near to the speed of light. It knocks a nucleus in the photographic emulsion sideways, and continues onwards with only a slight deflection in its flight. (This may not be easily apparent but the subtle deflection becomes more noticeable if you hold up the picture almost level with your eyes and look along the tracks at a glancing angle.) Even though the disturbance is very slight, it is enough to disrupt the carbon nucleus. After the collision we see not a single carbon nucleus, but three alpha particles flying

Figs. 3.13–3.14 *Alpha particles—helium nuclei—are particularly stable clusters of two protons and two neutrons, common among the fragments of nuclear collisions.*

Fig. 3.13 (*left*) *A cosmic ray carbon nucleus enters at the top of the picture and collides with a proton in emulsion. The carbon nucleus disintegrates into three alpha particles. The proton moves off to the left.*

Fig. 3.14 (*right*) *This time a colliding carbon nucleus breaks up into three lumps of different size—a proton, an alpha particle, and a lithium nucleus. (Hold the book up and look along the page horizontally, at the level of your eyes, to see this better.) The other faint tracks are probably due to pions produced in the collision while the heavy track to the right could be the struck nucleus.*

The scale bars in each picture show a distance of 50 micrometres (μ).

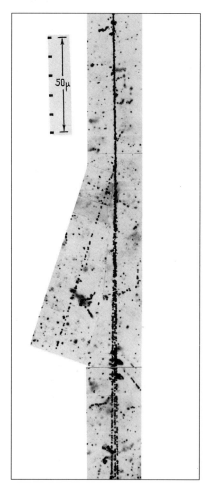

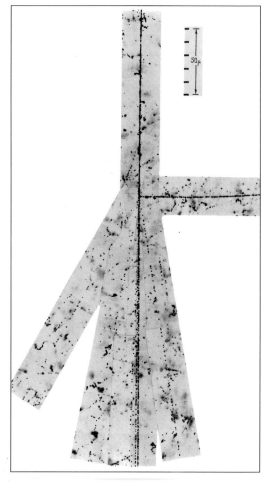

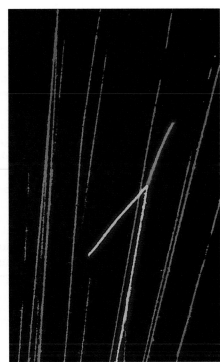

down the page. This shows that carbon's six neutrons and six protons behave just like three alpha particles bound loosely to each other.

In Fig. 3.14 we see the arrival of another carbon nucleus, but this time it splits up into one alpha, one proton, and a lithium nucleus (three protons and four neutrons). The six protons and six neutrons are preserved in total but they have grouped together differently. Notice how the quality of track varies for the different nuclei. The proton is almost invisible until you use the trick of viewing at a glancing angle.

Cosmic rays are one natural source of high-speed nuclei; natural radioactivity is another. But whereas cosmic rays contain the nuclei of a whole range of elements, natural radioactivity supplies only nuclei of helium—alpha particles—in any quantity. These were used in the early part of the century as tools to probe the structure of nuclei, as Chapter 2 describes. Figures 3.15-3.17 show some classic pictures of alpha-particle scattering, made visible with a cloud chamber.

Figure 3.15 shows what happens when an alpha particle collides with the nucleus of one of the hydrogen atoms filling a cloud chamber. With its two protons and two neutrons, the alpha is four times as massive as the single proton of the hydrogen nucleus. As a result the proton is propelled forwards, while the alpha is only slightly deflected. The angle between them is less than 90°. This all accords with experience if you have ever played shove-halfpenny: a big coin knocks a small one forwards. (Note also that the proton, with its single unit of electric charge, causes less ionization and so produces a less dense track than the doubly-charged alpha.)

What happens when identical coins collide, or two billiard balls (assuming that they are not given any spin or 'side')? They move off at 90° to one another. We see this in Fig. 3.16, where the cloud chamber is filled with helium instead of hydrogen. An alpha particle entering from below bounces off a helium nucleus, and the outgoing tracks produce a fork with an angle of 90°. This is because helium nuclei have the same mass as alphas; indeed, experiments of this kind helped to show that alpha particles are nothing more than the nuclei of helium atoms. (The tracks in fact appear to diverge at a little less than 90° because they are being viewed at a slight angle.)

In Fig. 3.17, the cloud chamber contains nitrogen. A nitrogen nucleus consists of seven protons and seven neutrons, and when an alpha particle entering from below meets·one head-on, it is turned back in its tracks, at the same time transferring most of its energy to the nitrogen nucleus, which moves a short distance forwards. The angle between their tracks is 142°.

Figs. 3.15–3.17 Patrick Blackett obtained clear evidence for the way that alpha particles scatter from nuclei of differing mass in a series of classic cloud chamber experiments in the late 1920s. The photographs have been coloured to identify particular tracks, with interacting alpha particles in yellow and non-interacting ones in green.

Fig. 3.15 *(left) An alpha particle collides with a proton in the hydrogen gas filling a cloud chamber. The proton (red) shoots off to the right, while the heavier alpha is deflected only slightly from its previous direction, so that the angle between the tracks is noticeably less than 90°.*

Fig. 3.16 *(centre) An alpha particle bounces off a nucleus in helium gas in a cloud chamber, the two particles moving off at right angles to each other. The 90° angle reveals that alpha particles and helium nuclei have the same mass; they are one and the same thing. (Note that the angle appears slightly less than 90° because the tracks have been photographed at an angle.)*

Fig. 3.17 *(right) An alpha particle bounces back from a much heavier nucleus in the cloud chamber gas, this time nitrogen. The alpha gives much of its energy to the nitrogen, which moves onwards, leaving a thick track (red) owing to its relatively high positive charge from seven protons. The deflected alpha also leaves a thick track because it is now moving much more slowly and therefore produces more ionization. Here the angle between the tracks is much greater than 90°.*

THE PROTON AND NEUTRON

BY THE 1930s, Rutherford and his team at the Cavendish Laboratory had shown that there are two types of particle that build up atomic nuclei: protons and neutrons. Electrically they are quite different—the proton is positively charged, while the neutron is neutral. But in other respects, the two particles are almost indistinguishable, and they are often regarded as charged and neutral versions of the same basic particle—the *nucleon*. We now know that the nucleons are not elementary in the sense that electrons are, but that they are themselves built from other particles—the quarks. This structure is reflected in the size of the nucleons, which are about 10^{-15} m across, at least 1000 times bigger than electrons. For many purposes, especially in studies of the behaviour of complex nuclei, it is useful still to regard the proton and neutron as the basic constituents of the nucleus, just as it is useful to consider atoms as simple objects when studying large-scale molecular behaviour.

In some senses the proton is more basic than the neutron, for free neutrons ultimately decay to protons—indeed, many of the particles that will be introduced later in this book decay ultimately to protons. With a mass of 1.6726×10^{-27} kg, the proton is the lightest member of a family of particles—the hadrons—built from three quarks. It carries a positive electric charge, which so precisely balances the negative charge of the electron that atoms and bulk matter are normally electrically neutral. This precise equality of charge, which is essential for the stability of matter, is profound because the electron and proton appear to be quite different forms of matter and yet conspire in this delicate way.

Protons are the nuclei of the simplest chemical element—hydrogen. This is made clear in Fig. 3.18, where an energetic proton enters a bubble chamber containing

Fig. 3.18 *A proton (red) enters a bubble chamber from bottom right and scatters from other protons (also red) in the liquid hydrogen filling the chamber. Each time the angle between the scattered particles is close to 90°, revealing the equality of their masses. The tracks of particles not involved in this game of subatomic billiards are coloured blue; the spirals are electrons knocked from hydrogen atoms in the liquid.*

liquid hydrogen. The proton eventually scatters from a hydrogen nucleus (a proton) in a billiard ball collision, and the two particles move off at 90°, revealing their equal masses. (Once again, the viewing angle makes this look like less than 90°). The 90° scattering angle occurs again and again as more and more protons are set in motion. It is also easy to see the difference between a proton and an alpha particle—the nucleus of helium. Both types of particle are positively charged and bend the same way in a magnetic field (Fig. 3.19), but although the alpha has double the charge of the proton, the proton has only a quarter of the alpha's mass. Consequently, the proton bends more easily in the magnetic field.

Until recently, physicists believed that protons are completely stable and live forever. The latest theories, however, suggest that they do decay after an enormously long period of time. The question is, how long? We know that they must live on average for more than 10^{17} years, otherwise our bodies would be highly radioactive. This is because particle lifetimes are only an average, and a human body contains so many protons—roughly 10^{27}—that many would decay even during a 70-year lifespan. Subtle experiments on proton decay, described in Chapter 10, show that protons must in fact live for at least 10^{32} years, which is 10^{22} times longer than the estimated age of the Universe.

The availability, stability, and electrical nature of protons makes them a favourite tool for particle physicists. Intense beams of protons can be accelerated and fired into matter and used to study nuclear behaviour under extreme conditions. This complements experiments using electrons because protons interact by both the electrical and strong forces, whereas electrons do not experience the latter.

Neutrons are about 0.1 per cent heavier than protons; they weigh 1.675×10^{-27} kg. The mass of a neutron in fact exceeds the total mass of a proton and an electron,

Fig. 3.19 *The tracks of a 1.6 MeV proton (red) and a 7 MeV alpha particle (yellow) curl round in a cloud chamber in a magnetic field. The curvature is proportional to charge and inversely proportional to momentum (the product of mass and velocity). So the alpha particle, with double the charge of the proton but four times the mass, curls less in the magnetic field. The particles were produced in collisions of a beam of 90 MeV neutrons from the 4.6 m (184 inch) cyclotron at Berkeley. The thin blue horizontal lines are wires which produce an electric field to remove old ions, while the thick black rectangle near the bottom of the picture is part of the structure of the chamber.*

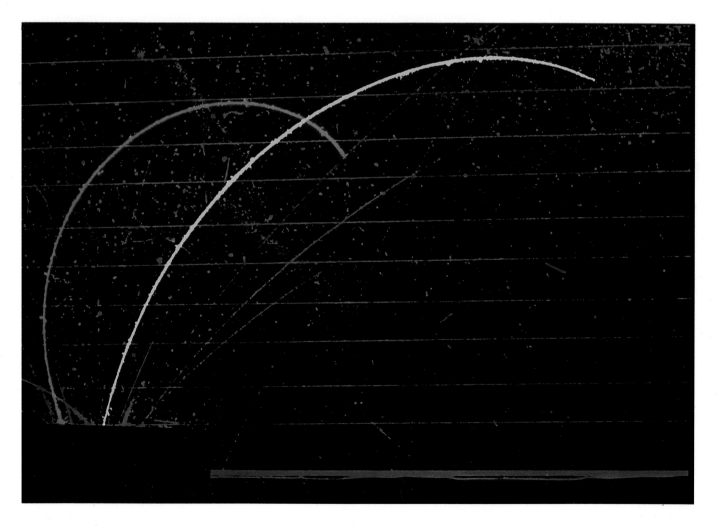

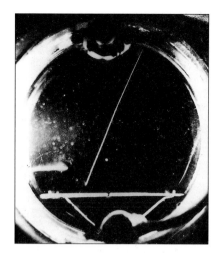

and this is sufficient in certain circumstances to make neutrons unstable. An isolated neutron decays into a proton and an electron on average after about 15 minutes. This is the basic process of beta radioactivity. In some nuclei, neutrons also decay this way, but in others, subtle nuclear effects tip the scales and enable the neutrons to survive, leading to a stable nucleus.

Whereas protons leave visible trails, neutrons are like H. G. Wells's invisible man, who gave away his presence indirectly—by jostling the visible crowd. If an invisible neutron bumps into a proton and sets it in motion we can detect the trail of the proton. An example is in Fig. 3.20, where a single neutron has struck a proton in a paraffin sheet. The proton shoots out across the cloud chamber. It is clear that something bulky has come in—protons do not spontaneously fly off without cause.

Figure 3.21 shows what happens when an intense beam of artificially energized neutrons, coming from the bottom of the picture, enters a cloud chamber. The

Fig. 3.20 (above) One of the first photographs of a proton recoiling after being struck by a neutron, taken by the Joliot-Curies in 1932. An invisible neutron, knocked out of beryllium by an alpha particle, enters a cloud chamber from below. It strikes a sheet of paraffin wax across the chamber (the white horizontal line) and knocks out a proton, which shoots up the picture. The small gap at the start of the proton's track, above the paraffin, occurred because drops did not condense on ions near the paraffin.

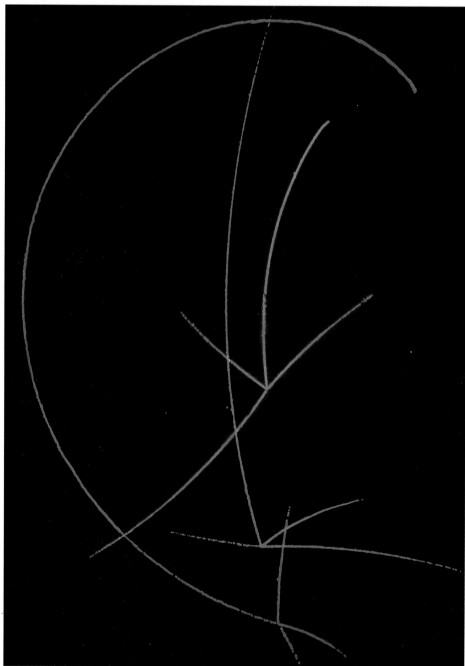

Fig. 3.21 A narrow beam of 90 MeV neutrons, produced by the 184 inch (4.6 m) cyclotron at the Lawrence Berkeley Laboratory, enters a cloud chamber from the bottom. The chamber is filled with hydrogen gas saturated with a mixture of ethyl alcohol and water. While the neutrons are invisible, their effects certainly are not. Neutron-induced transmutations of oxygen and carbon in the alcohol produce three 'stars', one above the other. Tracks not involved in these interactions have been removed; those remaining have been coloured according to which star they belong to.

hydrogen in the chamber has in this case been mixed with a little alcohol, which is basically a mixture of carbon (six protons and six neutrons) and oxygen (eight protons and eight neutrons). These nucleons like to clump together as alpha particles, making three alphas in the case of carbon, and four for oxygen. The energetic neutrons entering the chamber can break up these 'alpha clusters'. The four-pronged star at the centre of the picture is probably due to the four alpha particles produced by the break-up of an oxygen nucleus.

A second way of telling that neutrons are around is when they cluster with a proton to form an isotope of hydrogen such as deuterium (one proton and one neutron) or tritium (one proton and two neutrons). The neutrons add no charge to the proton but they do add mass, and therefore slow it down. An example appears in Fig. 3.22 where gamma rays enter a cloud chamber filled with helium. One of the gamma rays splits a helium nucleus (two protons and two neutrons). One proton flies off rapidly downwards; the other proton, bound to two neutrons in a tritium nucleus, moves up the picture. A proton has gone in each direction, but the extra bulk of the neutrons causes a slower flight, and therefore a thicker trail, for the tritium nucleus.

The neutron's neutrality can be an advantage. Whereas protons are repelled initially by the electrical forces surrounding nuclei, neutrons feel no such disruption. Slow neutrons can freely approach and enter nuclei, modifying their internal structure and creating new isotopes. This is the key to several technologies, such as the production of special radioactive isotopes for medicine.

Another consequence of the neutron's penetrating powers is its ability to split a uranium-235 nucleus into two fragments, releasing nuclear energy and two or three additional neutrons in the process. These neutrons can in their turn split further nuclei of uranium-235, releasing more energy and more neutrons. In a sufficiently large lump of uranium-235, a chain reaction will occur in which the multiplying neutrons cause the fission of ever more nuclei, leading to an explosive release of energy. This is how the atom bomb works.

A single fission of uranium (with no chain reaction!) has been captured on film in Fig. 3.23. Neutrons enter a cloud chamber containing a thin uranium-coated foil at the centre. Many protons and nuclei recoil. But there are also thick tracks due to the two fragments of the divided nucleus—the fission products—which leave the foil in opposite directions. Their high electrical charge produces strong tracks which end in characteristic branches where the fragments hit nuclei in the gas.

Fig. 3.22 *An invisible gamma ray, produced by the 100 MeV electron synchrotron at the University of Torino, enters a helium-filled cloud chamber from the right. It interacts with a helium nucleus to produce a proton and a nucleus of tritium. The proton and the tritium nucleus have equal charge, but different masses. The tritium contains two neutrons in addition to a proton and so is three times as heavy as a single proton. The tritium moves away, up the picture, relatively slowly and leaves a thicker trail than the lighter single proton. The distance between the central crosses is 16 cm.*

Fig. 3.23 *Neutrons entering a cloud chamber knock on many protons and nuclei, leaving short tracks. But one neutron has induced fission in a uranium nucleus in a thin layer coating a gold foil down the centre of the chamber. The two bulky, heavily ionizing fission fragments move out sideways, producing long tracks with short branches towards their ends. These are due to nuclei from the chamber gas, which have been knocked by the fission fragments. The chamber has a diameter of 25 cm.*

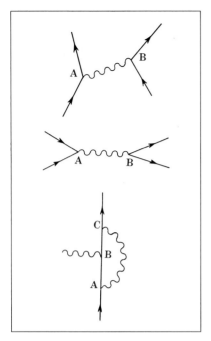

Fig. 3.24 *Richard Feynman invented diagrams like these to help in calculating the electromagnetic interactions of charged particles. The lines and vertices in the diagrams are a stylized shorthand for detailed mathematical expressions describing the behaviour of the particles. In general, charged particles are indicated by straight lines, photons by wiggly lines. The top diagram could represent a muon (on the left) kicking an electron (right) out of an atom by exchanging a photon (wiggly line). This is the kind of process that has occurred in Fig. 3.8 (see p. 52). The centre diagram shows an electron and a positron (the two straight lines on the left) annihilating at A and producing a photon (wiggly line) which rematerializes at B as new forms of matter and antimatter, such as an electron and a positron or a muon and an antimuon (compare Fig. 5.7, p. 89). The bottom diagram illustrates an electron emitting a photon at A, absorbing a second photon at B, and then reabsorbing the first photon at C. This kind of process makes a measurable contribution to the behaviour of an electron in a magnetic field.*

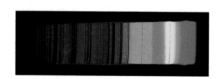

Fig. 3.25 *Photons of sunlight of specific energy—and therefore wavelength—are absorbed by elements in the Sun's atmosphere, and so do not reach Earth. When the Sun's spectrum is photographed, these missing wavelengths show up as black 'absorption lines'. There are some 20 000 of these lines in all, but this photograph shows only the major ones. The line in the yellow part is due to a pair of sodium lines, so close together that they are indistinguishable.*

THE PHOTON

OUR SENSES perceive the visible light and warm rays from the Sun as a continuous stream of radiation. So it comes as no surprise to learn that light and heat (infrared), and all the other forms of electromagnetic radiation, can be described in terms of vibrating waves—oscillations of intertwined electric and magnetic felds. Yet the quantum theories of the 20th century have provided us with another view of light, as a staccato burst of particles of zero mass called photons. This is true all the way across the electromagnetic spectrum, from high-frequency (short-wavelength) gamma rays through to low-frequency (long-wavelength) radio waves. Frequency and wavelength describe waves of light, but photons are described in terms of their energy and momentum. Gamma rays consist of high-energy photons, radio waves of low-energy photons.

Still more surprising is that photons are responsible for transmitting electromagnetic forces. The electric forces that hold atoms together are in effect 'carried' by photons, which flit back and forth between the atom's electric charges. These photons within the atom are transitory entities, existing on timescales too fine for us to perceive with our slow, macroscopic senses. Physicists call them 'virtual' photons.

This description of electromagnetic forces in terms of photons is a consequence of quantum theory. First defined by the British physicist Paul Dirac in 1928, it is now formulated in the theory known as quantum electrodynamics, or QED, which was worked out by two Americans, Richard Feynman and Julian Schwinger, and Sin-Itiro Tomonaga in Japan in the late 1940s. These three theorists shared the Nobel Prize for their work in 1965. Figure 3.24 shows examples of the diagrams that Feynman invented as an aid to calculating detailed electromagnetic effects through the exchange of photons between charged particles.

The predictions of QED have been verified many times and to a precision of better than one part in a billion. So precise and powerful is QED that it has become the template for subsequent theories of other fundamental forces, such as the weak and strong forces.

The photon belongs to a class of elementary particles, distinct both from the electron and its relatives, and from the quarks that make up nucleons and other subatomic particles. It is an example of what is now called a *gauge boson*. It acts as a mediator and force-carrier between particles. And just as the photon 'carries' the electromagnetic force, so other gauge bosons carry the other fundamental forces—the W and Z particles in the case of the weak force, the gluons in the case of the strong force between quarks, and the hypothetical graviton in the case of gravity.

If we shake an atom, by heating it or firing electrons or other charged particles at it, we can shake the photons loose from the atom's electric fields. When this happens photons emerge with energies characteristic of the parent atom, reflecting the pattern of electron energy-levels unique to each specific element. And because the wavelength (colour) of light is related to its energy, the resulting spectrum is an autograph of that element. Thus the light from sodium vapour glows yellow, that from neon is red, while copper colours a flame green. It was in attempting to explain the broad spectrum of radiation emitted from objects, that the German physicist Max Planck realized in 1900 that the radiation must be bundled into 'quanta', which we now call photons.

Atoms do not only emit photons, they can also absorb them. When a photon encounters an atom, the photon can give up its energy to an electron and raise it to a higher energy level. Absorption of this kind happens only if the energy of the photon exactly matches one of the energy steps within the atom, but it provides an important means of analysing the elements in a material. This is how various elements were first discovered in the outer regions of the Sun. Photons streaming out from the Sun cover a whole range of energies, some of which match the energy steps of particular elements. These elements absorb the photons and so we find dark shadows at these wavelengths in the Sun's spectrum, as in Fig. 3.25.

The energy of photons in the visible range can raise electrons from one energy level to another within atoms. But photons, particularly in higher-energy X-rays and gamma rays, can also knock electrons out of atoms entirely, or even shatter the

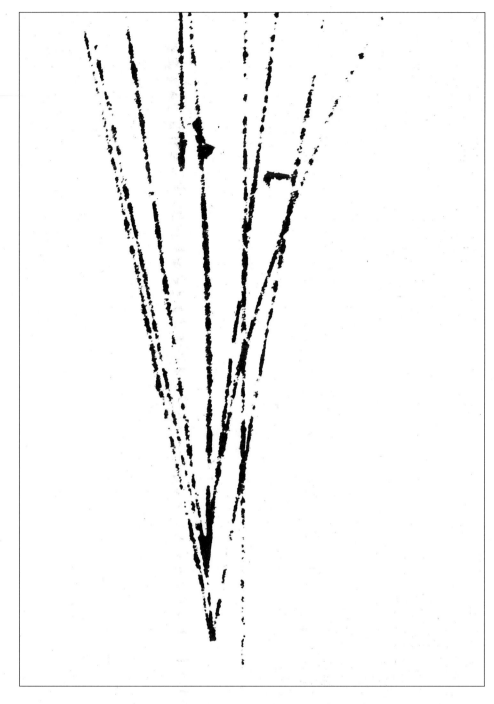

Fig. 3.26 *Tracks of particles appearing as if from nowhere betray the interaction of a high-energy photon with a proton in the 1 m hydrogen bubble chamber at the Stanford Linear Accelerator Center (SLAC). The photograph is about 10 times life size.*

nuclei of atoms into their constituent neutrons and protons.

The release of electrons from atoms by photons is known as the photoelectric effect. Working out the theory of the phenomenon won Albert Einstein the Nobel prize for physics in 1921. Nowadays the effect is exploited in many modern processes and gadgets such as solar cells and 'electric eyes'. The essential feature is that the photons liberate electrons, which can then flow and carry an electric current.

These devices work with photons at the relatively low energies of visible light. Figure 3.26 shows the effect of a photon at a much higher energy. In this case, a gamma ray photon has collided with a proton and given it so much energy that it has shaken loose new particles, spawned as the excited proton divests itself of its excess energy and returns to normal. Experiments of this kind helped to reveal the structure of the proton during the late 1960s and early 1970s, but that story belongs to later chapters of this book.

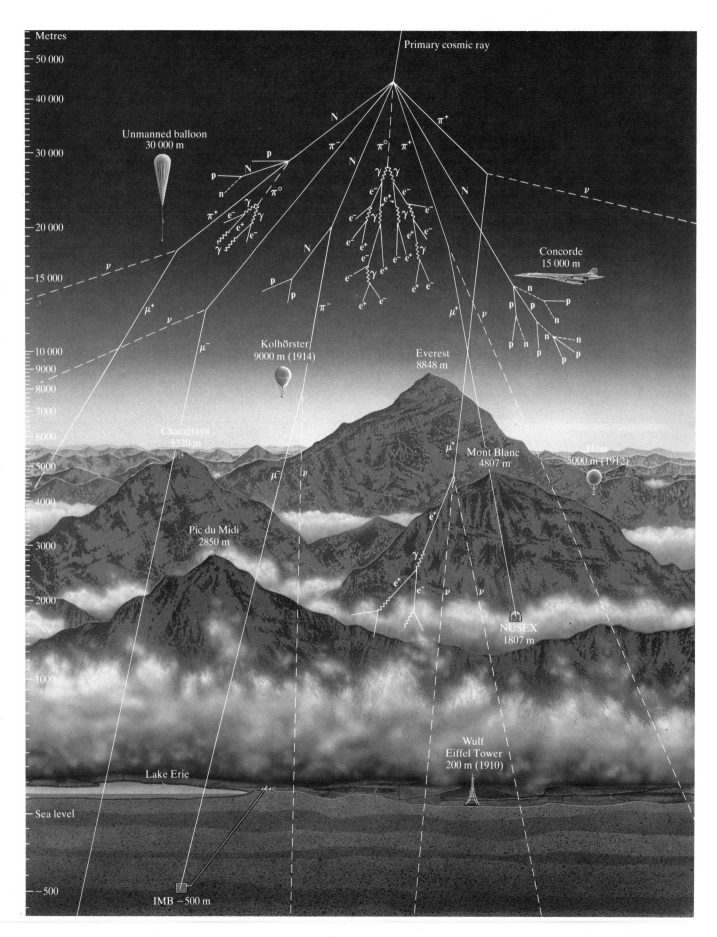

CHAPTER 4
THE EXTRATERRESTRIALS

THOUSANDS OF metres above the Earth's surface, the outer atmosphere experiences a continuous bombardment. The 'artillery' comes in two forms: photons and subatomic particles. The photons cover the whole spectrum of electromagnetic radiation—from radio waves through visible light to gamma rays. Most of the other particles are energetic nuclei, that is atoms stripped of all their electrons. The astronauts on board the Apollo spacecraft on their way to the Moon literally saw the effects of such energetic nuclei, as they tried to sleep. Whenever a heavy nucleus occasionally struck the retina in the closed eye of one of the dozing astronauts, he saw a tiny flash of light.

The high-energy rain of particles has become known as *cosmic radiation*, but in many respects it is quite different from the alpha and beta radiation emitted by radioactive nuclei, described in Chapter 2. Cosmic ray particles have much higher energies than alpha and beta rays, and are thinly spread. These two factors in particular have made the nature of cosmic rays much more difficult to discover.

The Apollo astronauts were out in space, exposed directly to the cosmic rays, but here on Earth we are shielded by the atmosphere. When the high-energy *primary* radiation from space hits the atoms and molecules of the upper atmosphere, it generates showers of subatomic particles, which form a rain of *secondary* radiation. Most of this is absorbed in the atmosphere before reaching the Earth's surface, leaving only a thin 'drizzle' of radiation to pass harmlessly through our bodies. We now know that this drizzle consists mainly of electrons, particles called *muons*—heavy electron-like particles that were discovered in studies of cosmic rays—and neutral particles called *neutrinos*. Roughly 20 particles per square centimetre arrive each second at the top of the atmosphere. At sea level this has been reduced to about one particle per square centimetre per minute. By comparison, a gram of a radioactive substance such as radium emits thousands of millions of particles a second.

Cosmic rays have very high energies, rising to 10 million million times the maximum energy of the radiation from radioactive sources. However, the highest-energy particles are very rare, with only a few arriving in a square kilometre over a period of 10 years! The majority of cosmic rays have energies up to 1000 times more than those produced by radioactivity. Most of these particles come from within our own galaxy, the Milky Way, and are accelerated by many different processes.

The ultra-high-energy cosmic rays may originate from beyond our galaxy. Though they are few and far between, they produce characteristic showers of millions of particles that spread over several square kilometres when they reach the Earth's surface. Studies of these 'extensive air showers' provide evidence that high-energy primaries come from the direction of the cluster of galaxies in the constellation of Virgo. This cluster includes a giant galaxy known as M87, which astronomers believe has a huge black hole at its centre.

Though we have a relatively poor understanding of where cosmic rays come from and how they reach the Earth, we do know in great detail what the radiation consists of and what it does to the atmosphere. Figure 3.12 (see p. 57) shows the tracks of different kinds of nuclei in the cosmic radiation, caught in photographic emulsions flown high in the atmosphere by balloon. The interactions of this primary radiation initiate a rain of secondary radiation, which develops through the atmosphere as Fig. 4.1 illustrates, and which can penetrate below ground.

Many scientists studied the cosmic radiation in great detail after its discovery in 1912. Their investigations progressed in parallel with the studies of the atom and the

Fig. 4.1 *A continuous rain of cosmic rays— energetic photons, protons, and other atomic nuclei—enters the Earth's atmosphere from outer space. This primary radiation collides with the nuclei of atoms in the upper atmosphere and produces showers of secondary particles. They include protons (p), neutrons (n), light nuclei (N), and many charged and neutral pions (π). The pions are relatively short-lived and decay in flight. The neutral pions decay to a pair of gamma ray photons (wiggly tracks), which in turn produce pairs of electrons (e^-) and positrons (e^+). If they have sufficient energy, these electrons and positrons can radiate more photons, which produce further electron–positron pairs, creating an avalanche of particles that can reach down to sea-level. The charged pions that are not absorbed by nuclei in the atmosphere also decay, transmuting into penetrating muons (μ) and neutrinos (v) which can continue their journey far underground.*

The exploration of cosmic rays began at ground level, and progressed gradually up through the atmosphere. Theodor Wulf only climbed the Eiffel Tower, but Victor Hess rose to 5000 m in his balloons, and Werner Kolhörster to 9000 m. Modern unmanned balloons can reach the edge of the atmosphere at 30 000 m or more. In the 1930s, mountain-climbing physicists established cosmic ray observatories on the Pic du Midi, in the French Pyrenees, and more recently on Bolivia's Mt Chacaltaya. Today, the penetrating muons and neutrinos are detected as a spin-off of underground proton decay experiments, such as NUSEX in the Mont Blanc road tunnel and the IMB detector beneath Lake Erie (see Chapter 10).

nucleus described in Chapter 2. The work involved similar experimental techniques, and indeed sometimes the same inquisitive minds. And there were surprises in store. By the end of the 1940s, the cosmic radiation had revealed unexpected particles of matter, different from the electrons, protons, and neutrons of atoms and nuclei. The discovery of these particles inspired the building of modern accelerator laboratories where we can produce high-energy particles to order—though still none as energetic as the highest energy cosmic ray particles.

Modern particle physics has thus grown out of the early studies both of radioactivity and of cosmic rays. Many of the cosmic ray physicists of the 1930s and 1940s were later to work at accelerators, bringing with them an armoury of techniques which, even today, underlie the complex technology of experiments in particle physics. The cosmic rays had in their turn been discovered by intrepid scientific explorers who went up in balloons at the turn of the century, to answer some of the questions raised in the study of radioactivity.

THE DISCOVERY OF COSMIC RAYS

RADIOACTIVITY ATTRACTED everyone's attention soon after its discovery. The radiation from radioactive bodies was easy to detect, for it splits the molecules of the air into positive and negative ions, and makes the air electrically conducting. In this way, the radiation reveals its presence in electrometers (see Fig. 2.15). But a puzzling phenomenon soon became apparent: even when no radioactive source was present, electrometers would indicate the presence of some other 'radiation' that ionized the air.

Armed with electrometers, scientists looked all over for the tell-tale indications of the mysterious emanations. The rays showed up everywhere, even out at sea, far from the radioactivity of rocks. But the most peculiar thing was that however much the researchers shielded their detectors, some radiation still penetrated. The rays from radioactive materials could not breach the shielding, so it seemed that another source of unknown rays of immense penetrating power must exist. But where?

The first clues emerged in 1910 when Father Theodor Wulf, a Jesuit priest, went up the Eiffel Tower and measured more radiation than he expected. Wulf guessed that the rays might have an extraterrestrial origin and he proposed going up in balloons to great heights to test this idea. But the spirit of adventure must have deserted him, since he seems to have been reluctant to do this himself! The risky exercise was undertaken instead by others, notably the Austrian, Victor Hess. During 1911–12 he made ten ascents in balloons complete with detecting apparatus, reaching heights of over 5000 m. These experiments showed that the intensity of the radiation increases rapidly above 1000 m, being some three to five times greater at 5000 m altitude than at sea level. Hess concluded that there must be a powerful radiation originating in outer space, entering the Earth's atmosphere, and diminishing as it passes through the air towards the ground.

Hess discovered the cosmic rays with the aid of instruments that required the personal attendance of the experimenter to watch and record the results. Robert Millikan's group at the California Institute of Technology (Caltech) in the mid-1920s developed an electrometer whose readings could be recorded on a moving film without need for anyone to be present. This extended the observational possibilities enormously. Unmanned balloons could take the recording equipment to very high altitudes and, at the other extreme, great depths of water could be plumbed for the presence of the rays.

At first Millikan had not believed Hess's claims that the radiation came from outer space. However, he made extensive investigations of his own and in 1926 he changed his mind, even going so far as to claim credit for the discovery himself! As the cosmic rays penetrate matter so easily, and gamma rays are the most penetrating form of radioactivity, Millikan believed, along with many others, that the cosmic rays are ultra-high-energy gamma radiation. He proposed that the primary gamma rays came from nuclear reactions occurring out in space, in which heavier elements were being

Fig. 4.2 *Victor Hess (1883–1964) in 1936, the year he won the Nobel prize for physics for his discovery of cosmic rays.*

Fig. 4.3 *Robert Millikan (1868–1953) assists in setting up instruments to measure cosmic rays on a balloon flight from Bismarck, North Dakota, in 1938.*

synthesized from lighter elements, ultimately from the lightest of all, hydrogen. He referred to the cosmic rays as the 'birth cries' of newly-born matter.

Hess was awarded the Nobel Prize for his work in 1936, and he is generally acknowledged as discovering the cosmic radiation. Millikan is honoured through the evocative name 'cosmic rays', which he coined and which is commonly used today. Wulf, whose ambitions reached only as high as the Eiffel Tower, is all but forgotten.

Once everyone had accepted the existence of the cosmic rays, there remained the problem of finding out exactly what they are. Hess had correctly identified outer space as the source of the rays. But the intensity of the cosmic radiation is so low, even at high altitudes, that with his relatively simple instruments Hess could do no more than recognize that cosmic rays exist; he could not identify their content. Also, having high energy and therefore high velocities, the cosmic rays are less ionizing than the lower-energy rays from radioactive sources. The more energetic cosmic ray particles pass atoms too quickly to have much effect, and so knock electrons out of fewer atoms; cosmic rays are not only thinly spread, they are also elusive. In addition, there is the problem of not knowing precisely from which direction the next cosmic ray will come. This is utterly different from radioactivity. Whereas a radioactive source set at the end of a collimating tube produces a narrow, well-defined beam of radiation, cosmic rays arrive at the Earth from all directions.

A big step forward came in 1928, when Hans Geiger and Walther Müller, at the Physics Institute in Kiel, developed what we now call the 'Geiger counter'. This was an improved version of the cylindrical counter with a wire down the centre that Geiger and Rutherford had used in 1908 to count alpha particles (see p. 32). With the Geiger–Müller counter, the electric field at the wire is so high that a single electron anywhere in the counter can trigger an avalanche of ionization. As many as 10 thousand million electrons would be released along the whole length of the wire. The tiniest amount of ionization would produce a signal from the new Geiger–Müller tube.

This ability to detect low levels of ionization makes the Geiger counter ideal for studying the high-energy cosmic rays. But the tubes form a still more powerful tool when two or more are used together. Soon after Geiger and Müller had demonstrated their new detector, Walther Bothe and Werner Kolhörster, both in Geiger's old laboratory in Berlin, used two tubes one above the other to make a 'telescope'—a device to define the path of a cosmic ray. With this they found the first conclusive evidence about the nature of the cosmic rays.

Fig. 4.4 *Hans Geiger (1882–1945) in 1930.*

Fig. 4.5 *A basic Geiger–Müller tube, or Geiger counter as it is more commonly known, consists of a metal cylinder, plugged at both ends, with a wire held at a high positive voltage (around 1000 volts) running along its axis. The tube is filled with a gas at low pressure. A charged particle passing through the gas will ionize atoms, releasing electrons which are attracted by the positive wire. As the electrons approach the wire they are accelerated in the electric field and become energetic enough to knock out more electrons. An 'avalanche' soon develops. The electric field around the wire is so intense that this avalanche propagates along the length of the wire, and the wire collects a large pulse of electric charge. This pulse can be used to activate a small loudspeaker, to produce the familiar 'click' associated with modern Geiger counters.*

In 1929, Bothe and Kolhörster used two of the newly invented Geiger–Müller tubes, one vertically above the other, to show that individual cosmic rays could penetrate a gold block 4 cm thick. A simultaneous discharge of the two electrometers attached to the tubes revealed the passage of a single cosmic ray.

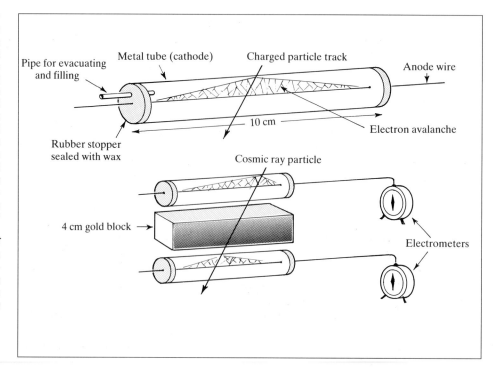

Fig. 4.6 *Bruno Rossi (b.1905) working in his laboratory at the Physics Institute of the University of Florence, around 1930. The horizontal tubes just visible in the centre of the picture are Geiger-Müller counters. Banks of dry cells (batteries) provided the high voltage for the counters.*

The common belief at that time, that cosmic rays are high-energy gamma radiation, implied that there should be charged particles associated with the cosmic rays, because the rays would tend to knock electrons out of atoms in the atmosphere. The Geiger counter telescope was the perfect tool to test these ideas.

Bothe and Kolhörster connected the two Geiger counters to electrometers and immediately observed simultaneous deflections of the fibre needles in the two electrometers. The surprise was that there were so many 'coincidences', each of which implied the passage of a cosmic ray through both tubes. A gamma ray fires a Geiger counter only if it first knocks an electron out of an atom; it is in fact the electron that triggers the counter. So the observation of coincident signals suggested that a cosmic gamma ray had either fortuitously produced two separate electrons—which was very unlikely—or that a single electron had fired both counters.

To test the latter possibility, Bothe and Kolhörster inserted material between the counters, to absorb electrons knocked from atoms. But they found that 75 per cent of the rays passing through the telescope were impervious even to gold blocks 4 cm thick! In fact, the particles triggering the Geiger counters were as penetrating as the cosmic radiation. The researchers were forced to conclude that these particles *are* the cosmic radiation; that the cosmic rays are highly penetrating electrically charged particles, and not gamma rays as had been supposed.

These results inspired, among others, Bruno Rossi at the University of Florence. He saw a way of using electronic valves—the predecessors of today's transistors—to register coincident pulses from the Geiger counters. This did away with the cumbersome arrangement of electrometers viewed by a self-winding camera, which Bothe and Kolhörster had used. Rossi used his technique to detect coincidences between counters that were not in a vertical line, but arranged in a triangle so that a single particle could not traverse all three counters. With a shield of lead over the counters, Rossi detected many coincident signals from all the counters. This showed for the first time the production of showers of secondary particles. Rossi went on to perform many key experiments in the detailed study of cosmic rays. More important still, his coincidence circuitry forms the basis of all the electronic counter experiments that today record the creation of many particles in man-made high-energy collisions.

Results from 'coincidence experiments' with Geiger counters, particularly those by Rossi, showed just how penetrating the cosmic rays are. They can pass through metre-thick lead plates and have even been detected deep underground, thousands of metres below the surface of the Earth. But because of their high energies and consequent low ionizations, identifying the exact nature of the particles present in cosmic rays was difficult or nearly impossible; that is, until Wilson's cloud chamber was used to study the cosmic radiation. The technique provided beautiful images of the cosmic ray tracks, and the real excitement began.

THE FIRST NEW PARTICLES

IN 1923, a young physicist called Dmitry Skobeltzyn began to investigate gamma rays in his father's laboratory in Leningrad. Gamma rays from a radioactive source knock electrons from atoms. Skobeltzyn hoped to detect the tracks of such electrons in a cloud chamber, but he encountered a problem: the gamma rays also knocked electrons out of the walls of the chamber, and these interfered with his measurements of the electrons produced from the gas in his chamber. To overcome this, Skobeltzyn decided to place the chamber between the poles of a large magnet, so that the magnetic field would deflect away the unwanted electrons.

A magnetic field exerts a force on an electrically charged particle so that its path becomes curved. The curvature depends on the strength of the field, and on the particle's momentum—the product of its mass and its velocity. The curvature is greater for slow or lightweight particles (low momentum) than for fast or heavy ones (high momentum). In this way, the magnetic field can distinguish different particles.

On some of the photographs he took in 1927–8, Skobeltzyn noticed a few tracks that were almost straight. These indicated values of momentum and energy much greater than for electrons from any source known at the time. Skobeltzyn supposed

Fig. 4.7 *Dmitry Skobeltzyn (b.1892) in his laboratory in Leningrad in 1924, with the cloud chamber that revealed for the first time the tracks of cosmic rays.*

Fig. 4.8 *One of the first photographs showing the track of a cosmic ray in Skobeltzyn's cloud chamber. He was studying the tracks of electrons recoiling under the influence of energetic gamma rays when he decided to subject the chamber to a magnetic field, so as to deflect unwanted tracks emanating from the walls of the chamber. In 1927 he noticed a few straight tracks, such as the vertical one at the centre of this picture, amid the curling tracks of the recoiling electrons. These straight tracks must have been made by very energetic particles, otherwise they would have been deflected by the magnetic field; it seemed they must be due to cosmic rays.*

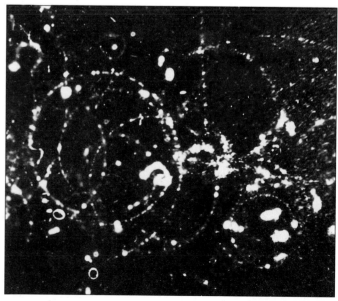

Fig. 4.9 *Carl Anderson (b.1905) working with the electromagnet for his cloud chamber in the Aeronautics Laboratory at Caltech, where a generator could supply 600 kW to power the magnet. The two coils of the magnet are wound with copper tubing, and cooled by tap water.*

that the tracks were due to fast-moving electrons knocked from atoms by cosmic gamma rays. (This was the year before Bothe and Kolhörster set up their cosmic ray 'telescope' and discovered that the cosmic rays actually are charged particles.) Although he did not realize it then, Skobeltzyn had become the first person to observe directly the tracks of the cosmic rays themselves.

Skobeltzyn did not follow up his discovery, but two years later, in 1930 at Caltech, Robert Millikan instructed one of his research students, Carl Anderson, to build a cloud chamber to study the energies of the cosmic ray particles. With the assistance of engineers at the nearby aeronautical laboratory, Anderson built a powerful water-cooled electromagnet, which could produce fields more than 10 times stronger than Skobeltzyn had used. His first dramatic results showed that the cosmic rays contain both positively and negatively charged particles in about equal numbers.

Millikan still believed that the cosmic ray particles were electrons knocked from atoms by primary gamma rays, so this observation came as something of a surprise. He insisted that the positive particles must be protons, also knocked from atoms by the high-energy gammas. To produce tracks of similar curvature to the electrons, the much heavier protons would have to be moving much more slowly. However, few of the tracks had the dense ionization expected for slow particles. Anderson, on the other hand, thought that the tracks could be due to electrons moving upwards through the chamber, rather than downward-moving positive particles. But Millikan did not like that idea at all, and stuck to his belief that the tracks were due to protons.

To settle the debate, Anderson inserted a lead plate across the chamber. Particles traversing the plate would lose energy and be curved more by the magnetic field on emerging into the chamber again. In this way he could tell the directions of the particles, and so be certain of their charges. The modification soon revealed that both Millikan and Anderson had been wrong. Anderson found a beautiful example of a positive particle that was clearly much lighter than a proton; its ionization and curvature suggested a mass similar to the electron's. He had discovered the positron — the 'anti-electron' predicted by theorist Paul Dirac (see pp. 84-6). This was the first observation of an 'elementary' particle that does not reside within the atom.

Positrons are not contained within atoms, so where do the positrons in the cosmic rays come from? Anderson was uncertain and it was an experiment by two physicists in Britain in the same year that provided the answer—positrons are created by the cosmic radiation itself. Patrick Blackett and Giuseppe Occhialini, in the Cavendish Laboratory at Cambridge University, had developed an improved version of the cloud chamber which was to provide an exciting new window on nature.

Up to this time, cosmic ray research with cloud chambers was in some senses rather hit and miss. The cloud chamber was expanded at some random instant and more often than not no cosmic ray happened conveniently to pass through. How could this

Fig. 4.10 *The first evidence for the positron, obtained in a cloud chamber photograph taken by Anderson. The particle must be moving up the picture, because it loses energy as it crosses a 6 mm thick lead plate at the centre, and curls round more in the top half of the chamber. The anticlockwise direction in which the track bends shows that the particle is positive, but its track is too faint for it to be due to a proton or an alpha particle. Anderson had earlier evidence for tracks from particles that appeared to be positive, because they were curling the opposite way to electrons, but which were too light to be protons. With hindsight we can say that they were positrons, but at the time Anderson believed them to be electrons moving upwards through the chamber, which would curl the opposite way in the magnetic field to electrons moving down. Millikan, Anderson's professor, argued that they must be protons. As Anderson has since recalled, 'Curiously enough, despite the strong admonitions of Dr Millikan that upward-moving cosmic ray particles were rare, this indeed was an example of one of those very rare upward-moving cosmic ray particles.'*

Fig. 4.11 *This picture, taken by Anderson on top of Pike's Peak in Colorado, shows a 'shower' of three electrons and three positrons produced by a cosmic ray which has interacted in the wall of the cloud chamber. The electrons bend to the left, and the positrons to the right, in the chamber's magnetic field.*

Fig. 4.12 *Giuseppe Occhialini (b.1907), on the far left, and Patrick Blackett, third from right, during a visit to the observatory on the Pic du Midi in the French Pyrenees in 1949.*

success rate—about 1 in 20—be improved upon? Blackett had been working on cloud chambers, producing images of nuclear transmutations, when he was joined by Occhialini, a young Italian physicist. Together they set about devising a way whereby cosmic rays would announce their presence. Occhialini had been a student in Florence under Rossi, and was able to bring Rossi's work on coincident signals from Geiger counters to blend with Blackett's gift for gadgets and expertise with cloud chambers.

Their idea was brilliantly simple. Put one Geiger counter above and another below the cloud chamber, then if both fire simultaneously it is very likely that a cosmic ray has passed through them and, by implication, through the chamber. Blackett and Occhialini connected the Geiger counters up to a relay mechanism so that the electrical impulse from their coincident discharges actuated the expansion of the cloud chamber and a flash of light to allow the tracks to be captured on film. Notice how it is the knowledge of the passage of the cosmic ray that is crucial—not an instantaneous knowledge of its presence. By the time the triggering signal from the Geiger counters is formed, the ray has already passed through the chamber. But the ionized track remains and the all-important droplets form on the ions. The cosmic rays are like jet-planes: their trails remain for a while and show where they have been.

Instead of 1 cosmic ray in 20 or more photographs, now the success rate was 4 out of 5! Blackett and Occhialini took their first photographs by this method during June of 1932, and then accumulated nearly a thousand photographs of cosmic rays during the late autumn of 1932.

Anderson was the first to report the observation of a positron, but the experiments of Blackett and Occhialini confirmed its existence without a doubt. Many of the pictures showed as many as 20 particle tracks diverging from some point in a copper plate just above the chamber, like water from a shower. The strong magnetic field throughout the chamber curved the tracks, showing roughly half the particles to be negatively charged and half positively charged. The pictures provided dramatic evidence that positrons are produced in the collisions of cosmic rays; Anderson's particle was not a peculiar 'extraterrestrial' object that entered the atmosphere with the primary cosmic radiation.

The explanation of the 'showers' begins with an energetic electron entering the copper plate; there the electric fields of the positive charges of the copper nuclei cause the electron to radiate photons—gamma rays. Provided they have sufficient energy, these gamma rays can in turn produce pairs of electrons and positrons, again under the influence of the nuclear electric fields. These are the electrons and positrons that Blackett and Occhialini saw, spawned from gamma radiation, itself produced by the cosmic rays in the copper plate above the chamber. Albert Einstein's equation, $E = mc^2$, implies that energy (E) can be converted into mass (m)—radiation into matter. Blackett and Occhialini had for the first time captured the process on film.

So, by the early 1930s, it was clear that cosmic rays, at least near ground level, contain electrons and positrons. But this did not seem to be the whole story. Among

Fig. 4.13 *Carl Anderson, left, and Seth Neddermeyer (b.1907).*

Fig. 4.14 *A Japanese dinner party in a Kyoto restaurant in 1956, with (left to right) Hideki Yukawa (1907–1981), Cecil Powell, Mrs Powell, and Mrs Yukawa.*

the tracks showing up in the cloud chambers were many that occurred with positive or with negative charge, but which were far more penetrating than electrons and positrons and which did not create showers. At Caltech, Anderson and his colleague Seth Neddermeyer at first favoured the idea that there might be two types of electron; they referred to the penetrating particles as 'green' electrons and the shower-producing particles as 'red' electrons. By 1936, however, they had convinced themselves that the penetrating particles were something new. In November of that year, they presented publicly their evidence that the particles have a mass between that of the electron and the proton.

The new particles were originally named 'mesotrons' from the Greek for 'middle', though this was later shortened to mesons. And as with the positron, there already existed a theory that could account for the mesotron—or so it seemed. In 1935 Hideki Yukawa, a Japanese theorist working in Kyoto, had built a theory of the powerful forces that bind the atomic nucleus. One of its consequences was that a new particle should exist, weighing some 250 times more than the mass of an electron and about 1/7 as much as a proton. Yukawa's work was not widely known outside Japan, but when Anderson and the others announced a new particle with a mass only 20 per cent away from his prediction, Yukawa claimed the particle as his own. Moreover, Robert Oppenheimer and Robert Serber at Berkeley in California had made the same connection, and promoted Yukawa's theory in the West.

But they were wrong to do so. Yukawa's prediction was fulfilled a decade later, with the discovery of the particle now known as the pi-meson, or pion (see pp. 90–1). The original 'mesotron' observed by Anderson and Neddermeyer, which is now known as the muon, is something quite different and uncalled for. Half a century later, it is still an enigma (see pp. 87–9).

STRANGE PARTICLES

THE DISCOVERIES of the positron and the muon were the first of a series showing that Earth-bound physics had sensed only a small corner of nature's rich pageant. By 1950 cloud chambers exposed to cosmic rays had revealed yet more new particles, inexplicable by the existing theories. Their discovery owed much to the improvements in experimental techniques immediately following the Second World War.

In 1937, Patrick Blackett moved to Manchester University from Birkbeck College, London, where he had been a professor since leaving Cambridge in 1933. He immediately began to gather about him a strong team of cosmic ray physicists, but their research was short-lived, being brought virtually to a standstill by the Second World War. Blackett became science advisor to the RAF and anti-aircraft defence and later director of the British naval operations research and anti-U-boat strategy. Many of his recently gathered team went off to other duties, but a few remained to carry on the training of physics students. So a little research on the cloud chamber studies of cosmic rays continued at Manchester—with Blackett's permission and the proviso that it cost nothing! George Rochester and the Hungarian Lajos Janossy spent their spare time investigating a particular interest of Janossy's—the so-called 'penetrating showers', in which extremely energetic particles enter a cloud chamber and produce a cascade of many tracks.

After the war Janossy moved to Dublin. Blackett, however, was sufficiently impressed by the wartime studies to encourage Rochester, together with Clifford Butler, to continue the work, this time with a cloud chamber in a magnetic field. (Electromagnets use electricity, so it had been out of the question to employ a magnetic field for the research during the war!) When Blackett came to Manchester he had brought with him the large electromagnet he had built in 1935 especially for work with counter-controlled cloud chambers. Rochester and Butler set about building an entirely new cloud chamber, which they placed in Blackett's magnet. They arranged Geiger counters between sheets of lead above and below the chamber, so that it would expand only when a penetrating shower had occurred.

The pictures collected with this arrangement during 1946 and 1947 revealed a major surprise—the first examples of the so-called 'strange' particles. Cosmic rays hit the lead above the chamber and created penetrating showers. Among the many tracks were two curious pronged, or V-shaped, patterns. Rochester and Butler pointed out that both these 'vees' could be explained in terms of the spontaneous decay of new unstable particles. In one case the new particle had to be neutral, in the other a

Fig. 4.15 (below) *George Rochester (b.1908) in 1958.*

Fig. 4.16 (right) *Clifford Butler (b.1922), in about 1947, with a system for projecting photographs from a cloud chamber. The image of a particular track was projected through a prism on to the white screen, and the prism was turned until the track became straight on the projected image. The curvature of the track could then be read from the position of the prism, which was calibrated in terms of radius of curvature.*

charged particle was called for; and both particles would have to weigh about half as much as a proton.

Such a discovery was totally unexpected and, on such sparse evidence, not without some controversy. The tension rose for Rochester and Butler as two years passed and no more examples turned up. To improve their prospects of observing penetrating showers, Blackett gave Butler the task of setting up a magnet and cloud chamber at altitude.

After some deliberations, Blackett agreed to the choice of the 2850 m high Pic du Midi de Bigorre in the French Pyrenees. An astronomical observatory already existed at the summit, and by November 1949 a team from Manchester had successfully installed Blackett's 11 tonne magnet there. Then, over on White Mountain in California, Carl Anderson reported that he and Eugene Cowan were obtaining one of the 'vees' per day—a total of 28 in all. He said that 'to interpret our photos we require the same remarkable conclusion as Rochester and Butler: spontaneous decay of neutral and charged unstable particles of a new type.'

Between July 1950 and March 1951, the chamber on the Pic du Midi produced 10 000 photographs of high-energy showers. Among these were 67 'vees' of which 51 were due to neutral decays and 12 to charged decays. This was decisive proof that the 'vee' particles were genuine. But what was novel was that four of these cases indicated the presence of a neutral particle *heavier* than the proton. The first 'vees' found were from particles with a mass about half that of the proton; they have since become known as charged and neutral *kaons* (see pp. 92–3). The neutral particle heavier than the proton is called the *lambda* (see pp. 94–7). Together, the kaons and the lambda became known as the 'strange' particles because their behaviour was unexpected.

We now know that these particles carry a new property—'strangeness'—which is akin to electric charge, but which does not exist in ordinary matter. No one had predicted the existence of any of these particles, and their discovery created great excitement. Here was a whole new family of particles that might lead to some novel ideas about the laws of the Universe. The kaons and the lambda and still heavier relations took some years to understand, as Chapter 5 describes. Researchers were helped in the 1950s by the advent of new powerful particle accelerators which allowed them to simulate the cosmic ray collisions under controlled conditions. Meanwhile, another technique for tracking particles was bringing still more exciting information about the cosmic rays.

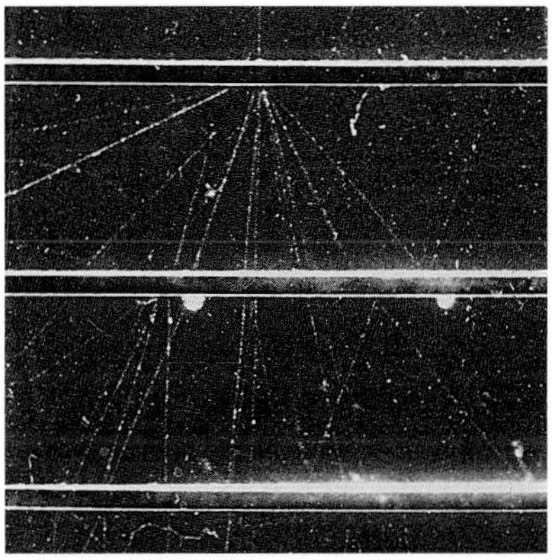

Fig. 4.17 *A cosmic ray interacts in a 0.6 cm lead plate across a cloud chamber and produces several particles, including a neutral one that subsequently reveals itself by a 'V' (top left) when it decays into two charged particles. Pictures like this, taken by W. B. Fretter at Berkeley, demonstrated that the 'V' particles were created in high-energy nuclear collisions.*

Fig. 4.18 *The Pic du Midi Observatory, in the French Pyrenees, where the cloud chamber from Manchester University was moved in 1949 to improve its chances of observing 'V' particles.*

POWELL, PIONS, AND EMULSIONS

THE CLOUD chamber had helped immensely in unravelling the content of the cosmic rays at ground level during the 1930s and 40s. But at the same time, it had become clear that the electrons, positrons, and muons are secondary particles created from very energetic primary radiation as it strikes the atmosphere at high altitudes. Several experiments revealed that the intensity of cosmic rays varies with the Earth's latitude. This can be explained only if the primary radiation is influenced by the Earth's magnetic field, which also varies with latitude; and it demonstrates that the cosmic rays carry positive charge and must consist of particles. But what kind of particles?

The extremely high energies of the primary cosmic rays at first made it difficult to ascertain their nature. However, in the late 1940s, the development of special photographic emulsions, which could easily be carried aloft by balloons, brought physicists their first beautiful images of the interactions of high-altitude cosmic rays. These emulsions were especially sensitive to high energy particles; just as intense light darkens photographic plates, so can the passage of charged particles. We can detect the path of a single particle by the line of dark specks that it forms on the developed emulsion. The particle literally takes its own photograph.

Photographic plates had figured in the very earliest work on radioactivity; indeed, it was through the darkening of plates that both X-rays and radioactivity were discovered. The essential feature of photography is that paper, or a glass plate, covered

Fig. 4.19 *Cecil Powell (1903-1969).*

Fig. 4.20 *An example of Powell's work with emulsions in accelerator beams. This shows the collision of a proton, coming in from the bottom of the picture, with a proton in the emulsion. Notice the 90° angle between the tracks of the two scattered protons (compare Fig. 3.16, p.59). The shorter proton track is about 0.04 mm long.*

with a thin layer of silver bromide responds to light. (The bromide is usually in the form of an 'emulsion' of crystals mixed with gelatin.) Light affects the silver bromide crystals in such a way that when they are treated chemically—developed—they release some pure silver. The more light landing at a point, the more silver is produced and the darker the image on the developed film or plate. In this way a 'negative' image forms, with dark regions corresponding to places that have received the most light, and which are the brightest on the scene being photographed.

X-rays are an energetic form of visible light, so it is perhaps not surprising that they register on photographs. What may seem more remarkable is that a speck of radium left on a photographic plate should produce an image, as in Fig. 2.12 (see p. 28). This happens because the alpha particles from the radium can ionize atoms in the emulsion and, as with light, this darkens the crystals of silver bromide after processing. Becquerel discovered radioactivity through blurred images from uranium salts and, in 1911, M. Reinganum became the first person to record the tracks of individual alpha particles in emulsion. This achievement was due to the fact that Reinganum ensured that the alpha particles travelled within the thin layer of emulsion for some distance, parallel to the surface of the plate.

Despite these early successes, the photographic technique for detecting radiation appeared to suffer from two serious drawbacks. First, the emulsions had to be in thin layers, only fractions of a millimetre thick, so that they could be developed properly. As a result, only particles travelling within the plane of the wafer-thin emulsion left any significant track; particles passing through the emulsion left an almost invisible spot. Secondly, the emulsions available at the turn of the century were sensitive only to slow particles. Faster particles passed through so quickly that they did not affect enough light-sensitive grains to form a visible track. To produce denser tracks the emulsions needed to contain more of the active ingredient—silver bromide—but this was technically difficult to achieve.

One person who persevered with the use of emulsions to record the tracks of low-energy particles was Cecil Powell, who had been a student at the Cavendish Laboratory under Charles Wilson, the inventor of the cloud chamber. Powell moved to Bristol University in 1928, and in 1935 his team began to build a Cockcroft-Walton accelerator. They initially used a cloud chamber to study the interactions of the particles produced, but Walter Heitler, one of the theorists at Bristol at the time, drew Powell's attention to the work of two Viennese researchers, M. Blau and H. Wambacher. They had been using photographic emulsions to detect cosmic rays; in particular, they had shown that emulsions could be sensitive to protons, not just the heavier alpha particles.

For Powell, the photographic technique had the advantages over the cloud chamber of greater simplicity and a far greater accuracy in measurements of a particle's range. A set of emulsion-covered plates is sufficient to collect particle tracks; a cloud chamber, on the other hand, is a complex piece of apparatus, needing moving parts so that the chamber can be continually expanded and recompressed. Powell's team tested emulsions by detecting cosmic rays high on mountain-tops. Then they turned to detecting the collisions of particles from the accelerator at Bristol.

The technique proved worthwhile, and in the late 1930s, and to a lesser extent during the Second World War, Powell's group studied nuclear collisions not only at Bristol but at a more powerful accelerator at Liverpool University. In some ways they were lucky, for their emulsions were of unusually high quality. Moreover, the accelerator provided a copious supply of particles travelling in a well-defined direction, so it was not too difficult to arrange the plates in suitable positions for recording tracks. However, after the end of the war, Powell was to extend the technique to make detailed studies of cosmic rays at high altitudes.

In 1945, the newly-elected Labour government set up an important scientific committee at the Ministry of Supply in London, chaired by Patrick Blackett. One of the committee's decisions was to encourage nuclear research outside the immediate concerns of national defence. To this end it formed two panels, one to develop accelerators and the other (which included Powell) to investigate special 'nuclear' emulsions, particularly sensitive to energetic subatomic particles. With support from the Ministry of Supply, a research team at Ilford Ltd had by May 1946 produced an emulsion incorporating about eight times as much silver bromide as normal. The improved sensitivity provided a medium that rivalled the cloud chamber in the visual beauty of its images.

These better emulsions were immediately exploited in cosmic ray research in the skilled hands of Powell and of Occhialini, who had come to Bristol in 1945 on Blackett's recommendation. Occhialini took some plates to the French Observatory on the Pic du Midi. The results stunned the physicists by revealing a whole new world of nuclear interactions, produced by the primary cosmic radiation entering the atmosphere. 'It was as if, suddenly, we had broken into a walled orchard, where protected trees had flourished and all kinds of exotic fruits had ripened in great profusion,' recalled Powell in later years.

The fruits turned out to be more exotic than the physicists had anticipated, for trapped in the emulsions examined in 1947 was evidence for a new type of particle. This new particle, like the 'mesotron' Anderson had found 10 years previously, had a mass between that of the electron and the proton. But it was slightly heavier than Anderson's particle, and in the tracks revealed by the high-sensitivity emulsions the new particle could be seen to decay after a few tenths of a millimetre into a particle like the 'mesotron'.

Powell and his team had at last really discovered the particle that Yukawa had predicted back in 1935 as the carrier of the strong force—the particle we now call the pion. Anderson's particle, now known as the muon, results from the decay of the pion, but for ten years the pion had remained hidden, and the muon had been presumed to be Yukawa's particle. Only with the use of sensitive emulsions at high altitude did the pion's brief life become visible for the first time. Yukawa received the Nobel prize shortly after the pion's discovery, in 1949; Powell was honoured in the following year.

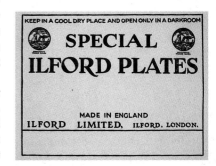

Fig. 4.21 *A packet of the new emulsions from Ilford Ltd.*

Fig. 4.22 *The interaction of a cosmic ray in the emulsion known as Kodak NT4, which was first produced in 1948 in response to the demand for emulsions of greater sensitivity. This was the first emulsion to be completely sensitive to electrons, and it gave recognizable tracks for particles at all velocities—notice how all these tracks are solid lines.*

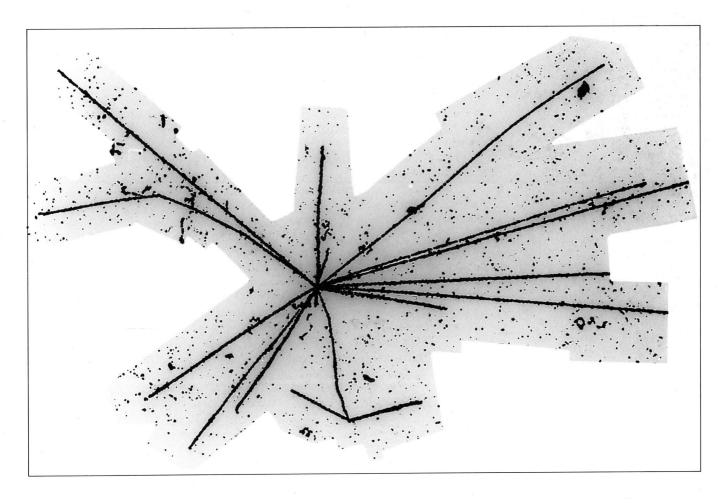

PARTICLES FROM OUTER SPACE

Fig. 4.23 *A balloon to carry emulsions up into the primary cosmic radiation in the high atmosphere awaits launch from Cardington in Bedfordshire, c.1950s.*

THE NEW emulsions at last enabled physicists to identify the nature of the primary cosmic radiation. Polyethylene balloons, their skins fractions of a millimetre (25 micrometres!) thick ascended to great altitudes, carrying emulsion. On return to Earth, and after developing, the emulsion revealed the primary rays to be atomic nuclei, moving at nearly the speed of light. Heavy nuclei, with large positive charge, gave thick tracks; lighter nuclei, with smaller net charge, gave finer tracks. It was possible to identify the nuclei of different elements with ease (see Fig. 3.12, p. 57). By 1950, it was clear that the primary rays are mainly protons (86 per cent), with a proportion of helium nuclei (12 per cent), and carbon and oxygen at the level of about half a per cent. Rarer nuclei were identified as more emulsions were exposed, and now many nuclei, up to uranium, have been found.

An important breakthrough in this work was the development of techniques to expose larger amounts of emulsion. Occhialini, while at Bristol, devised a means of successfully processing thick emulsions, up to a millimetre deep. The way to expose a greater depth of emulsion to cosmic rays is to pack many layers together in a stack. This may sound obvious, but recall that the emulsion is held on a backing, generally of glass to keep it as flat as possible. In the early 1950s, Powell and other researchers discovered how to peel unexposed emulsions away from the glass plates on which they were made—a process known as 'stripping'—so that as many as 100 layers of emulsion can be stacked together. After exposure, the layers of emulsions are carefully separated and returned one by one to the glass plates, so that they can be developed and studied in the usual way.

As with the cloud chamber, the tracks in emulsions provide sufficient clues to allow particles to be identified. But in this case the pictures must be studied through a microscope, for individual tracks are invisible to the naked eye. Moreover, because emulsion is much denser than the gas of a cloud chamber, particles do not travel so far, even at high energies.

One clue to a particle's nature comes from its 'range'—the distance it travels through the emulsion before stopping. The more energetic a particle, the greater its range. In addition, the person looking through the microscope can count the darkened grains; the more dark spots in a given length of track—usually 100 micrometres—the greater the rate of ionization. Heavy ionization—many dark spots—can occur because the particle has a large electric charge; or it can arise because the particle is moving slowly, and is near the end of its range. A third clue comes from the scattering that occurs, when the dense material of the emulsion deflects a particle from its straight path. Detailed measurement of the change in angle from one section of track to the next reveals information about a particle's mass.

It is worth pointing out that one problem with examining emulsion through a microscope is that the depth of focus is typically only 0.5 micrometres, one thousandth the thickness of the material; after all, the darkened grains are only a few tenths of a micrometre across. The scanner can look at tracks dipping through the emulsion only one portion at a time. Most of the pictures of tracks in emulsion are not exactly what one sees through a microscope; they are collages of the view into different layers as the focus is slowly changed.

Examining emulsions, especially those that have been stacked in a sandwich of many layers, can be very time-consuming. In the late 1940s, Powell built up a team of women—known as 'Cecil's beauty chorus'—to help his group in this task; and in publications, he and his fellow physicists were careful to credit the particular person who had found an interesting event. In this way, Powell anticipated the teams of scanners needed for more modern detectors, such as bubble chambers. He also established the precedent for international teams of scientists from a number of institutions, who collaborated on collecting the data—exposing the emulsions—and then divided the spoils for analysis.

Stacks of emulsion carried by balloons completed the final part of a picture that Hess had begun on his flights in hot-air balloons. Atomic nuclei from outer space fragment on collision with atoms in the upper atmosphere. The fragments consist for

the most part of protons, neutrons, and light nuclei, many of which are clearly visible in the photographs. But the nuclear maelstrom also includes pions, which can carry positive or negative electrical charge or can even be uncharged.

The uncharged pions decay rapidly to gamma rays, which produce showers of electrons and positrons as they travel through the atmosphere. The charged pions that are not first absorbed by nuclei in the atmosphere decay in flight, transmuting to muons. These muons traverse the atmosphere with ease and can even penetrate far underground. Although muons are much longer-lived than pions, they also often decay in flight. Powell photographed examples of a pion decaying into a muon, which in turn decayed into an electron (Fig. 4.25). The abrupt changes in direction at each stage result from the simultaneous emission of an unobserved lightweight particle called the neutrino, which is also very penetrating and can even travel right through the Earth.

Thus by the early 1950s, a picture of the whole sequence of processes involved in the cosmic radiation, from the upper atmosphere to below ground, had emerged. But it had also revealed rather unexpected things, such as the muon, the 'vees' discovered by Rochester and Butler, and the pion finally pinned down by Powell and Occhialini. The suspicion arose that a weird world of exotic particle varieties lay undiscovered, and this spurred scientists to build their own particle accelerators so that 'cosmic rays' could be generated with high intensity and to order under controlled conditions. The heyday of cosmic rays was past; it was now to be the turn of the accelerators, as Chapter 6 describes.

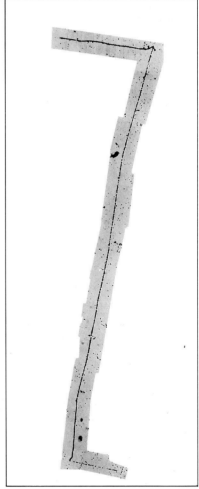

Fig. 4.25 *With the development of electron-sensitive emulsions, Powell was able to record the complete decay chain of a charged pion in images such as this one taken in October 1948. The pion comes in to the top of the picture from the left, leaving a strong track. It decays to a muon and an invisible neutrino. The muon proceeds down the page and then itself decays into an electron and a second neutrino. Again the neutrino remains invisible; the electron, however, leaves a faint but clear track. Notice how the muon's track, which is about 0.6 mm long, becomes denser as it slows down before it decays.*

Chapter 5
The Cosmic Rain

When physicists in the late 1920s proved once and for all that cosmic rays are penetrating high-energy particles, they opened up a new means for studying matter. Radioactivity, the phenomenon that had revealed the contents of the once 'indivisible' atom to Rutherford and his contemporaries, had now become a tool used by other scientists, including chemists and biologists. Cosmic rays became the new mystery for physicists to understand.

The cloud chamber was automated in the early 1930s, so that Geiger counters could trigger the expansion of a chamber only when specific patterns of tracks passed through, and it became a valuable tool for studying cosmic rays. Then in the late 1940s, the improved photographic emulsions brought to life an amazing wealth of detail in the tracks of particles captured high on mountain tops, or still higher in balloon-borne experiments. These new techniques revealed for the first time particles unseen in the study of atoms and nuclei at the relatively low energies associated with radioactivity. The new particles did not appear as part of the stable matter of the world about us; rather they proved to be transient objects, formed in the high-energy maelstrom of cosmic ray collisions in the upper atmosphere. That they were discovered at all is due to the ability of cloud chambers and emulsions to provide time-exposures of a particle's path, which sometimes recorded the birth and death of a particle on the same image.

Cosmic ray research in the 1930s and 40s led to the discovery of several new particles, and in a sense gave a preview of things to come later in experiments at particle accelerators. Some of the new particles had been predicted by theory. For example, the positive electron, or positron, the first example of an antiparticle, fitted in with the theory of the electron that Dirac had put forward in 1928. The pion also had been predicted, this time by Yukawa, as the carrier of the strong force. We now know that it was the first example of the group of particles that are classifed as 'mesons'. Other new particles were, however, entirely unexpected. The muon was at first confused with the pion, and it was only in the 1950s that physicists came to recognize it as a heavy relative of the electron. Most mysterious of all were the so-called 'strange' particles: the kaon, the lambda, the sigma, and the xi.

A better appreciation of the the role of the muon and the strange particles did not come until particle accelerators could mimic the actions of cosmic rays and produce plentiful supplies of the 'extraterrestrials'. With hindsight, we can see that the new particles were the first hints of a deeper complexity in nature, which is still not fully understood.

The following portraits of particles discovered in the cosmic rays include pictures from these early experiments at accelerators, along with others from more modern machines. In particular, there are a number of images from bubble chambers, which superseded the cloud chamber in the mid-1950s. Instead of revealing tracks of ionization as chains of droplets in a vapour, the bubble chamber shows them as strings of bubbles in a superheated liquid.

The images show how our knowledge of the particles has increased over the years, as the detectors to study them have increased in sophistication. Though we do not yet understand everything about the particles portrayed in this chapter, they no longer seem exotic. Nowadays physicists make them all routinely at modern high-energy particle accelerators. The particles found first in the cosmic radiation have become useful tools right here on Earth, helping to solve the problems raised by their very existence.

Fig. 5.1 *The collisions of cosmic rays provided physicists with their first glimpses of new subatomic particles, such as pions and kaons. Here, in a colour-coded positive print of a photograph taken by Powell in 1950, a cosmic ray sulphur nucleus (red) collides in emulsion and produces a spray of particles. They include a fluorine nucleus (green) and other nuclear fragments (blue), as well as about 16 pions (yellow). The length of the sulphur track is about 0.11 mm.*

THE POSITRON

ALL THE atoms of matter contain negatively-charged electrons and positive protons; the total negative and positive charges are equal, making matter electrically neutral overall. But we can imagine a world in which electrons are positive and protons negative. After all, the definitions of 'positive' and 'negative' are purely arbitrary; what is important for atoms is that electrons and protons have opposite charges and are bound together by the electric force. Nature, however, seems to have made a choice, because all electrons have the same charge, as do all protons. Or do they?

In 1928 a theorist at Cambridge University, Paul Dirac, had been attempting to combine Einstein's theory of special relativity with the equations governing the behaviour of electrons in electromagnetic fields. In so doing, he was led eventually to the remarkable conclusion that particles of the same mass as the electron but of opposite charge must exist. At that time none had been seen and Dirac proposed that there might exist parts of the Universe where positive and negative charges were reversed. Then in 1932, Carl Anderson at Caltech observed a new kind of particle in the cosmic radiation passing through his cloud chamber (see Fig. 4.10, p. 73). The new object was similar in mass to the electron, but it was positively charged. He had discovered the positive electron, or 'positron'. This was the first example of 'antimatter'—an 'antiparticle' with properties opposite to those of the familiar particles.

The French physicists Irène and Frédéric Joliot-Curie later discovered, in 1934, that certain nuclei can spontaneously emit positrons, in a form of radioactivity akin to the emission of electrons. In the familiar form of beta decay, a neutron converts to a proton by emitting an electron (and an antineutrino). However, some nuclei can become more stable by converting a proton to a neutron while at the same time emitting a positron (and a neutrino).

Positrons are also created together with electrons when pure energy 'freezes' into matter. One of the most common ways in which this happens is when an energetic gamma-ray photon produces an electron-positron pair. Figure 5.2 is a bubble chamber photograph that has been 'cleaned up' to show only the relevant tracks, and then coloured. Two electron-positron pairs have been produced simultaneously by separate gamma rays. The bottom pair is relatively energetic and the two tracks curve only a little in the bubble chamber's magnetic field. The top pair, on the other hand, have less energy and curl round to form spirals; the reason they are less energetic is that the gamma ray spent much of its own energy in knocking an electron (the long green track) out of an atom in the bubble chamber's liquid. The image shows clearly the transformation of energy into matter, in accordance with Einstein's equation $E = mc^2$. This is how positrons are formed in the cosmic radiation, from gamma rays with sufficient energy to create the total mass of an electron-positron pair.

Figure 5.3 shows positrons entering a cloud chamber and curving to the right in the magnetic field. One of them comes close to an electron in the chamber's gas, and knocks it out of its parent atom. The electron curls off to the left, as befits its negative charge. The 90° angle between the tracks of the electron and the deflected positron reveals the equality of their masses. (Note that that the electron and positron do not 'annihilate' in the manner described below because they do not come close enough.)

Energy in the Universe is constantly being changed from one form into another, though the total energy remains constant. We are familiar with the conversion of electrical energy into light (as in lamps), and chemical energy into heat (as in fires). Less familiar forms of energy conversion fuel the Sun, which in turn provides the energy for the cycle of life on Earth to continue. Matter is, in a sense, 'frozen energy', and the discovery of the positron revealed a new form of energy transformation— radiant energy converting into matter (electrons) and antimatter (positrons). This may be the way that much of the matter in the Universe formed, freezing out from the radiation of the hot Big Bang with which the Universe began.

Dirac's original theory applied to electrons, and predicted the existence of the positron. Now we know that it applies equally well to the proton, neutron, and many

Fig. 5.2 *Invisible gamma ray photons produce pairs of electrons (green) and positrons (red) in a bubble chamber at the Lawrence Berkeley Laboratory. The photons come in at the top of the picture. In the upper pair, some of the photon's energy is taken up in displacing an atomic electron, which shoots off towards bottom left. In the lower example, all the photon's energy goes into the production of the electron-positron pair. As a result, these particles are more energetic than the upper pair, and their tracks do not curve so tightly in the chamber's magnetic field. (Like many of the bubble chamber photographs that follow, this picture has been 'cleaned up' to remove all but the relevant tracks, and then colour-coded to identify the tracks.)*

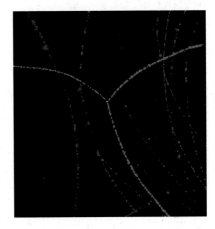

Fig. 5.3 *Positrons enter a cloud chamber, from the bottom of the picture, and curl clockwise in the magnetic field. One of the positrons (red) comes close enough to an atomic electron to knock it out of the atom so that it produces its own track curling anticlockwise (green). Although they interact, the positron and electron do not come close enough in this case to annihilate. The 90° angle between their two tracks reveals the equality of the particles' masses. The blue non-interacting tracks are due to other positrons and one other electron.*

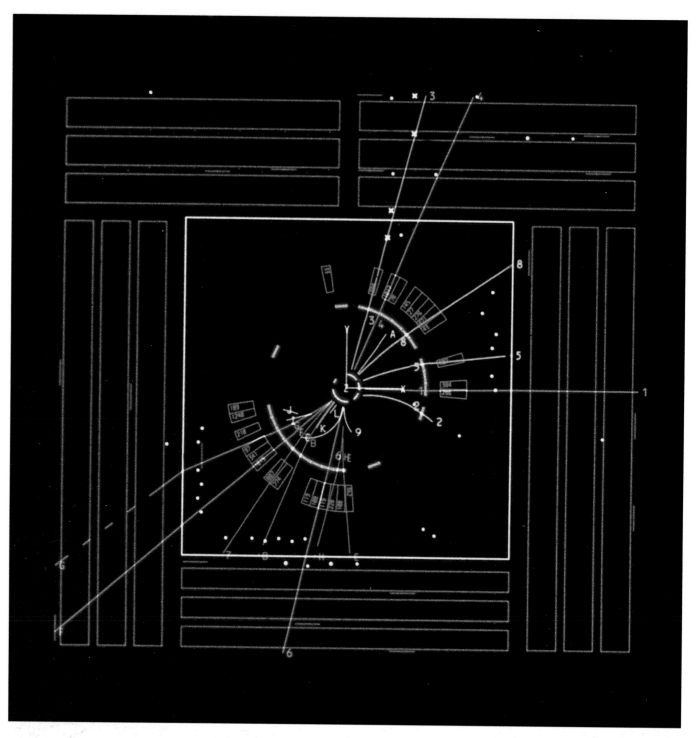

Fig. 5.4 *A computer display of the tracks of charged particles flying out from an electron-positron annihilation at the centre of JADE, a detector at the PETRA collider in Hamburg. The picture shows a cross-section through the detector, with various parts outlined in different colours. The electron and positron have come in at right angles to the image, from in front of and behind the page. The pink arcs represent scintillators 95 cm from the centre of the detector.*

other particles of matter, all of which have antimatter equivalents (see pp. 134–7). This knowledge enables physicists to use electrons and positrons together as tools to create new forms of matter and antimatter. Bringing a positron and electron sufficiently close together reverses the events seen in Fig. 5.2; the two particles mutually convert into energy, in a process called annihilation. The energy can then revert to matter again, but not necessarily as an electron–positron pair. Providing matter exactly balances antimatter, and the rule $E = mc^2$ is obeyed, any variety of particle can emerge; we are not limited to the common forms prevalent on Earth. Figure 5.4 shows the aftermath of the annihilation of an electron and a positron of very high energy. Their total energy materializes as a burst of particles, echoing the formation of matter from high-energy radiation in the early moments of the Universe.

THE MUON

'WHO ORDERED that?' physicist Isidore Rabi once remarked about the muon, which physicists now recognize as being a heavy version of the electron. It seems unnecessary for nature to have provided more than one variety of the same type of particle. Moreover, the early history of the muon was full of confusion. It was mistaken for an entirely different particle—the pion—which ironically turned out to be the very particle that gives birth to the muon in the first place.

Whereas electrons, protons, and neutrons are the stuff of ordinary matter, muons are typically the stuff of cosmic rays. Atomic nuclei from outer space hit the upper atmosphere and produce a debris of pions. These soon decay to produce showers of negative and positive muons, which rain down on the Earth continuously. But the muon is not a stable particle. Negative muons decay into an electron, a neutrino, and an antineutrino, positive muons into a positron and a neutrino and an antineutrino, and this occurs in 2.2 microseconds when the muons are at rest. However, when muons are in motion, like all particles they experience the subtle slowing of time inherent in Einstein's theory of relativity. A fast-moving, energetic muon has its life prolonged as measured by clocks on Earth. The majority of cosmic ray muons decay during their flight through the atmosphere, but the most energetic of them survive long enough to penetrate hundreds of metres underground.

Figure 5.5 shows the reconstructed tracks of two cosmic muons passing through an enormous tank of ultra-pure water located 600 metres underground, in a salt mine in Ohio. The tank is designed to search for the decay of protons, as described in

Fig. 5.5 *The penetrating quality of the muon is shown in this computer display of two high-energy cosmic ray muons (blue diagonal tracks) streaking through the IMB detector, a 20 m cube filled with ultrapure water located 600 m underground in a salt mine near Cleveland, Ohio (see pp. 201–202). Phototubes arrayed on the sides of the cube detect Čerenkov radiation—a shock-wave of light emitted as the muons travel through the water faster than light does. Colours indicate the times of arrival of the light at the phototubes, with the red end of the spectrum indicating those tubes that fired first. The tracks of the two muons have been calculated from the detected patterns of Čerenkov light and drawn in by computer.*

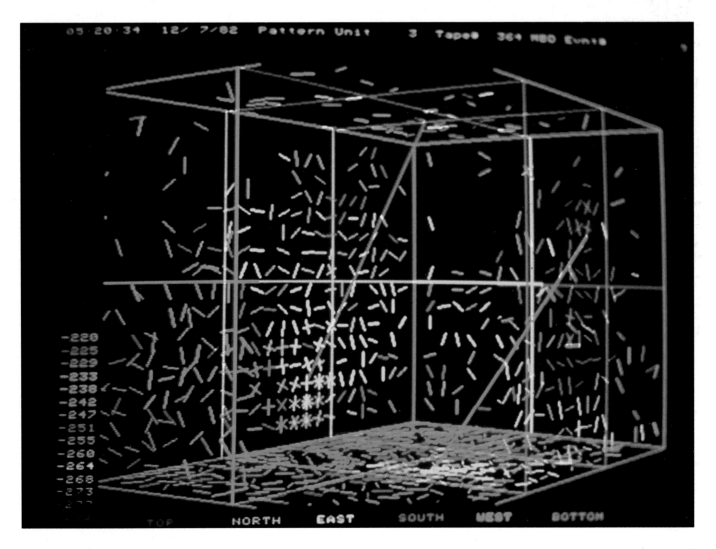

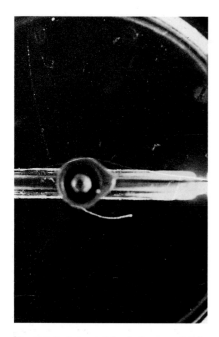

Fig. 5.6 *A photograph by Anderson and Neddermeyer of a positive muon coming to rest before it decays in a cloud chamber that was activated by a Geiger counter inside the chamber. (The counter lies horizontally across the centre of the picture; the circular structure is part of the counter.) The incoming muon leaves a faint track at the upper left of the picture. The track curls round and becomes thicker after the muon loses energy in traversing the glass walls and copper cylinder of the Geiger counter. The chamber was not sensitive enough to record the track of the positron produced in the muon's decay. The muon travels 2.9 cm after emerging from the counter.*

Fig. 5.7 *An electron and a positron annihilate and produce a positive and a negative muon, which travel out sideways from the colliding beams. The tracks of the muons are revealed here in a computer display of an event in the Mark-J detector at the PETRA collider in Hamburg. The yellow squares outline the iron of a magnet. We know the tracks are due to muons because they penetrate the iron, as the white dots beyond the yellow squares indicate; other charged particles would be absorbed. The muon counters are some 2 m from the centre of the detector.*

Chapter 10, but the physicists who run the experiment need to distinguish possible proton decays from the effects of penetrating cosmic ray particles such as muons. On average, three muons a second pass through the tank.

It is not only because muons have relatively long lifetimes that they can penetrate so far; it is also because they are heavy, weighing in at some 200 times the electron's mass. When electrons pass through matter, they radiate energy in the form of photons and are rapidly brought to a standstill. The heavier muons have much less tendency to radiate and slow down.

Despite these differences between muons and electrons, the two particles are so similar in other respects that when Carl Anderson and his colleague Seth Neddermeyer first observed cosmic ray muon tracks in their cloud chamber in the early 1930s, they believed them to be due to ultra-high-energy electrons obeying new laws of physics. But by careful measurement of the tracks, Anderson, Neddermeyer, and others showed in 1936 that the tracks had to be made by a new particle with a mass somewhere between that of the electron and the proton—hence their name for it, the 'mesotron', or middle particle. Figure 5.6 shows a muon coming to a halt in Anderson's cloud chamber after passing through a Geiger counter placed across the centre. The positively charged muon decays to a positron, although the chamber was not sensitive enough to show the latter's track. Analysis of this picture gave one of the first determinations of the mass of the muon—now known to be 210 times that of the electron.

It was this mass that caused the muon to be confused with the pion. Only the year before the discovery of the mesotron, Yukawa had put forward his theory of the strong force, which predicted the existence of a particle weighing about 250 times the mass of the electron. When the mass of the new mesotron was found to be so close to this value, an obvious conclusion was that it must be the predicted particle. But this never seemed quite right. Yukawa had invented his particle in order to explain the strong force, and it should therefore have had a great affinity for nuclear matter. The muon, however, displays quite the opposite characteristic, and can easily penetrate matter.

The puzzle was finally resolved by a remarkable experiment begun secretly in Rome during the Second World War by three young Italians, Marcello Conversi, Ettore Pancini, and Oreste Piccioni. To begin with, their makeshift laboratory was in a basement near the Vatican City, where they were hiding from the occupying Germans. There they set up their apparatus of Geiger counters, some material to slow down the cosmic ray particles, and some magnetized iron bars, which acted as lenses to concentrate particles of the same electric charge. Their aim was to discover what happened when the mesotrons decayed.

At first, using iron to slow down the particles, the researchers found that the mesotrons behaved as expected. The positive particles were seen to decay; the negative particles were not, presumably because they were more readily captured around the positively-charged atomic nuclei and absorbed by the strong force before they could decay. The surprise came when the Italians changed the absorber to a lighter material, carbon: the negative mesotrons were no longer absorbed, but were also seen to decay. This behaviour completely ruled out the mesotron's identification with Yukawa's particle, which should still have been absorbed by the carbon before the end of the mesotron's life. The work of the three Italians was confirmed in 1947 when a particle was finally discovered which *did* fit Yukawa's description completely—the pion. And soon afterwards the mesotron was renamed the muon.

Study of the muon's behaviour has repeatedly affirmed that it is like a heavy electron and is not influenced by the strong force. In Fig. 5.7 we see the symmetry of a reaction in which an electron and a positron have annihilated, their energy rematerializing as a positive and a negative muon. Through studying reactions like these, physicists hope to solve the mystery of the muon's existence. In the mid-1970s fresh clues came with the discovery of another particle resembling the electron and the muon, the tau (see pp. 184–5), which weighs in at 20 times the muon's mass. Why should there be three kinds of 'electron', and are there heavier ones still undiscovered? These are among the questions still challenging particle physicists.

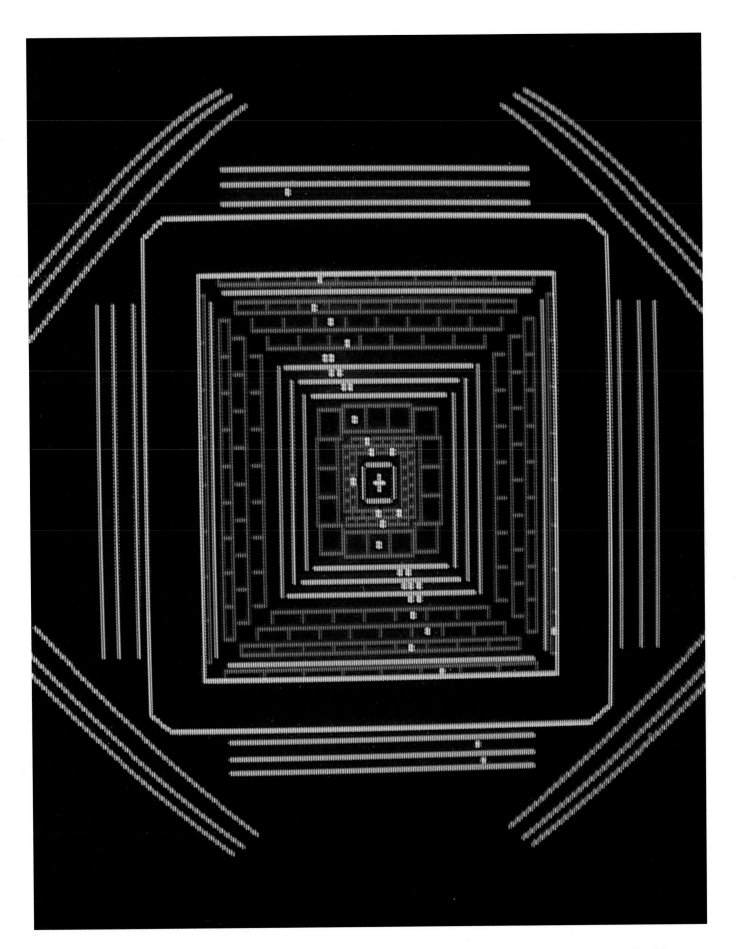

THE PION

ONE WAY to learn about an object is to pummel it about. Shake electrons about, for example, and they emit electromagnetic radiation; this is how radio waves are produced, when electrons are made to oscillate at particular frequencies. The burst of radiation is released by the disruption of the electric field surrounding the charged electron. What happens if instead we pummel protons about? Again we find that in disturbing the tranquillity of a proton at rest, we release a burst of radiation. But this time it is not simply photons of electromagnetic radiation; instead, it consists mainly of particles known as pions. The pions are set free when the *nuclear* force field associated with the proton is disturbed, and the more energetically we disturb it, the more pions are produced. Figure 7.24 (see p. 147), for example, shows many pions released in the violent collision between an energetic proton and one at rest.

Collisions between primary cosmic rays and the upper atmosphere produce positive, negative, and neutral pions in vast numbers. They are unstable, however, and decay

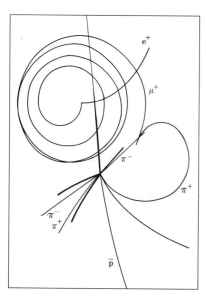

Fig. 5.8 *The decay of a positive pion is captured in a false colour streamer chamber image of an antiproton annihilating in the neon gas filling the chamber. The antiproton (p̄) comes in from the bottom of the picture and interacts to produce a typical starburst of tracks. One of the positive pions (π⁺) curls round on the right in the chamber's magnetic field, before decaying to a muon (μ⁺) which forms a beautiful spiral. Eventually, the muon decays to a positron (e⁺). At each decay the tracks change direction abruptly, indicating the simultaneous emission of an undetected neutrino. The thick tracks not identified in the diagram are nuclear fragments. The distance between the central crosses is 28 cm.*

rapidly. It is their children and grandchildren that form the bulk of the cosmic radiation near sea level. Particles decay via one or more of the fundamental forces, and they always decay into particles lighter than themselves. Since pions are the lightest particles subject to the strong force, they cannot decay into lighter particles under the influence of this force. Pions would therefore be stable particles if they were not also subject to the electromagnetic and weak forces. These step in and cause the pions to decay as follows.

The positive and negative pions decay within 10^{-8} s to positive and negative muons, which decay in their turn into positrons or electrons. The neutral pion, however, decays very much more rapidly—within 10^{-16} s—to gamma ray photons, which then spawn the electron-positron pairs in the cosmic radiation. Partly because of the great rapidity of its decay, the neutral pion was not discovered until many years after its charged siblings and it is therefore described separately on pp. 130-1.

In Fig. 5.8 the decay of a positive pion is captured by a device called a streamer chamber, in which tiny luminous streamers form along the trails of ionization in a gas in a high electric field. The successive steps of the decay from pion to muon to positron are clearly visible. At each step the new particle deviates markedly from the path of its parent, indicating that invisible neutrinos are also released in the decay.

The pion had been 'on order' for more than a decade before it was discovered in 1947 in emulsions exposed on the Pic du Midi by Powell's group from Bristol. In 1935, Hideki Yukawa had proposed a new particle, necessary to convey the force between the protons and neutrons of the nucleus. This idea was not entirely new, for Yukawa was building on the early quantum theory of electromagnetic forces developed by Dirac in 1928. In this theory, electrically charged particles—electrons, protons—interact by exchanging bundles of light, or photons.

By analogy, Yukawa argued that protons and neutrons in a nucleus must exchange some particle. He reasoned that because the effect of the strong force is limited to the tiny dimensions of the nucleus, the carrier particle must be heavy—unlike the massless photon. Yukawa calculated that the particle must have a mass between that of the proton and the electron—roughly 15 per cent of the proton's mass. Many physicists, including Yukawa, were at first misled by the discovery of the muon, in 1937, with almost the right mass. Powell's discovery of the pion in 1947 put the record straight.

Within a nucleus, pions form an invisible, evanescent web between protons and neutrons, binding them together. It is only when nuclear particles collide at high energies that we can see the pions liberated. This is how the pions are formed in cosmic rays, and in Fig. 5.9 we see an instance where both the birth and the death of a cosmic ray pion have been captured in emulsion. This picture, taken in 1947, was one of the first observations of a pion and demonstrated its strong affinity for nuclei. It had no sooner been produced at A than it shattered a nucleus at B.

Today, tailored beams of pions of chosen energy and charge are readily produced at particle accelerators throughout the world. They can be aimed at nuclei and used to probe the details of nuclear behaviour. The first observations of pions created by an accelerator came in 1948, a year after Powell's discovery, at the 184 inch cyclotron at the Lawrence Berkeley Laboratory in California. The cyclotron had been in operation since late 1946, but it was only in the autumn of 1947 that the emulsion group at Berkeley began a search for pions. At first they had no luck, but help was at hand in the person of the Brazilian physicist Cesare Lattes, who arrived in February 1948 with a fund of knowledge from working with Powell's group. He also brought some of the new, high-sensitivity plates made by Ilford. These plates were exposed to the debris from collisions between alpha particles accelerated in the cyclotron and a carbon target, and then developed according to Lattes's recipe. The pictures revealed the first examples of the decays of man-made pions.

Although the pions have been described as the transmitters of the strong force, they are not classed with the other force-transmitting particles, such as the photon or the W and Z particles. This is because they are now known not to be elementary particles, but composites made up of quarks. The strong force is transmitted by the pions only at the larger, nuclear level; at the deeper level of the quarks, it is transmitted by particles called gluons (see pp. 189-91).

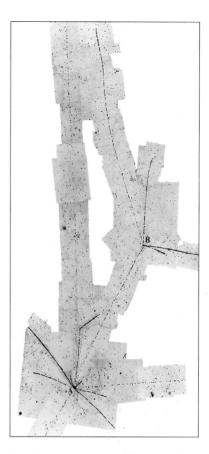

Fig. 5.9 *The birth (A) and death (B) of a pion are recorded in this photograph taken by Lattes, Occhialini, and Powell in 1947. It was one of the first observations of the creation of a pion. The distance between points A and B is about 0.11 mm.*

THE KAON

ON 15 October 1946, George Rochester and Clifford Butler observed something unusual in their cloud chamber at Manchester University. Two tracks appeared from a single point beneath a lead plate, as if from nowhere (Fig. 5.10). Their stereo views of the chamber showed that the tracks indeed originated from the same point and did not merely appear coincident from a particular perspective. Nor were the tracks caused by protons knocked out of the gas; the ionization and curvature showed them to be due to much less massive particles. In the following months, Rochester and Butler calculated that the two particles could be the decay products of a neutral particle with a mass some 800 times that of the electron—unlike anything they had seen before. Seven months later, on 23 May 1947, they found a similar occurrence—this time a track with an unusual kink, which calculations of angles and energies showed could not possibly be due to scattering (Fig. 5.11). With these two pictures, Rochester and Butler had found the first examples of the decays of particles we now call kaons.

With hindsight we can say that Fig. 5.10 shows the decay of a neutral kaon to two pions (one positive, one negative), while Fig. 5.11 reveals the decay of a positive kaon to a muon, accompanied by an invisible neutrino. A clearer example of a charged kaon appeared in a new extra-sensitive emulsion exposed by Powell's group from Bristol, at a laboratory high in the Jungfraujoch in 1948 (Fig. 5.12). The emulsion revealed the decay of a particle into three pions. The scattering and density of grains along the track showed that the particle is about 1000 times more massive than an electron—about half the mass of a proton.

The decay to a muon and a neutrino is the charged kaon's most common mode of decay, occurring 63 per cent of the time. Less frequently—21 per cent of the time—a charged kaon will transmute into a charged pion together with a neutral pion, as shown in a beautiful example in Fig. 7.2 on p. 130. The decay of a charged kaon to three charged pions leaves a particularly distinctive 'signature', however. This kind of decay is so easy to spot that it was discovered early on (Fig. 5.12) despite the fact that only 5 per cent or so of charged kaons decay this way. The neutral kaon decays most frequently to two charged pions—one positive, one negative—though it too can

Figs. 5.10–5.11 *The original observations of 'V' particles, recorded in 1946 and 1947 by Rochester and Butler at Manchester.*

Fig. 5.10 *(left) This image of the first 'V' particle found shows a pair of tracks forming a pronounced fork (A) just below the lead plate across the centre of the chamber. This was probably due to a neutral kaon, produced in an interaction in the lead, which decayed into a negative and a positive pion.*

Fig. 5.11 *(right) The second 'V' is the wide-angled fork (B) near the top right-hand side of this image. In this case a positive kaon has probably come in from the top of the picture and decayed into a muon and a neutrino; the neutrino leaves no track, but the visible track carries on down the picture and easily penetrates the 3 cm lead plate, a characteristic behaviour of the muon.*

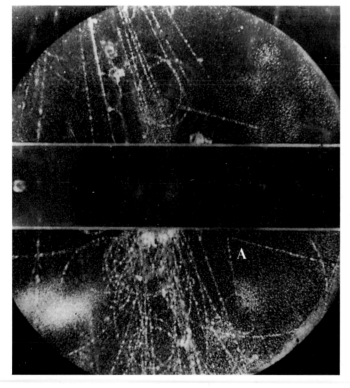

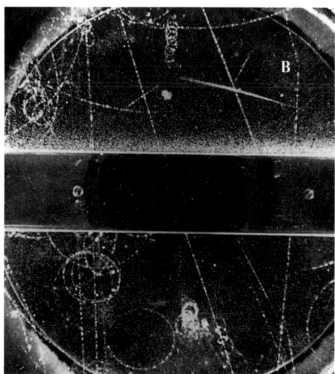

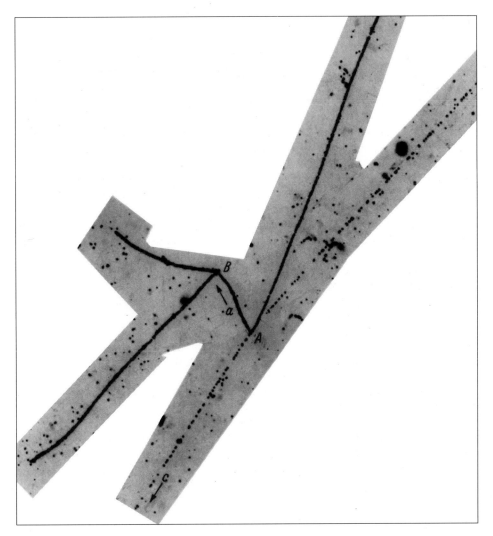

Fig. 5.12 *The first observation of the decay of a kaon into three pions, recorded by Powell's group in 1948 in some of the early electron-sensitive emulsion. The kaon produces the strong track coming in from the top right and decays at A into three pions. One of these moves away slowly, leaving a thick track, and then interacts at B. The other two pions are faster and leave only faint tracks as they move off in opposite directions towards the top right and bottom left of the picture. The distance from A to B is 25 micrometres.*

decay to neutral pions or to combinations of charged pions with muons or electrons and neutrinos.

The many different decays of the kaon posed problems for physicists studying the particles in cosmic rays. From the few events observed in emulsions and cloud chambers it was not entirely clear whether they were dealing with several particles of similar mass, or a single type of particle that could end its life in a variety of ways. Only with the advent of particle accelerators, which produced large numbers of kaons under controlled conditions, did it finally become clear that there is one type of particle and that it can be positive, negative, or neutral.

The kaon was the first of a number of particles found in cosmic rays that were dubbed 'strange'. This name arose because the particles all live for a surprisingly long time—in the case of the kaon, about 10^{-8} s, which is a million billion times longer than expected. The kaon, like the pion, is produced by the strong force; unlike the pion, it should be able to decay via the strong force. As we have seen, the pion cannot decay 'strongly' because it is the lightest particle subject to the strong force. The heavier kaon, on the other hand, should be able to decay strongly to produce pions, in a mere 10^{-23} s. Instead, with kaons and certain other particles, the strong force seems to have been cut off by something, and this is what at first seemed strange. To illustrate by how much its effects are postponed, one scientist said, 'It is as if Cleopatra fell off her barge in 40 BC and hasn't hit the water yet.'

The mystery of this postponed death began to be explained in the early 1950s when a whole family of strange particles first became apparent from studies of cosmic rays. The discovery of the kaon was soon followed by that of the lambda, and as the following pages describe, it provided the first clues as to just what strangeness is.

THE LAMBDA

THE LAMBDA particle leaves one of the most distinctive signatures in particle track detectors: it writes its own Greek name—Λ! In Fig. 5.13 we see an inverted 'lambda' formed by two tracks emanating from the decay of a neutral particle, produced in a high-energy collision along with 16 charged particles. This decay into a proton and a negative pion is the most common decay of the lambda, occurring 64 per cent of the time, though often the particle vexes physicists trying to track it by decaying into two neutral particles, a neutron and a neutral pion. Like the kaon, the lambda is a 'strange' particle—it lives for 10^{-10} s; but unlike the kaon it is *heavier* than the proton and neutron. Indeed, it was the first such 'hyperon', or heavy particle, to be found.

The first images of decaying lambda particles came in 1951 from a number of cloud chambers: from Manchester University's device, once it had been raised up to the lofty heights of the Pic du Midi; from Anderson's cloud chamber on White Mountain in California; and from Robert Thompson's chamber in Indiana. These images showed clearly tracks of different quality emerging from the 'V' of the neutral decay, but precise identification of the particles was often difficult. The team from Manchester established a neat method of analysing the motions of the particles, and used it to show that their neutral Vs corresponded to two kinds of particle, one some 950 times the mass of the electron, and the other 2250 times more massive than the electron, or some 20 per cent heavier than the proton. The former was the neutral kaon; the latter the lambda particle.

So in the early 1950s, physicists were faced with two varieties of strange particle, which lived far longer than expected. What were they to make of them? Soon after the discoveries, Kazuhito Nishijima and other theorists in Japan, as well as Abraham Pais in the US, began the process of unravelling the strange code. They proposed that these particles are produced by the strong force *in pairs*, and that they can be disrupted by the strong force only *in pairs*. When a pair of strange particles separate from one another, the strong force can no longer act on them as a pair, and their deaths are postponed. Instead, they decay by the much feebler electromagnetic and weak forces that are responsible for the decay of the pion or neutron.

According to the theory of production in pairs, or 'associated production', the lambda is produced along with another strange particle, such as the kaon. Confirmation of this came in 1954 with accelerators that could produce particle beams of high enough energy to create these particles. At the Brookhaven National Laboratory on Long Island, New York, experimenters found that a lambda and a kaon were often produced together. Figure 5.14 overleaf is a later bubble chamber photograph from Berkeley, which clearly shows the associated production of a lambda and a kaon, and their subsequent separate deaths.

Associated production was the first step towards solving the puzzle of the strange particles. The next step came in 1954 when American theorist Murray Gell-Mann, and independently Nishijima and T. Nakone in Japan, proposed that 'strangeness' is a new property of matter, akin to electric charge. Just as electric charge is conserved, so is strangeness conserved when the strong force is at work.

A pion and a proton have no strangeness. If they collide, as in Fig. 5.14, and produce a neutral kaon, with strangeness +1, then they must balance the books by producing a particle with strangeness −1, the lambda. This is why strange particles are always produced in pairs. (The allocation of positive strangeness to certain particles and negative strangeness to others is of course arbitrary, just as the allocation of negative electric charge to electrons and positive charge to protons is arbitrary; the fact is that protons and electrons have opposite electric charges, and the positive and neutral kaons have opposite strangeness to the negative kaon and the lambda.)

There is a major difference, however, between strangeness and electric charge. The latter is conserved, as far as we know, under all circumstances—an electron can only disappear from the Universe by annihilating with a positron. Strangeness, on the other hand, is conserved only in interactions via the strong force. Once created, two strange particles go their separate ways and usually decay via the weak force. The heavier strange particles, the xi and the sigmas (see pp. 98–9), can decay to lighter

Fig. 5.13 *A high-energy proton (yellow) enters from the bottom and collides with a proton at rest in the liquid hydrogen of the '80 inch' (200 cm) bubble chamber at the Brookhaven National Laboratory. The small electron spiral (green) shows that negative particles curl anticlockwise and positive particles clockwise. The collision produces seven negative pions (blue); nine positive particles (red), which include a proton and a positive kaon as well as seven positive pions; and a neutral particle—a lambda. The lambda travels up the picture leaving no track, but betrays its existence when it decays into a proton (yellow) and a negative pion (purple), which curls rapidly to the left. (This picture has been 'cleaned up' to show only the relevant tracks.)*

Fig. 5.14 *An early photograph of the associated production of two strange particles, taken in a bubble chamber at the Lawrence Berkeley Laboratory. A pi-minus (green) with zero strangeness (S=0) interacts with a proton (S=0) in the chamber liquid, producing a lambda (S=-1) and a neutral kaon (S= + 1). The neutral kaon and lambda leave no tracks but are revealed when they decay; the lambda decays into a proton (red) and a pi-minus (green), while the neutral kaon decays into a pi-plus (yellow) and a pi-minus (green). Notice how strangeness is conserved when the strange particles are produced, being zero both before and after the initial interaction; this is because their production occurs via the strong force. When each strange particle decays, however, strangeness changes; both the lambda and the kaon decay into particles with zero strangeness. The decays occur via the weak force, and this does allow strangeness to change one unit at a time. (The blue tracks are particles not involved in the interaction.)*

strange particles as long as overall strangeness is conserved. But the two lightest strange particles, the kaon and the lambda, cannot decay into lighter strange particles; instead they decay separately into non-strange particles. And since on average the kaon decays in 10^{-8} s and the lambda in 10^{-10} s, there is a time—however brief—when there is an imbalance of strangeness. Whereas electric charge is conserved always, strangeness leaks away when the weak force acts. Physicists do not yet fully understand why this should be.

The lambda behaves in many ways like a heavy neutron—a neutron imbued with strangeness. For instance, the similarity between the decay of the lambda and that of the neutron is illustrated in Fig. 5.15. Rather as the neutron decays into a proton, an electron, and a neutrino, so can the lambda decay into a proton, a muon—the heavy version of the electron—and a neutrino.

And just as the neutron can be incorporated into nuclei, so can the lambda. It is only because of the lambda's brief lifetime that we do not see strange nuclei around us. In Fig. 5.16, a cosmic ray at A has hit a nucleus in an emulsion and broken it up. In the process a lambda is produced which binds to protons and neutrons to form a lambda nucleus. Its heavily ionizing track heads downwards at 5 o'clock and explodes when the lambda decays at B.

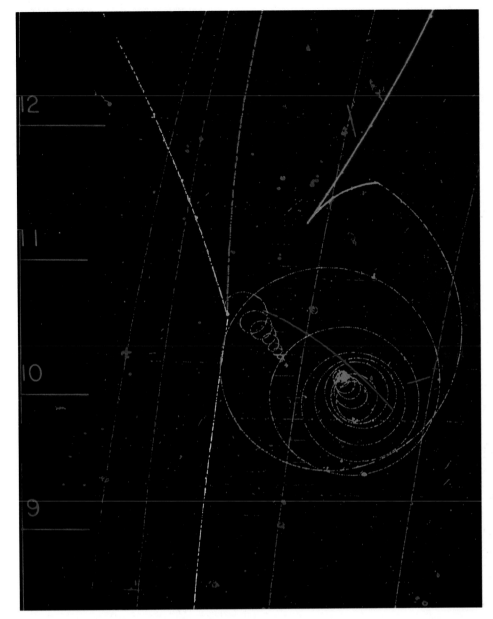

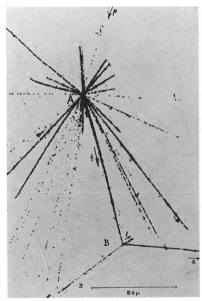

Fig. 5.15 *A rare decay of a lambda into a negative muon, a proton, and a neutrino is photographed in the '72 inch' (180 cm) hydrogen bubble chamber at the Lawrence Berkeley Laboratory. The lambda is produced, together with a positive and negative pion (salmon pink and purple), when a negative kaon (yellow) coming in from the bottom of the picture strikes a proton. The neutral lambda leaves no track but is revealed by the 'V' formed when it decays. The proton (red) shoots off to the top of the picture, while the muon (pale blue) decays, producing an electron (the larger of the two green spirals) and two neutrinos, their invisible presence betrayed by the kink at the start of the electron spiral. Note how the positive pion produced in conjunction with the lambda knocks out an atomic electron, which forms the tighter of the two electron spirals. Background tracks are dark blue. The scale up the side is in units of about 9 cm.*

Fig. 5.16 *The first observation of the formation and disintegration of a 'hypernucleus', photographed in emulsion in 1953 by two Polish physicists, Marian Danysz and Jerzy Pniewski. A hypernucleus is a nucleus in which a lambda particle replaces one of the neutrons. In this instance it is produced in the collision at A and disintegrates at B, and it has about the same charge as a boron nucleus.*

A single lambda is electrically neutral and leaves no trail. In Fig. 5.16, the track is visible because of all the charged protons in the nucleus containing the lambda. One interesting feature is that the density of the track decreases as it travels towards the bottom of the page. This is because the positive charges of the protons attract electrons. By the end of the trail the nucleus has grown into a 'strange atom', with several electrons and correspondingly less net positive charge to darken the emulsion. It has too little energy left to make a star by shattering another nucleus. The star here is due to its internal explosion caused by the decay of the lambda.

The discovery of 'lambda nuclei' was an important result because it demonstrated that the lambda is an additional, and unexpected, member of the family containing the proton and neutron. What is more, it suggested the possibility of a Universe that includes not just the novelty of strange particles but even strange nuclei ('hypernuclei') and strange atoms. A whole 'strange universe' could exist in principle. The reason that we are unaware of this in daily life is that strange particles, and hence strange matter, are unstable. We become aware of them only when we study matter on timescales of millionths of a second. In these ephemeral intervals the Universe exhibits a richer, more complex structure.

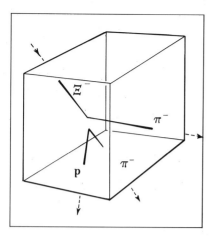

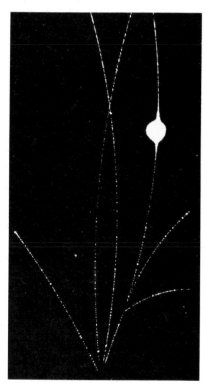

THE XI AND THE SIGMA

TWO MORE strange particles—the negative xi or cascade particle, and the sigma—were discovered shortly after the lambda, and they helped to confirm the picture of strangeness that theorists were developing. In 1952, the cosmic ray group at Manchester chanced upon a startling image recorded by their cloud chamber on the Pic du Midi. By good fortune a particle never seen before had entered the chamber and decayed within it. The original stereoscopic pictures are too poor to reproduce, so in Fig. 5.17 we have reconstructed the event in three dimensions. There is no doubt about it: a sharp kink, with a 'vee' pointing back at it. We now know that the kink is due to the new particle, the negatively charged xi, decaying to a negative pion and an invisible lambda. The vee is the result of the lambda's decay into a proton and a negative pion.

This was first time that a particle had been seen to descend to a proton in a sequence, or cascade, of decay steps. This simile led the physicists of the time to name it the cascade, today summarized by the Greek letter 'xi', or Ξ. The xi decays ultimately to a proton by cascading through a lambda. Here we begin to see strangeness as a property carried by particles in discrete amounts like electric charge. The xi has two units of negative strangeness and descends to a proton by shedding them one at a time. First it sheds one unit by decaying to the singly-strange lambda, and then the lambda sheds its own unit when it decays to a proton and a pion in the 'vee'.

This is not, of course, the whole story, since the conservation of strangeness appears to have been violated. But by going back in time, to the production of a xi, we can see how the strangeness books are balanced. Figure 5.18 is a streamer chamber photograph in which a negative kaon, carrying one unit of negative strangeness, enters from below and strikes a proton in the chamber's gas. The collision gives rise to two new strange particles, a xi and a positive kaon. The xi has strangeness -2 and the positive kaon strangeness $+1$, giving the same total of -1 strangeness brought into the interaction by the original negative kaon. At the next stage, the doubly-strange xi loses one unit of negative strangeness when it decays into a singly-strange lambda and a pion. The lambda decays in its turn into a proton and a pion, shedding its strangeness of -1. The positive kaon escapes the image before it too decays, and strangeness finally 'leaks away'.

In 1953, the year after the discovery of the xi, a group of Italian physicists identified a new strange particle in emulsion exposed to cosmic rays, and a similar object was also observed in a cloud chamber by a team from Caltech. The particle was positively charged, decayed to a proton, and analysis of the tracks showed that it was 30 per cent heavier than the proton; as a result it became known at first as the 'superproton'. Later in the same year, a negatively charged version of the particle was found in accelerator experiments, and in 1956 a neutral version was identified in a bubble chamber experiment at Brookhaven's Cosmotron accelerator. The three particles are today known as the positive, negative, and neutral sigma particles, after the 's' of superproton, and they each carry one unit of negative strangeness.

Figure 5.19 shows the separate production and decay of both a positive and a negative sigma in a bubble chamber exposed to a beam of negative kaons. In the lower half of the photograph, one of the kaons collides with a proton in the bubble chamber liquid to produce a positive sigma (the short track) and a negative pion. This positive sigma takes a different decay path from that first noted by the Italians in 1953, transmuting to a positive pion and an invisible neutron.

In the upper half of the photograph, another kaon interacts with a proton and produces a negative sigma, together with a negative pion and two positive pions. This sigma decays to a negative pion and another invisible neutron. The white curlicues that bedeck the picture are spiralling low-energy electrons knocked out of atoms in the bubble chamber liquid.

The theory of associated production and the concept of strangeness developed by Gell-Mann, Nishijima, and Nakone served to explain the observed behaviour of the strange particles discovered between 1947 and the end of the 1950s. It led to the prediction of the existence of the neutral sigma, and of the neutral xi, whose discovery

Fig. 5.17 *In this three-dimensional reconstruction of the discovery of the xi-minus, the Manchester cloud chamber is notionally represented as the rectangular box. The xi-minus (Ξ^-) enters from the front and decays in the middle of the chamber to a pi-minus (π^-) and a neutral lambda. The lambda betrays its presence when it also decays, into a proton (p) and another pi-minus, the 'V' they form pointing back towards the xi's decay.*

Fig. 5.18 *A negative kaon interacts with a proton in the gas in a streamer chamber at the Lawrence Berkeley Laboratory. It produces, from left to right, a pi-plus, a pi-minus, a positive kaon, and a negative xi particle. The xi decays to a lambda and a pi-minus, which veers right. The invisible lambda decays after a very short distance in a typical 'V', formed by a proton and a pi-minus, which again veers right. Tracks not involved in the interaction have been removed from this photograph.*

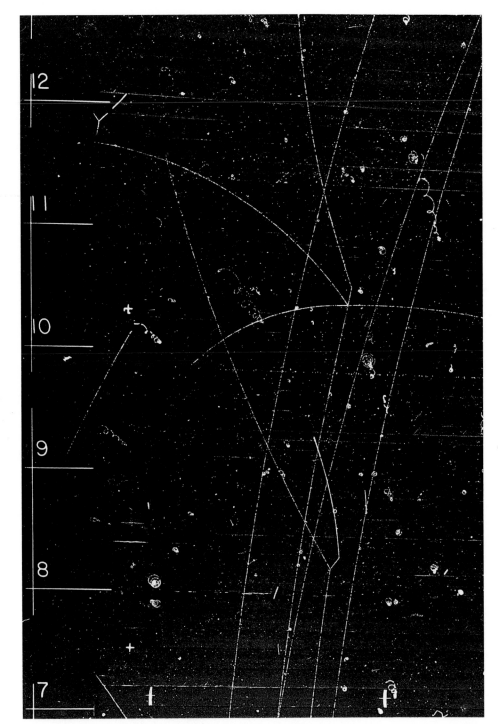

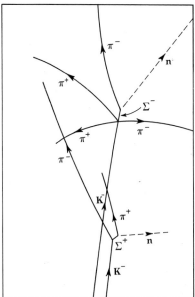

Fig. 5.19 *This picture from the 72 inch bubble chamber at the Lawrence Berkeley Laboratory shows the characteristic kinked tracks due to both a negative and a positive sigma. The particles are produced in the interactions of negative kaons* (K^-). *The sigma-plus* (Σ^+) *decays to a pi-plus* (π^+) *and a neutron* (n). *The sigma-minus* (Σ^-), *on the other hand, decays to a pi-minus* (π^-) *and a neutron. The scale is in units of about 9 cm.*

is described in a separate portrait (see pp. 132–3). But how and why strangeness occurs in the first place remained a mystery.

The first steps towards its solution came in the following decade, when Gell-Mann and the Israeli physicist Yuval Ne'eman developed the classification of particles that became known as the Eightfold Way, and used it to predict successfully the existence of a particle with three units of strangeness, the omega minus (see pp. 141–2). Shortly afterwards Gell-Mann went further and proposed the existence of a new level of elementary particles, the quarks, one of which is the strange quark. These theoretical developments were made possible only by the construction of increasingly powerful particle accelerators and sophisticated new techniques for detecting the subatomic debris produced in high-energy collisions.

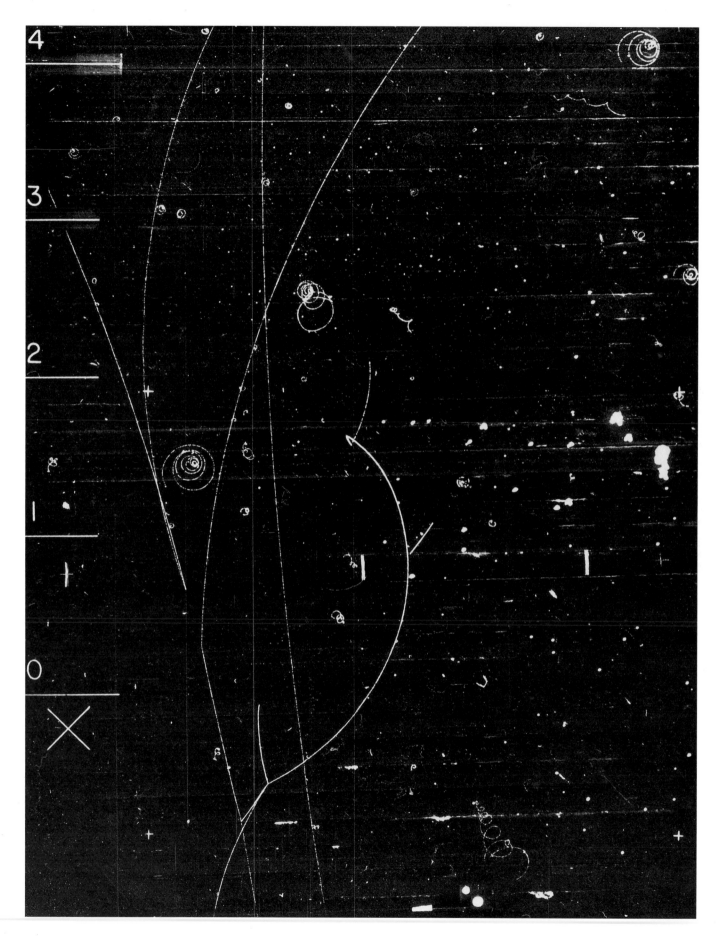

READING THE PICTURES

SOME IMAGES of subatomic particles are difficult to read. You must work out the momentum of the particles from the amount of bending in the magnetic field, measure the scattering angles, and then balance the books; only occasionally can you deduce which particle is which. However, sometimes it is much easier to identify the particles and interpret the picture. The particles sign their autographs distinctly and provide enough clues to fill in the missing links.

This chapter has presented images chosen to demonstrate specific particles. Now in Fig. 5.20, we can see many of them in a single picture that records a continuous ebb and flow of different particles. What is more, in this remarkable image, the particles display their most noticeable characteristics, which makes them easy to identify without recourse to detailed measurements.

The picture was taken in a bubble chamber containing hydrogen—electrons and protons—exposed to a beam of negative kaons. Two kaons enter at the bottom of the picture. The one on the left hits a proton and makes new particles, whereas the other passes clean through the chamber. The gradual curve of the right-hand kaon shows that negative charges are bending clockwise in this picture (and hence by inference, positives bend anticlockwise). Armed with this knowledge we can interpret the sequence initiated by the other kaon.

The negative kaon hits a proton and produces two charged particles forming a 'Y' near the bottom of the picture. The particle ejected to the right is clearly a kaon, because it shows the characteristic decay into three pions (compare Fig. 5.12). We can tell that these are pions because one of them exhibits the famous 'pion-muon-electron' decay (to be exact, pion-muon-positron in this case) at the top of its trail; it shows the two sudden kinks as the pion decays to a muon and an invisible neutrino, followed by the muon's decay (compare Fig. 5.8). The other tracks have the same quality and bend in the same way as this identified pion, so it is most likely that this is the decay of a positive kaon; positive because two of the pions curl anticlockwise (positive charge) and the other clockwise (negative charge).

A proton in the hydrogen attracts the negative pion and absorbs it. Both their charges are neutralized and no further trail is seen. One positive pion escapes from the chamber and the other shows the characteristic decay chain. The end product of the pion's decay chain is a positron. Being antimatter, this does not survive long in the hostile matter surroundings, and it annihilates with an electron in the hydrogen in the chamber. The annihilation produces two photons. These leave no overt tracks but we can sense them as they kick electrons from atoms in their path. These are the characteristic electron spirals, roughly symmetric at 45° to either side of the positron trail.

So much for the action on the right-hand side of the picture. Now look at the other half of the story, the particle produced along with the positive kaon at the foot of the picture. A negative kaon has hit a proton and produced a positive kaon and something else, 'X'. Balancing electrical charges shows that X is negatively charged. Balancing strangeness tells us more about it.

The negative kaon brings in one unit of *negative* strangeness. The positive kaon takes out one unit of *positive* strangeness and the proton has none at all. So to balance the books X must have two units of negative strangeness. Strangeness minus two and negative charge: this is a xi-minus. Indeed, we see the famous autograph of the xi-minus's cascade involving a neutral lambda and a pion: the 'kink with vee arrow' (compare Fig. 5.17).

The track curving off to the right is a fast negative pion. The neutral lambda leaves no track but it soon decays into a proton and a pion, yielding the characteristic 'V'. Here again the negative pion flies off curving clockwise while the more sluggish proton leaves a denser trail with a gradual curve anticlockwise.

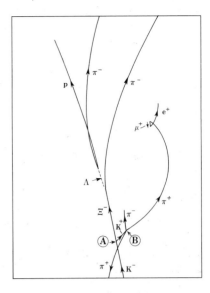

Fig. 5.20 *A negative kaon* (K$^-$) *interacts at A with a proton in a bubble chamber filled with hydrogen and produces a xi-minus* (Ξ^-) *and a positive kaon* (K$^+$). *This kaon travels only a short distance before it decays at B into three pions—one negative* (π^-) *and two positive* (π^+). *One of the positive pions then exhibits the characteristic decay sequence to a muon* (μ^+) *and then a positron* (e$^+$). *The xi-minus travels further up the picture before it too decays, to a negative pion and a lambda* (Λ) *which, being neutral, leaves no track. The lambda betrays its presence, however, when it decays to a proton* (p) *and a negative pion.*

CHAPTER 6
THE CHALLENGE OF THE BIG MACHINES

FIFTY KILOMETRES west of Chicago a ring so huge that it is visible from space is etched on the Illinois prairie. A modern temple to science, it is the site of the Fermi National Accelerator Laboratory, which produces the highest energy particle beams in the world. Sprawled over 6800 acres, the laboratory employs over 2000 people, and in addition caters for about 1000 visiting particle physicists from more than 200 institutions both in the US and abroad. When fully operational, its machines consume some 60 megawatts of electricity, roughly the amount used by the nearby city of St Charles with a population of 175 000. The laboratory incorporates its own mid-western village to house visiting scientists and their families, its own herd of buffalo, and a number of fascinating examples of modern architecture. This is Fermilab, a prime example of a modern particle physics laboratory—a far cry indeed from Rutherford's small empire at the Cavendish Laboratory.

Fig. 6.1 *The 2 km diameter ring of the Tevatron, the big accelerator at the Fermi National Accelerator Laboratory (Fermilab), is marked out by the lights of a car circling the service road above the underground machine. The land within the circle has been restored to natural prairie by volunteers from the laboratory. The glow of Chicago is visible in the distance.*

Fermilab's centrepiece is the Tevatron, the 2 km diameter accelerator ring. The machine itself lies buried in a tunnel, but it is marked out above ground by a service road. The ring contains a total of 2000 electromagnets, designed to steer protons on a circular course until they reach an energy which can be as high as 1000 GeV, or 1 tera electronvolt (TeV). The protons travel through a narrow stainless steel pipe, about 10 cm across. A vacuum inside the pipe ensures that the particles are not deflected from their path by collisions with extraneous molecules. The Tevatron can deliver bursts of 20 million million (2×10^{13}) protons every 60 seconds to serve a variety of experiments. The protons are often used to create secondary bursts of particles such as pions, kaons, and neutrinos; it's all in a minute's work for the Tevatron.

The Tevatron actually consists of two accelerators, one on top of the other within

Fig. 6.2 *The Cockcroft–Walton generator at Fermilab provides the first stage in the acceleration of the protons. Electrons are added to hydrogen atoms in the cube-shaped structure (an ion source) at the top of the picture. The resulting negative ions—each consisting of one proton and two electrons—are accelerated to 750 keV in the pipe to the upper left, before being passed to the second stage, the linear accelerator (linac). Negative ions, rather than protons, are used in the first stages of acceleration for technical reasons. They are easier to inject from the linac into the Booster, the third stage, and they allow the machine physicists to get more protons into the Booster. The dome-topped structure to the right contains components of the 'Cockcroft ladder' that builds up a high direct voltage (750 kilovolts) from the alternating mains. The dark columns are supports with smoothly curving 'corona rings' (the shiny collars) which help to prevent unwanted discharges.*

the same 3 m high tunnel. Surprisingly, it is the ring of smaller magnets, tucked beneath the larger ones, that carries the protons on the final part of their journey. The magnets steer the electrically charged protons on many thousands of circuits around the ring. On each circuit, the protons receive a small accelerating 'kick' from electric fields. By the end of their journey, they are travelling almost at the velocity of light, 300 000 kilometres per second.

In most particle accelerators today, including the Tevatron, the electric fields are set up by radio waves in hollow copper vessels called 'cavities'. Radio waves, like all forms of electromagnetic radiation, are coupled vibrating electric and magnetic fields. When pumped in to a copper cavity of the correct size and shape they will form a 'standing wave' rather like a sound wave in an organ pipe, but one that varies between regions of positive and negative electric field. Providing the field is in the right direction, protons entering a cavity absorb energy from the radio waves and are accelerated.

The purpose of an accelerator like the Tevatron is to take stationary protons and boost them almost to the speed of light. This cannot be done in a single step. The protons must be accelerated through different stages, rather like going up through the gears when accelerating a car from rest.

First gear at the Tevatron is a Cockroft–Walton generator, a machine that looks like a science fiction fantasy. Here the protons are accelerated to an energy of 750 keV (0.00075 GeV), or 4 per cent of the speed of light. They then move on to second gear in the 150 metre long linear accelerator or 'linac'. This consists of a series of copper cylinders in which electric fields accelerate the protons to 200 MeV (0.2 GeV), or 55 per cent of the speed of light.

From the linac, the protons move into third gear in a small accelerator ring, only 150 metres in diameter, known as the Booster. This whirls them up to 8 GeV, or 99 per cent of the speed of light, before injecting them into the uppermost ring of magnets in the main accelerator. Here in the top ring, the fourth gear, the protons make 70 000 circuits to reach 150 GeV or 99.998 per cent of the speed of light. They then move into overdrive in the lower ring of magnets, where a further 400 000 circuits hurtle them to their final energy. At the design energy of 1000 GeV, they are moving at an incredible 99.99995 per cent of light speed. From rest in the Cockroft–Walton, the protons have been accelerated over a distance of 3 million kilometres in under 20 seconds.

Figure 6.3 shows the Tevatron's two rings of magnets. The larger magnets on top form the original ring completed in 1972. They consist of coils of copper wire, which create a magnetic field when a current flows through the wire The higher the current, the stronger the field—but only up to a point. The magnetic field 'saturates' at a value of about 2 teslas—roughly 100 000 times the strength of Earth's magnetic field. Increasing the current beyond this point heats up the magnet rather than strengthening the field.

On its own, this ring of copper-coiled magnets was able to accelerate protons up to 500 GeV. To double this energy to its current maximum required a different technology for building magnets, which is where the second ring comes in. These magnets have coils formed from *superconducting* cables of niobium–titanium alloy.

A superconductor is a material in which electric currents can flow with practically no resistance, provided the temperature is extremely low—typically a few degrees above the absolute zero of temperature at −273°C. Superconducting magnets have two major advantages over conventional electromagnets. First, they produce higher magnetic fields—in the Tevatron, double the strength of the upper ring of conventional magnets. And secondly, because the current meets so little resistance, they can achieve these stronger fields with less electrical power—an important factor when your electricity bill comes to $16 million a year.

Once they are at maximum energy, the protons are deflected from their orbit by electrostatic fields that direct them to a 'switchyard'. Here the beam is split into a spray of several individual beams which hurtle off in different directions to serve detectors in three experimental areas. One of these is the Meson Area, where the incoming protons collide with a metal target to produce secondary showers of mesons—pions and kaons. Combinations of electric and magnetic fields then select particles of a given type and energy and deliver beams of them to individual experiments. In all, the Tevatron's secondary beams supply about 13 different experiments simultaneously.

The Tevatron is a wonder of the modern world, its operation dependent on split-second timing and the reliable functioning of thousands of individual components,

Fig. 6.3 (*left*) *The big ring at Fermilab contains two accelerators, built one on top of the other in the same 3 m wide tunnel. The 6 m long electromagnets in the upper ring—red and blue—guide the protons as they are accelerated to 150 GeV. The yellow and red superconducting magnets in the lower ring then take the protons on the last part of their journey to 1000 GeV (1 TeV).*

Fig. 6.4 (*right*) *The protons receive kicks of energy supplied by radio waves in copper 'cavities'. The ones shown here, in a picture taken before the superconducting ring was installed, are now used in accelerating the protons on the fourth stage of their journey, from 8 GeV to 150 GeV.*

Fig. 6.5 *The main control room at Fermilab is crammed with displays that monitor the status of thousands of components in the complex accelerator network.*

each of which can cause a breakdown if it fails. Every aspect of the system is monitored by banks of powerful computers. In the main control room, the accelerator physicists can call up colour displays to show the state of the vacuum in the beam pipe, the position of the beam itself, and dozens of other parameters.

Similarly, each of the experiments incorporates microprocessors to control the simpler aspects of the apparatus, as well as larger computers to take overall charge of the operation of the experiment. The researchers are not quite redundant, however! They must be on hand night and day while the experiment is running to see that nothing untoward happens. Like a continual industrial process, the average experiment at a modern particle accelerator requires a team of experts and technicians to work shifts and keep a 24 hour watch.

Such complexity is very different from the experiments that Rutherford performed 70 years previously. It is even a long way from the cloud chamber experiments that first revealed new particles in the cosmic radiation. Yet the forerunners of the Tevatron and the other modern giant particle accelerators were invented in the early 1930s, just as cosmic ray research was entering its heyday.

THE WHIRLING DEVICE

IN NOVEMBER 1927, in a presidential address to the Royal Society, Rutherford wished for 'a copious supply of atoms and electrons which have an individual energy far transcending that of the alpha and beta particles from radioactive bodies'. His words inspired physicists and engineers both in America and in Britain. In Rutherford's own Cavendish Laboratory, Cockroft and Walton built a machine that produced the first nuclear disintegrations from artificially accelerated particles in 1932 (see pp. 41–3). But the invention that was to lead directly to today's giant accelerators was a different type of machine, the *cyclotron*. It was the inspiration of one man, Ernest Orlando Lawrence, who arrived at Berkeley in 1928 to be associate professor of physics.

The 27 year old Lawrence had originally intended to continue his researches on photoelectricity, but in 1929 he came across the doctoral thesis of Rolf Wideröe, a Norwegian engineer working in Germany. Wideröe had put into practice an idea for accelerating particles that had been suggested five years earlier by a Swedish physicist, Gustaf Ising. Lawrence immediately saw a way to improve Wideröe's device still further, and thereupon changed the course of not only his own future but that of particle physics.

Ising and Wideröe had considered accelerating particles to high energy through a series of small pushes from relatively low accelerating voltages. In Wideröe's design,

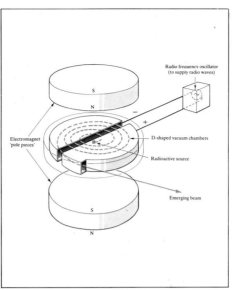

Fig. 6.6 (*left*) *Lawrence's first successful cyclotron, built in 1930, was only 13 cm in diameter and accelerated protons to 80 keV.*

Fig. 6.7 (*right*) *The basic components of a cyclotron are an electric field to accelerate the particles and a magnetic field to curve them round on a cycling path. In practice, the magnetic field is supplied by two electromagnetic 'pole pieces'. These generate a vertical north–south field through the path of the particles, which are contained in a horizontal plane. The electric field is provided across a gap between two hollow D-shaped metal vacuum chambers. Particles emanating from a radioactive source at the centre of the device are accelerated when they cross the gap between the 'dees'. By applying an electric field oscillating at radio frequencies, the direction of the field can be changed so that it is in the correct direction to accelerate the particles each time they cross the gap. The particles curl round in the cyclotron's magnetic field, but as they increase in energy, they curl less and less, so that they spiral outwards until they eventually emerge from the machine.*

the particles travel through a series of separate metal cylinders in an evacuated tube. Within the cylinders there is no electric field and the particles simply coast along. But across the gaps between the cylinders Wideröe set up electric fields by means of alternating voltages, which switch between positive and negative values. He matched the frequency of the alternating voltage with the length of the cylinders, so that the particles would always feel a kick, not a brake, as they emerged into a gap. In this way, the particles could be accelerated every time they crossed between one cylinder and the next. This is the basis of the operation of the modern linear accelerators, or linacs, used in the preliminary stages of acceleration at today's big rings.

Lawrence's inspired idea was to use a magnetic field to bend the particles into a circular orbit. Then they could pass across the same accelerating gap many times, rather than travel through a succession of gaps as Wideröe's scheme required. Lawrence saw that if the particles are accelerated on each circuit, they must spiral outwards as they increase in energy and become more resistant to the bending influence of the magnetic field. But he also realized that as the radius of the orbit increases, so does the particle's speed, so that the time taken for each circuit remains constant. Despite their spiralling orbit, the particles can still cross a gap at equal intervals of time and remain in step with an alternating accelerating voltage.

The principle underlying Lawrence's 'whirling device' was to place between the circular north and south poles of an electromagnet two hollow semi-circular metal cavities, or 'Ds'. A gap separates the two Ds and an electric field across the gap accelerates the particles as they cross it on the first half of their circuit. On the second half of the circuit, the particles cross the gap again, but in the opposite direction; so if the particles are to be accelerated again, the electric field must have changed direction. To accelerate the particles continuously, the electric field in the gap must switch back and forth. All Lawrence needed to do was to match the frequency at which the electric field switched with the time taken for the particles to complete the circuit. Then particles issuing from a source at the centre of the whirling device would spiral out to the edge and emerge with a greatly increased energy.

Lawrence announced the successful operation of his first 'cyclotron', as the whirling device had become officially known, to the American Physical Society in January 1931. Together with his research student Stanley Livingston, he had built a machine that accelerated protons to an energy of 80 keV. As a result, Lawrence received a grant of $500 from the National Research Council for work on a larger, useful machine.

A year later, Lawrence and Livingston, together with a new research student, David Sloan, successfully operated a cyclotron with a diameter of 28 cm—the '11-inch'—and reached the magical figure of 1 MeV. But in their zeal to improve the design of their accelerator, the team at Berkeley had neglected to exploit its appli-

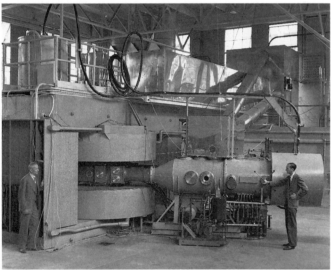

Figs. 6.8 (*top left*) *Lawrence's cyclotrons grew in rapid succession. The '11 inch' (28 cm), was the first to exceed 1 MeV, in January 1932.*

Fig. 6.9 (*top right*) *The next, the '27 inch' (69 cm), reached 4.8 MeV in December 1932. It was based on an unwanted magnet that Lawrence obtained free from the Federal Telegraph Company in Palo Alto. This picture shows Lawrence, kneeling in white shirt, and his team trying to coax a beam in the device. Edwin McMillan is on the right of the group standing.*

Fig. 6.10 (*bottom left*) *The '60 inch' (1.5 m) bears more resemblance to the cyclotrons used today, for example in hospitals. It delivered its first beam, of 19 MeV deuterons (nuclei of 'heavy hydrogen' consisting of a proton and a neutron), in October 1939, two months after the picture shown here was taken.*

Fig. 6.11 (*bottom right*) *Lawrence's final masterpiece, the '184 inch', as it appeared in 1957. It is still at work today.*

cations. Instead, Cockroft and Walton observed the first artificially induced nuclear transmutations at Rutherford's Cavendish Laboratory.

Berkeley did however become the Mecca for accelerator builders, and spawned cyclotrons at other laboratories not only in the US but across the world. Places such as Cornell, Princeton, Chicago and Michigan, Liverpool and Birmingham, Paris, Stockholm, and Copenhagen were soon to have cyclotrons of their own. Nor were these machines used only to do physics. Work at Berkeley had shown the importance of making radioactive isotopes for use in medicine, biology, and chemistry. The cyclotron gradually became a tool of 'nuclear science'.

By 1939, the diameter of the largest cyclotron at the Radiation Laboratory at Berkeley had risen to 1.5 metres—the '60-inch'. During the early 1930s, while the US was sunk in a deep economic depression, Lawrence had, amazingly enough, created 'big science'—research on an industrial scale that involved the collaboration of scientists and engineers, many technicians and support staff, and large sums of money. Lawrence's talents lay not only in daring to reach for goals that were increasingly challenging, but also in a remarkable ability to raise the funds to support enterprises that were increasingly costly.

In November 1939, Lawrence was awarded the Nobel prize for physics for his invention and development of the cyclotron. Five months later he had a promise of $1.4 million from the Rockefeller Foundation to build a giant 100 MeV cyclotron, based on an enormous magnet with poles 4.6 metres (184 inches) in diameter.

Lawrence wanted to produce the supposed carrier of the strong force, later called the pion, and he believed that bombarding nuclei with alpha particles accelerated in his proposed machine would do the trick. With double the charge of protons, alphas would be accelerated to double the energy, or 200 MeV, and Lawrence calculated that 150 MeV alphas would be energetic enough to release the pions from the clutches of the strong force. But the Second World War intervened, and although Lawrence got his 4.6 m magnet, it was as a 'mechanism of warfare' which was used in a method he had devised to separate the fissile isotope uranium-235 from the much more common uranium-238.

The pause for war had one fortuitous consequence. The original '184-inch' design would probably never have produced the desired beam of 150 MeV alpha particles. One effect of Einstein's theory of special relativity is that as objects approach the speed of light they become increasingly heavy. The cyclotron works on the principle that the particles always take the same time to complete a circuit, but this ceases to be true when special relativity applies. The heavier a particle becomes, the longer it takes to complete a circuit; eventually it will arrive too late at the gap between the cyclotron's Ds to catch the alternating voltage during the accelerating part of its cycle.

In the smaller cyclotrons made before the war, this effect was insignificant. But for protons of about 25 MeV, at about one fifth the velocity of light, the increase in mass of about 2 per cent is enough to begin to make itself felt. This is really the practical limit for a proton cyclotron. At 100 MeV, and approaching half the speed of light, protons are over 10 per cent heavier than they are at rest. But Lawrence, in characteristic spirit, was not to be deterred by relativity. In 1939, he had hoped to beat the increase in mass by using the brute force of a very high accelerating voltage, taking protons to 100 MeV within a few turns. By the end of the war a more subtle technique had come to light, and one that could go far beyond the limit of 25 MeV.

Ed McMillan, 'conscripted' during the war from Berkeley to work on the atomic bomb at Los Alamos, and Vladimir Veksler in the Soviet Union, independently thought of the same idea to enable the cyclic accelerator to break free from the constraints of relativity. They proposed adjusting the frequency of the applied voltage so that it remains in step with the particles as they take longer to circulate.

A machine operating at variable frequency could no longer accelerate a continuous stream of particles, as the cyclotron had done. Changing the frequency to keep in time with higher-energy particles would mean that any particles still at lower energies

Fig. 6.12 *Members of Lawrence's 'Rad Lab' relax at a party in 1939. Lawrence, fork in hand, is seated at the head of the table to the left; McMillan is between the two women with polka dot dresses at the same table.*

would become out of step. Instead the 'synchronized' cyclotron, or *synchrocyclotron*, would take particles from the source a bunch at a time, and accelerate these bunches out to the edge of the magnet. The frequency of the accelerating voltage would meanwhile decrease to compensate for the particles' increasing mass. The final energy of the particles is then limited only by the strength and size of the magnet.

When McMillan returned to Berkeley after the war, his idea for varying the cyclotron frequency was applied to the design of the 184-inch. The great magnet was relieved of its uranium enrichment duties and could at last be incorporated in a particle accelerator. At the beginning of November 1946, the new synchrocyclotron produced its first beam—deuterons with an energy of 195 MeV. But before the physicists at Berkeley began to search for pions, they were overtaken by events in cosmic ray research. Powell and his colleagues found the charged pion early in 1947. However, as a 'consolation prize', Berkeley was rewarded with the discovery of the neutral pion two years later.

MAN-MADE COSMIC RAYS

LAWRENCE'S 4.6 metre synchrocyclotron complemented cosmic ray studies by producing copious supplies of pions to order. But even as it did so, in 1947, the cosmic ray physicists found the first of a series of exotic new particles. The kaon and its fellow strange particles were significantly heavier than the pion; some were even heavier than the proton!

Lawrence's machine was not powerful enough to produce these heavy particles. It was limited by the strength of the magnetic field and the diameter of the magnet's poles: once the accelerated particles reached a certain energy their orbits could no longer be contained between the poles. As so often in the history of accelerators, the cry went up for 'more energy'. But Lawrence's 4.6 metre magnet was as large as it was practical to make, so how could higher energies be reached?

The solution was to alter not only the frequency of the accelerating voltage to match the increasing energy of the particles, but also to increase the magnetic field. If the magnetic field is strengthened continuously as the circling particles gain energy, they can be kept on more or less the same orbit instead of spiralling outwards. Moreover, the enormous single magnet of the cyclotron can be replaced by a doughnut-like ring of smaller magnets, each with a profile like a 'C'. The particles travel through a circular evacuated pipe held in the embrace of the magnets; they are accelerated during each circuit by an alternating voltage of varying frequency, which is applied at one or more places around the ring; and they are held on their circular course through the pipe by the steadily increasing strength of the magnetic field. Such a machine is called a *synchrotron*, and it is still the basis of large modern accelerators, such as the Tevatron at Fermilab.

When McMillan returned to Berkeley after the war, he set about building a prototype electron synchrotron. It was easier for technical reasons to begin with an electron machine rather than a proton device. But although electron synchrotrons were to play an important role during the next 30 years, it was proton synchrotons that became the order of the day, both at Berkeley and at other laboratories around the world.

In 1947, the US Atomic Energy Commission approved the building of proton synchrotrons at two competing sites—Berkeley on the West Coast and the Brookhaven National Laboratory on Long Island, New York. The machine at Brookhaven was designed to reach 3 GeV, so that its beam of protons would produce pions in profusion after colliding with a suitable target. Berkeley's preliminary goal was to find the antimatter counterpart of the proton, the negatively-charged antiproton. The antielectron, or positron, had been discovered in cosmic rays by Carl Anderson in 1932. Detecting the antiproton would provide the missing link in establishing that the laws of physics are symmetrical between matter and antimatter. Theory suggested that an energy of just over 6 GeV would be necessary to produce antiprotons from the collisions of protons with a target, so Berkeley aimed for this higher energy

The 3 GeV machine at Brookhaven, the Cosmotron, became the first proton synchrotron to operate, in 1952, and it led the field for two years. Early experiments there complemented well the work done on strange particles with cosmic rays. It discovered the negatively charged partner of the positive sigma particle found in the cosmic radiation. And, more important, it provided the first concrete evidence that the 'vees' formed by the decays of two kinds of strange particle—the kaon and the lambda—always emerge together. This did much to strengthen the theory of *associated production*, which had predicted that strange particles are always produced in pairs.

Meanwhile, in California, the Bevatron was nearing completion at Berkeley. By November 1954, it was delivering 10^{10} protons per pulse at 6.2 GeV, and in 1955 a number of teams began the hunt for the antiproton. There were already faint indications that such an object might have been found in cosmic ray experiments in Europe; Berkeley did not want to be eclipsed by the cosmic radiation yet again.

The first antiproton searches at Berkeley used the tools of the cosmic ray physicists—emulsions and cloud chambers. But because the antiprotons were rare, the photographs revealed no signs of the anticipated nuclear starburst that would result from a proton–antiproton annihilation. The collisions between the accelerated protons and protons at rest in a target would produce only one antiproton for every 50 000 pions. What the physicists needed were techniques that would automatically sift out the occasional antiprotons from the large 'background' of pions *before* the information about the particles was recorded.

Two teams, led by Edward Lofgren and Emilio Segrè, planned to seek out antiprotons in this way. They designed a series of detectors to determine the momentum and velocity of the particles created in the collisions. If you know a particle's momentum and velocity, you can calculate its mass; and if you find a particle with the same mass as the proton, but with negative instead of positive charge, you can be fairly certain that you have found an antiproton.

The hunt began with the selection of negatively charged particles out of the debris produced by collisions of protons with a target inside the Bevatron's magnet ring. This was the easy part. The Bevatron's magnetic field bends positive and negative particles in opposite directions, so a beam of negative particles was selected by suitably aligning a hole in the accelerator's shielding with the internal target. A more delicate problem was to pick out from this beam the particles with the same mass as the proton, while ignoring the light pions and the slightly heavier kaons.

The first step was to use a magnetic field to spread the particles out according to their momentum, much as a prism spreads out visible light according to wavelength. A magnet bends particles with high momentum less than particles with lower momentum, and so a suitably placed slit or collimator will allow through a narrow beam of particles, all of which have more or less the same momentum. Now 'all' that was needed was to measure the velocity of each particle, and thereby calculate its mass.

Fig. 6.13 (*left*) The Cosmotron at the Brookhaven National Laboratory was the first proton synchrotron to come into operation, in 1952. The protons were accelerated first to 3 MeV in a Van de Graaff generator—the cylindrical tank in the foreground. Then they moved through the slim pipe to the 'doughnut' of the main accelerator, where they were raised to an energy of 3 GeV. The four sections of the magnet ring (three are visible here) each consisted of 72 steel blocks, about 2.5 m × 2.5 m, with an aperture of 15 cm × 35 cm for the beam to pass through. The machine ceased operations in 1966.

Fig. 6.14 (*right*) The Bevatron at the Lawrence Berkeley Laboratory began to accelerate protons in 1954, up to an energy of 6 GeV. A Cockcroft-Walton generator, to the right of the picture, fed protons down the linac (the tube at the centre) for injection into the main magnet ring at 10 MeV. The Bevatron's magnet weighs 10 000 tonnes, fives times greater than the magnet in the Cosmotron. The Bevatron is still operational today, routinely accelerating heavy ions—even as heavy as uranium—which are fed into it from a special linear accelerator callled the Super-Hilac. Together the two machines are known as the Bevalac, one of the world's leading facilities for the study of heavy-ion collisions.

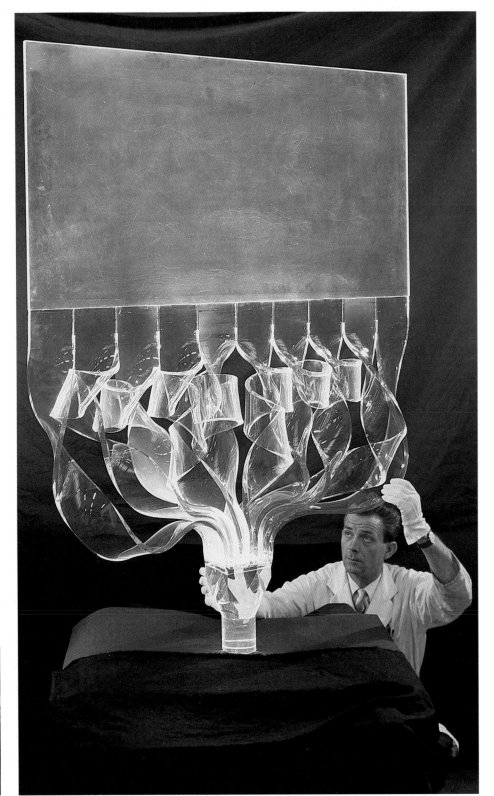

Fig. 6.15 *A large rectangular sheet of plastic scintillator is prepared for an experiment at CERN. The curly structures are acrylic 'light guides' which collect the light emitted in the scintillator and concentrate it in the circular 'pipe' at the bottom. This pipe will fit against the surface of a photomultiplier tube, which converts the light into small electrical signals. The whole structure will be carefully wrapped, first in reflecting foil and then with layers of black tape, to make it completely light tight. Notice how the light guides are designed to be all of the same length, so that the light from different parts of the scintillator arrives at the phototube at nominally the same time.*

Fig. 6.16 *Members of the team who discovered the antiproton surround Edward Lofgren, leader of the other team of antiproton hunters at Berkeley. Left to right: Emilio Segrè, Clyde Wiegand, Lofgren, Owen Chamberlain, and Thomas Ypsilantis.*

Segrè, together with Owen Chamberlain, Clyde Wiegand, and Tom Ypsilantis, chose to use two ways of pinpointing the velocity so as to be doubly sure that they had indeed captured an antiproton. One method used two scintillation 'counters', which produced a flash of light each time a charged particle passed through. Modern plastic scintillators are the descendants of the scintillating materials Rutherford used in his scattering experiments. But whereas Rutherford and his colleagues had to use their own eyes to see and count the flashes, by the 1950s electronic components made the process automatic. Each tiny burst of light is converted to a pulse of electricity, which is then amplified to produce a signal suitable for feeding into co-incidence counting circuitry of the kind Bruno Rossi had invented in the 1930s. In this way, two or more scintillation counters can reveal the flight path of a particle as it produces flashes in each counter.

Segrè and his colleagues set up two scintillation counters 12 metres apart. At the particular momentum the physicists had selected, antiprotons would reach the second counter eleven thousandths of a microsecond later than the faster, lighter pions. By using lengths of cable to delay the signal from the first counter, the physicists could make it coincide in time with the second signal, but only if the signals came from antiprotons. When a pion passed between the counters, it would make the journey too quickly, and the signals would no longer coincide. In this way, coincident signals revealed the passage of the elusive antiprotons.

The second method the team used to measure the velocity of their particles was based on the Čerenkov effect, named after the Russian physicist Pavel Čerenkov, who discovered the phenomenon in 1934. When a particle moves extremely swiftly through a material, it can create a kind of shock wave of visible light, known as Čerenkov radiation. The crucial factor is that the particle must be moving faster through the material than ordinary light does in the same substance. The Čerenkov radiation emerges at an angle to the particle's path, and the greater the particle's velocity, the larger this angle becomes. By choosing the right material, you can create a 'window' that reveals only particles above a certain velocity.

Segrè's team in fact used two Čerenkov counters, one containing an organic liquid ($C_8F_{16}O$), the other built from fused quartz. The liquid Čerenkov counter produced a signal for any particle moving faster than an antiproton, to identify the pions; the quartz counter, specially designed by Wiegand and Chamberlain, revealed only particles with a velocity close to that expected for antiprotons.

By early August 1955, Segrè and his colleagues had set up their apparatus at the Bevatron, and on September 21 they obtained their first evidence for antiprotons. Barely a month later, the team was confident enough to send a paper announcing the discovery to the *Physical Review*. They had sifted out 100 antiprotons from a background of 5 million pions. The Bevatron had done its stuff, and Segrè and Chamberlain shared the Nobel Prize in 1959. (See pp. 134–5 for antiproton portrait.)

Fig. 6.17 *Čerenkov radiation is responsible for the blue glow in the water surrounding the core of the nuclear reactor at the State University of North Carolina, Raleigh. Energetic charged particles travel through the water faster than light does, and as a result emit the Čerenkov radiation.*

GLASER AND THE BUBBLE CHAMBER

THE DISCOVERY of the antiproton confirmed the promise of the Bevatron. With twice as much energy as the Cosmotron, Berkeley's new accelerator could discover new heavy particles, as well as investigate the behaviour of known lighter particles at high energies. But the high energies also made it more difficult to detect the particles. Energetic particles could hurtle all too easily through a cloud chamber without decaying or interacting with the atoms in the chamber's thin gas. To record the whole life of a strange particle, for example, from production to decay, at the Bevatron's energies would have required a cloud chamber 100 metres long! In addition, cloud chambers are slow devices. The cycle of recompression after an expansion can take up to a minute; the Bevatron, on the other hand, delivered pulses of protons every two seconds.

What was needed was a detector that would capture the long tracks of high-energy particles and operate quickly. Gases were much too tenuous for the job. Liquids, on the other hand, were a more promising alternative, because their much greater density means they contain far more nuclei with which the high-energy particles can interact. But how do you make particle tracks visible in a liquid? The cloud chamber had depended on the production of liquid droplets in a gas, but liquid droplets in a liquid are like the proverbial black cat in a coal heap. The problem was to turn the black cats white.

The solution came not from Berkeley, but from a young physicist at the University of Michigan—Donald Glaser. Glaser had done his PhD at Caltech, where he had come under the influence of Carl Anderson in the late 1940s, just when the strange particles were causing consternation in the physics community. In 1949, Glaser moved to Michigan to begin teaching and research, and three years later he had the brilliant idea of how to make visible the tracks of particles passing through a liquid. He had worked out how to make the cats in the coal heap white.

A homely example of the effect Glaser wanted to harness is the action of opening a bottle of beer. The fall in pressure as you release the bottle's cap causes bubbles to rise up through the liquid. Glaser's idea was to hold a liquid under pressure and very close to its boiling point. If you lower the pressure in these circumstances, the liquid begins to boil—an effect familiar to mountaineers, who can brew up a cup of tea on a mountain top at a lower temperature than is possible at sea level. But if you lower the pressure very suddenly, the liquid will remain liquid even though it is now above its boiling point. This state is known as 'superheated liquid' and because it is unstable, it can be maintained only so long as no disturbance occurs in the liquid.

Glaser realized that charged particles shooting through a superheated liquid will create a disturbance and trigger the boiling process as they ionize the atoms of the liquid along their paths. For a fraction of a second, a trail of bubbles will form where a particle has passed, and this trail can be photographed. But you must act quickly, or the whole liquid will begin to boil violently. Glaser therefore planned to release the pressure and then immediately restore it. Particles entering the liquid during the critical moments of low pressure would leave trails that could be photographed. The immediate restoration of pressure would mean that the liquid was once again just below boiling point, and the whole process could be repeated.

In the autumn of 1952, Glaser began experiments to discover if his 'bubble chamber' would work. After thoroughly considering possible liquids, he chose to use diethyl ether. With a small glass vessel holding just 3 centilitres of the liquid, he successfully photographed the tracks of cosmic rays. But he faced an uphill battle in developing his invention. He was refused support by the US Atomic Energy Commission and the National Science Foundation. They said his scheme was too speculative. And his first paper on the subject was rejected on the grounds that it used the word 'bubblet', which was not in the dictionary. But his luck changed in 1953, when a chance meeting brought the bubble chamber to fruition.

Glaser's first talk on his idea was to be given on the last day of the American Physical Society's meeting in Washington DC in April 1953. Among the participants at the meeting was Luis Alvarez, a distinguished physicist with a long record of important discoveries. He was involved with the Bevatron, which was still under construction at the time, and was concerned about the problem of how to detect the high-energy particles that the machine would produce.

On the first day of the meeting, Alvarez was sitting at lunch with colleagues from his wartime days at Los Alamos. On his left was a young man who had not experienced those times and was missing out on the reminiscences. Alvarez started talking to him about physics and the current ideas. The young man was Glaser, who complained to Alvarez that his ten-minute talk had been allocated to the final slot on the Saturday, by which time most people would be on their way home. Alvarez admitted that he too would be unable to attend the talk for that very reason. Sheepishly, he asked Glaser what he was going to report. Glaser explained how he had invented the bubble chamber and built a small version 2 cm in diameter. Alvarez was impressed; he realized immediately that this was the breakthrough he had been looking for.

That night Alvarez told his colleagues from Berkeley what he had learnt and suggested that it might be possible to build a big chamber filled with liquid hydrogen. This makes an ideal target for nuclear collisions because hydrogen is the simplest form of matter. Alvarez's colleagues were won over as dramatically as he had been. They all agreed that this was the way to proceed, and on their return to California they set about designing a large hydrogen-filled bubble chamber.

The idea of using hydrogen instead of ether made the work more difficult. Hydrogen becomes a liquid only when cooled to a chilly 20 degrees above absolute zero, or −253°C! But by the end of the year, only eight months after Alvarez had talked to

Fig. 6.18 The track of a cosmic ray passes through Donald Glaser's first bubble chamber, a small glass phial holding a mere 3 cl of diethyl ether.

Glaser, one of the group at Berkeley, John Wood, had observed tracks in a hydrogen-filled bubble chamber. It was only 3.5 cm in diameter, but it proved that the idea worked. Moreover, Wood made the important discovery that he could obtain clear images of tracks despite accidental boiling of the hydrogen induced by 'flaws' in the chamber's walls.

Glaser, and everyone else, had originally thought that ultra-smooth walls were needed, and had therefore concentrated on glass chambers. Now that it was clear that smoothness was not so crucial, Alvarez's team turned to a construction based on metal walls with glass windows. They first built a chamber 6 cm in diameter and then a 10 cm chamber, which was ready for testing on the Bevatron in November 1954. The team then designed a 'big' chamber, 25 cm in diameter, which began regular work at the Bevatron in 1955. But Alvarez was already thinking much bigger.

In early 1955, before the 25 cm chamber was even complete, he proposed building a 75 cm chamber. Like Topsy, this grew in Alvarez's imagination until he eventually settled for a monster, 180 cm long, 50 cm wide, and nearly 40 cm deep. It would hold 17 cubic litres of liquid hydrogen and the window would contain 800 square centimetres of glass—which of course had to be thick enough to withstand the pressures inside. Even Lawrence, director of the laboratory at Berkeley and daring pioneer of the cyclotrons, was amazed at Alvarez's audacity. 'I don't believe in your machine,' he told Alvarez, 'but I do believe in you, and I will help you to obtain the money.'

The monster was not cheap. By the time it was completed in 1959, the '72-inch' had cost over $2 million. It was a far cry from Glaser's first tiny chamber. It occupied its own building, complete with crane, compressors, and a magnet drawing 3 MW of power. Detecting the smallest fragments of matter had become big business.

As in the case of the cyclotron, other laboratories followed Berkeley's lead and built bubble chambers of various sizes and filled with a variety of liquids. One that was to gain fame in the 1960s was the '80 inch' (200 cm) hydrogen bubble chamber at the Brookhaven National Laboratory. This was fed particles from the successor to the Cosmotron, an accelerator known as the Alternating Gradient Synchrotron or AGS.

The operation of a bubble chamber is always intimately tied to the operating cycle of the accelerator that feeds it. In the case of the 80-inch, the expansion of the chamber began some 15 milliseconds before the burst of particles from the AGS was due. The expansion was accomplished by the withdrawal of a large piston, 90 cm in

Fig. 6.19 (*left*) *Donald Glaser (b.1926) inspects a xenon bubble chamber at the Lawrence Berkeley Laboratory in the early 1960s. Xenon is useful because it forms a dense liquid in which gamma rays readily become 'visible' by converting to electron–positron pairs.*

Fig. 6.20 (*right*) *Luis Alvarez (b.1911) in 1954.*

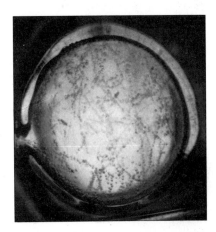

Fig. 6.21 *The first tracks observed in liquid hydrogen, in John Wood's 3.5 cm bubble chamber in 1954.*

Fig. 6.22 *The '80 inch' (200 cm) liquid hydrogen bubble chamber at the Brookhaven National Laboratory in 1965. The stainless-steel chamber is almost totally obscured by the surrounding magnet coil, the huge steel magnet yoke, and the equipment to expand the chamber and to keep the liquid hydrogen cool. Together the assembly weighed some 450 tonnes and stood about 7.2 m high, but it could be moved up and down, from side to side, and even rotated on a turntable, according to the desires of the experimenters. The hydraulic ram to move the apparatus sideways is visible at the lower left. The man on the platform just above the ram is removing one of the three automatic cameras that photographed tracks in the chamber. The particle beam entered the chamber through the vertical rectangular 'window', seen just to the right of centre. The pictures opposite show stages in the assembly of the chamber, which took 250 man-years to design and build, between 1959 and 1963, and which cost in the region of $6 million.*

diameter and 80 cm high, located in the cylinder above the wide neck of the chamber visible in Fig. 6.24. The withdrawal of the piston through just 1 cm reduced the pressure inside the chamber from over 5 atmospheres to 2 atmospheres.

The particles entered the chamber when the piston was fully withdrawn, the pressure at its minimum, and the liquid superheated. Then, about one millisecond later, an arc light flashed on for a fraction of a millisecond. The flash illuminated the trails of bubbles formed by charged particles, and exposed the film in three or four cameras viewing the chamber. The delay between minimum pressure and the flash allowed the bubbles to grow to a diameter of about 10 millionths of a metre, large enough to show up on the photographs. Meanwhile, the piston moved back in towards the chamber, increasing the pressure again, and the film in the cameras was automatically wound on to the next frame. It then took about a second for the chamber to 'recover' and be ready for the next expansion.

One of the 80-inch's most famous discoveries occurred in 1964 when researchers found one photograph—out of a batch of some 80 000—with the tell-tale pattern of tracks that betrayed the birth and death of the omega-minus particle (see Fig. 7.15 and pp. 141–2). In a typical bubble chamber experiment, over a million photographs

Fig. 6.23 *The thick glass window through which the cameras viewed the 80 inch bubble chamber at Brookhaven. The chamber itself was shaped like a squat oval cylinder, with its faces vertical. The window, made in Germany from borosilicate crown glass, had to be strong enough to withstand pressures of over five times atmospheric. It was 200 cm long, 75 cm high, 16.5 cm thick and weighed nearly 700 kg. It took 8 hours to grind and polish the window so that no surface roughness could set off unwanted boiling.*

Fig. 6.24 *The uninsulated body of the chamber is being manoeuvred between the two halves of the magnet. A rectangular structure extends upwards from the top of the chamber. This is the 'neck' through which the chamber was expanded, the 12 kg piston being housed in the cylinder above. To the left of the bubble chamber, the vacuum chamber that surrounded it is visible, with the vertical rectangular beam 'window'. This outer chamber was like a huge 'vacuum flask', designed to help keep the liquid hydrogen cool.*

Fig. 6.25 *Here the 'safety chamber' (in the middle of the picture) is being moved into place. This attaches to the 'front' of the bubble chamber, next to the glass window. Its purpose is to catch glass fragments and hydrogen should the window break, and to shield the bubble chamber from thermal radiation. The right hand face of the safety chamber contains apertures for the four cameras to view the bubble chamber, and a fifth larger aperture for light to illuminate the tracks. The turntable for rotating the whole assembly is visible towards the bottom of the picture.*

Fig. 6.26 *Scanners at work inspecting film from bubble chambers at Brookhaven in 1964. Film was projected onto a table to give a life-size image, which could then be measured.*

may be taken, occupying hundreds of reels of film. How do the researchers cope with all this information, and find rare events like the production of the omega-minus?

The first task in examining a bubble chamber photograph is to spot the interactions that look interesting and to identify the particles that have produced the tracks. Sometimes certain tracks are instantly recognizable, like the tight spirals formed by low-energy electrons, but generally only careful measurement of the tracks gives the correct identification. The techniques used are basically the same as in the interpretation of cloud chamber or emulsion pictures.

For instance, the curvature of a track in a magnetic field reveals a particle's charge and its momentum. But this is not usually sufficient to label a track correctly because two particles of different mass and energy can have the same momentum. Often the only way is to assign identities to the different tracks, and then to add up the energy and momentum of all the particles emerging from an interaction. If they do not balance the known values before the interaction, the assumed identities must be wrong, and others must be tested, until finally a consistent picture is found.

Identifying particles through such trial-and-error calculations is the kind of repetitive job at which computers excel. The first machine devised at Berkeley in the late 1950s to scan bubble chamber photographs was nicknamed the Franckenstein, after its creator Jack Franck. The device projected an image of a stereo pair of photographs. A human operator then set a photosensitive 'scanner' at the beginning of each track. The scanner would follow the tracks, bright against the dark background, and under the operator's control would punch out information about the trajectories on cards that a computer could read. The computer used this data to reconstruct the tracks in its 'brain' and compare them with pre-programmed patterns that the physicists thought might occur. Programming the computer to do this took two years. Even so, Franckenstein could analyse only 100 or so interactions a day, while the Berkeley 72-inch was photographing thousands of interesting interactions in the same time. By the late 1960s, however, the machines had improved to the point where more than 100 photographs could be analysed per hour.

STRONG FOCUSING

THE BUBBLE chamber, invented in 1952, was to be the workhorse of particle physics for the best part of 30 years, displaying the tracks of particles at ever increasing energies. But the early synchrotrons were soon to be superseded, and even as the Bevatron and the Cosmotron spewed forth their first protons, the ideas for a new breed of synchrotron were already on paper.

The basic concept of the synchrotron had reduced the huge magnet with circular poles, which the synchrocyclotron required, to a magnetic ring formed from smaller sectors. Even in the best behaved synchrotron beams, however, the particles cannot all crowd exactly on to the ideal orbit. There is an initial spread to the beam as it is injected into the ring; collisions with residual air molecules in the vacuum chamber can deflect the paths slightly; and the particles will have slightly differing energies. To keep a hold on as many particles as possible, the poles of the C-shaped magnetic sectors in the first synchrotrons were gently shaped to provide a weak magnetic focusing, so that stray particles returned to the ideal orbit. The result was that the particles oscillated about the perfect path, weaving from outside the main orbit to within it, as they were nudged first this way and then that by the magnetic forces. But the focusing was weak; to keep hold of as many particles as possible, the 'race-track' they whirled around had to be broad.

In the Bevatron, for example, the vacuum chamber through which the beam travels is 30 cm high and 120 cm wide; this forms the 'filling' for a magnetic doughnut weighing 10 000 tonnes. So the Bevatron gives nearly ten times the energy of the 4.6 m synchrocyclotron for two and a half times the weight, but 10 000 tonnes of iron is still a large amount. An even larger machine, which started up at Dubna in the USSR in 1957, accelerates protons to 10 GeV, but its racetrack is 40 cm by 150 cm, and it weighs a colossal 36 000 tonnes. At 10 GeV the weak-focusing synchrotron was in danger of becoming a dinosaur; how could it be saved?

Stanley Livingston and his colleagues at Brookhaven, Ernest Courant and Hartland Snyder, already had the rescue package for the synchrotron in 1952. They proposed a method for focusing the particle beams strongly, so that they swung less far from the ideal orbit. Unbeknown to the physicists at Brookhaven, a Greek engineer, Nicholas Christofilos, had patented the same idea in 1950; he later joined the rival establishment at Berkeley. The principle behind strong focusing is to shape the magnet pole faces so that they guide a deviant particle quickly back towards the middle of

Fig. 6.27 The 10 GeV proton synchrotron at the Dubna Laboratory in Moscow is the biggest of its kind, relying as it does on the principle of 'weak focusing'. The magnet weighs a total of 36 000 tonnes, and must accommodate a beam up to 1.5 m wide.

Fig. 6.28 *CERN's proton synchrotron—known simply as the 'PS'—accelerated its first protons to 24 GeV in 1959. Now the machine can run up to 28 GeV and forms a vital part of the accelerator complex that ultimately collides protons with antiprotons at very high energies.*

the vacuum chamber; such a particle will naturally swing across to the other side of the ideal orbit, but it will be swiftly directed back again, and criss-cross the vacuum chamber many times on its way round the machine. But there is one catch; a magnet shaped to focus the beam in the horizontal plane tends to *defocus* the beam vertically. To avoid losing as many particles in one direction as they were hoping to save in the other, Livingston and his colleagues, and Christofilos, realized that they needed to alternate two shapes of magnet.

The first would focus horizontally, but defocus vertically; the second would have the opposite effect, focusing vertically, but defocusing horizontally. With this combination, the net result would be a tightly-controlled beam that could be kept in a smaller vacuum chamber, and which could be guided by smaller magnets with smaller jaws. The concept of strong focusing was first used in an *electron* synchrotron at Cornell University, in Ithaca, New York. There, Robert Wilson, who had worked with Lawrence at Berkeley, built a 1.5 GeV machine which started up in 1954.

Meanwhile in Europe, a number of nations had come together under the auspices of UNESCO with the idea of rebuilding the shattered remnants of European unity in some project that no country could afford on its own. Particle physics, with its huge accelerators, was a natural choice. This was the origin of CERN, the 'Conseil Européen pour la Recherche Nucléaire'; and as early as May 1951, a board of consultants chosen personally by Pierre Auger, UNESCO's Director of Natural Science, suggested building not only a modest synchrocyclotron, but also an accelerator that would be the biggest and best in the world.

In 1952, when a 'Provisional CERN' was officially set up, the task of exploring the possibilities for the big machine was given to one of four special study groups. The still untested potential of strong focusing was not lost on the experts involved; indeed, the European plans for a giant accelerator had spurred Livingston and his colleagues to devise a means of strong focusing. So a team began to design a strong-focusing synchrotron that could reach up to 25 GeV, four times higher than the Bevatron.

By the end of September 1954, CERN officially came into being, with a permanent

Convention ratified at first by nine European countries, and in the following five months by three more. Now it had become the European Organization for Nuclear Research, but the acronym CERN stuck, and has stayed with it ever since. A site for the laboratory had been chosen on the outskirts of Geneva, and the synchrotron designers had already moved there in the previous October. By then the team had proposals for the new machine fit for the rest of the world to approve at an international conference that was also attended by representatives from Brookhaven. They too had plans for a 25 GeV machine, which the US Atomic Energy Commission approved shortly afterwards. The race was on.

CERN crossed the finishing line first. Its proton synchrotron—the 'PS'—accelerated protons to 24 GeV on 24 November, 1959. The machine had been completed on schedule, within six years of the Convention's being signed, and to cost—around £10 million. John Adams, who had led the accelerator team to its triumph, celebrated by opening a bottle of vodka given him by Vladimir Nikitin from the Dubna Laboratory in Moscow. Until that night Dubna had held the energy record with its 10 GeV machine. The following day the bottle was sent back to Dubna, empty of vodka, but containing a photograph of the instrument screen that proved that 24 GeV had been reached. The 10 GeV record was well and truly beaten.

The PS contrasted completely with the Synchrophasotron—the name the Soviets used for their synchrocyclotron at the Dubna Laboratory. With 100 strong-focusing magnet sectors arranged around a ring of 100 m average radius, the total weight of iron in the PS is 3200 tonnes—less than one tenth the amount in the weak-focusing 10 GeV machine at Dubna. Moreover, in the PS, the vacuum 'tank' of the weak-focusing machines has shrunk to an elliptical 'pipe' 14.5 cm across and 7 cm high.

Fig. 6.29 *John Adams (1920–1984), on the day after CERN's new proton accelerator had successfully reached 24 GeV, beating the record of 10 GeV previously held by the laboratory at Dubna. In his left hand he holds a picture of a monitor display confirming the energy; in his right hand he holds the (empty!) vodka bottle given by Dubna's director to be drunk once the record was broken.*

SPARK CHAMBERS

CERN WAS first in the race to 25 GeV, but Brookhaven was not far behind, and in 1960 the AGS (Alternating Gradient Synchrotron) began operation. On July 29, CERN's record was broken, when the beam in the AGS reached 30 GeV. By the following December the physicists at Brookhaven had begun experiments. The American researchers had a tradition of designing and operating experiments at Berkeley, especially on the Bevatron, as well as on the Cosmotron at Brookhaven. For the Europeans, working on such a grand scale was new, and after the brilliant success of the new accelerator at CERN it took some time for them to develop large experiments. During the 1960s, Brookhaven claimed several notable 'firsts', but the work at CERN was equally important, especially in the development of new kinds of detector to explore the territory that the 30 GeV machines had opened up. In particular, new detectors emerged to challenge the supremacy of the bubble chamber.

A bubble chamber can provide a complete picture of an interaction, but it has some limitations. It is sensitive only when its contents are in the superheated state, after the rapid expansion. Particles must enter the chamber in this crucial period of a few milliseconds, before the pressure is reapplied to 'freeze' the bubble growth. But how do you tell which incoming particles will produce interesting reactions? The question echoes the earlier difficulties with cosmic rays in cloud chambers. In that case, the problem of deciding when to take pictures could be solved because the cloud chamber has a 'memory'. Its expansion can be triggered *after* particles have passed through, using a signal from external counters that indicate that something interesting might have happened.

A bubble chamber cannot be triggered in this way; the expansion must occur *before* the particles arrive. And because the whole cycle of expansion and recompression takes about 1 second, the collection of rare events can take a long time. To study large numbers of rare interactions requires a more selective technique. In the 1960s, the *spark chamber* proved the ideal compromise.

Like the technique of coincidence counting, which was so vital in the discovery of the antiproton, the spark chamber was spawned from work on cosmic rays. Marcello Conversi—one of the Italians who had helped to identify the muon during the Second

Fig. 6.30 *An excited team crowds the control room of the Alternating Gradient Synchrotron at the Brookhaven National Laboratory on 29 July 1960, the day it first accelerated protons, reaching 24 GeV. Ken Green, chairman of the accelerator department, sits in front of the oscilloscope at the centre of the picture.*

Fig. 6.31 *The control room of the Alternating Gradient Synchrotron at Brookhaven in 1966. Contrast its appearance with the modern 'computerized' control room at Fermilab, shown in Fig. 6.5 (see p. 106).*

World War—had invented 'flash tubes' in the mid-1950s, and these became widely used to study cosmic ray showers. Flash tubes are sealed glass tubes filled with neon, which are arranged in layers between metal plates, rather like rafts. Charged particles ionize the gas in the tubes, and if a high voltage is applied to the plates luminous discharges—sparks—occur in the tubes that the particles traverse. A photograph of the array of tubes end-on reveals the tracks of the particles. Around the same time, researchers in the UK and Japan independently developed a different way of putting the sparks to use. They did away with the glass tubes and applied the high voltage across plates with gas between them. They had invented the spark chamber.

The basic spark chamber consists of parallel sheets of metal separated by a few millimetres and immersed in an inert gas such as neon. When a charged particle passes through the chamber it leaves an ionized trail in the gas, just as in a cloud chamber. Once the particle has passed through, you apply a high voltage to alternate plates in the spark chamber. Under the stress of the electric field, sparks form along the ionized trails. The process is like lightning in an electric storm. The trails of sparks can be photographed, or their positions can even be recorded by timing the arrival of the accompanying crackles at electronic microphones. Either way, a picture of particle tracks for subsequent computer analysis can be built up.

The beauty of the spark chamber is that like the cloud chamber it has a 'memory' and can be triggered. Scintillation counters outside the chamber, which respond quickly, can be used to pinpoint charged particles passing through the chamber. Provided all this happens within a tenth of a microsecond, the ions in the spark chamber's gaps will still be there, and the high-voltage pulse will reveal the tracks. Any longer, and the ions will have been swept away by a low-voltage 'clearing' field that mops up unwanted ions.

A still better version of the spark chamber was invented in the 1960s by Frank Kriernen at CERN. His idea was to subdivide the plates of the spark chamber into sheets of parallel wires, a millimetre or so apart. As before, when a charged particle travels through the spark chamber's filling of inert gas it leaves an ionized trail; a high voltage applied to alternate planes of wires provokes sparks to form along the trail. But the pulse of current associated with each spark is sensed only by the wire or two nearest to the spark. So by recording which wires sensed the sparks you have a reasonably accurate (to within a millimetre) idea of where the particle has passed. Notice how there is no longer any need for one stage in the data analysis—the film scanning necessary to convert visual information into numbers. The wire spark chamber produces information ready for a computer to digest with little further processing.

Wire spark chambers became popular in the late 1960s, and several ways of recording the information from the wires were developed. As well as bypassing the need for film-scanning, the wire chambers offered the additional advantage of a faster response. This is because for electronic recording, the sparks do not have to grow as large as they do if they are to be photographed, and this in turn means that the chamber 'recovers' more quickly—in other words, the ions from one set of sparks can be mopped away more rapidly in preparation for the next trigger pulse. Wire spark chambers can be operated up to 1000 times per second—1000 times faster than most bubble chambers.

The wire spark chamber fitted in particularly well with the computer techniques for recording data that were developed in the 1960s. Signals from many detectors—scintillation counters, Čerenkov counters, wire chambers—could be fed into a small

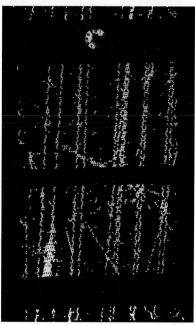

Fig. 6.32 *Fred Ashton (b.1935) at work on an array of flash tubes at Durham University. The glass tubes, seen end on, are a little more than 1.5 cm in diameter and are filled mainly with neon. They are stacked between metal plates, across which a high voltage is applied when a charged particle has passed through. The particle ionizes the gas in the tubes it crosses, and they 'flash' under the influence of the electric field between the plates, revealing the path of the particle.*

Fig. 6.33 *The tracks of a cosmic ray shower passing through an array of flash tubes are clearly visible when the tubes are viewed end on.*

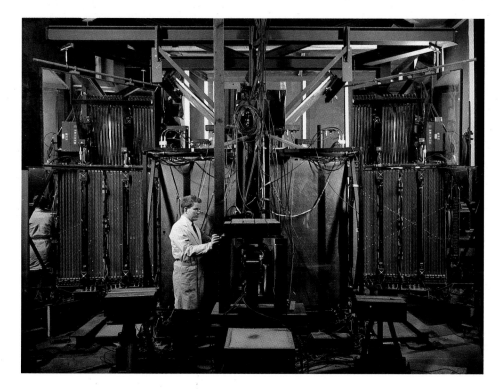

Fig. 6.34 *Spark chambers in operation at CERN in 1969. Sparks fly along the ionized trails of charged particles in layers of gas sandwiched between metal plates. A high voltage applied across adjacent plates in the spark chambers makes the gas flash where it has been ionized.*

Fig. 6.35 *Murray Gell-Mann (b.1929).*

Fig. 6.36 *Yuval Ne'eman (b.1925) in 1966.*

'on-line' computer. The computer would not only record the data on magnetic tape for further analysis 'off-line', but could also feed back information to the physicists while the experiment was in progress. Sets of chambers with wires running in three different directions provided enough information to build up a three-dimensional picture of the particle tracks. And the computer could calculate the energy and momentum of the particles and check their identification, exactly as in bubble-chamber analysis.

Experiments based on spark chambers, scintillation counters, and Čerenkov counters proved a useful complement in the 1960s to the bubble chambers at CERN and Brookhaven. The spark chambers allowed the rapid collection of data on specific interactions; bubble chambers, on the other hand, gave a far more complete picture of events, including the point of interaction or 'vertex'. The 'electronic' and 'visual' detectors were complementary, and together they proved a happy hunting ground for the seekers of previously unknown particles. First in cosmic rays, then at the accelerators, physicists were discovering an ever-growing 'zoo' of particles. This apparent complexity of nature at the subatomic level was compounded by the discovery in the 1950s of the 'resonances' (see pp. 138–40), extremely short-lived energetic states of the common proton and heavier particles.

The confusion began to be resolved in 1962. Theorists Murray Gell-Mann and Yuval Ne'eman had realized independently that the known particles, including the resonances, could be fitted symmetrically into a series of 'families'. Gell-Mann called this beautiful symmetry the 'Eightfold Way' after the Buddha's 'Eightfold path to truth'. At a meeting at CERN in 1962, Gell-Mann proposed the existence of a new particle that would make one of the Eightfold Way's 'families' complete. He called it the omega, for obvious reasons, and the race was on to discover it. In February 1964, a team at Brookhaven studying the interactions of kaons in the 80 inch bubble chamber found the vital evidence to clinch the theory—the decay of a particle that could only be the predicted omega (see pp. 141–2). Within weeks, similar evidence at CERN confirmed Brookhaven's discovery.

THE SUPERSYNCHROTRONS

WITH THE discovery of the omega particle, the 'Eightfold Way' symmetry proposed by Gell-Mann and Ne'eman seemed on firm ground. But it did not tackle the fundamental problem of the early 1960s. Why is nature so complex? Why is there this great diversity of particles? Gell-Mann's answer was that the observed particles are not in fact fundamental objects, but are built from more basic building blocks, which he called 'quarks'.

The idea of quarks took some time to catch on, the more so because there was little evidence that quarks could be knocked out of protons, say, in the same way that protons can be knocked out of nuclei, or electrons out of atoms. But the search for simplicity, and the possibility of discovering 'free' quarks at higher energies, drove particle physicists in the early 1960s to consider still bigger machines, which would reach out to energies far beyond those of Brookhaven's AGS and CERN's PS.

In 1967, the Soviet Union again became world leader, with a proton synchrotron that could reach 70 GeV. This machine, at the Institute for High Energy Physics at Serpukhov, outside Moscow, was for five years the world's biggest particle accelerator. Physicists from Europe and the US were keen to join in and investigate the new energy region; détente was the order of the day, at least in particle physics. Then the big machine at Fermilab started up in 1972, accelerating protons to 200 GeV. By 1976 it had successfully reached 500 GeV, but the route had by no means been easy.

Although the machine's design originated at the Lawrence Berkeley Laboratory, it was in reality the baby of Robert Wilson, one of the pioneers of electron synchrotrons at Cornell University. In 1967, Wilson became head of the project to build the new machine at a site known as Coon Hollow, to the west of Chicago. He faced the unenviable task of being allocated $250 million for its construction, rather than the $350 million the designers at Berkeley had said it would need.

Wilson decided not only to accept the financial challenge but to exceed it by building a machine that would reach 500 GeV—two and a half times the original design energy! He has described this decision as 'close to bravado', but it was a calculated attempt to attract the right sort of people to the project. Wilson's 'band of stouthearted men' put together a design for a machine with a ring exactly 2 km in diameter. Even with this large circumference, in order to reach 500 GeV the team still needed to design electromagnets that would produce magnetic fields nearly 20 per cent higher than had been achieved before. To help do this, they used separate magnets to bend and to focus the beam—a new concept in synchrotrons.

In an attempt to speed up sluggish funding, Wilson announced in 1969 that the machine would be ready a year early, in July 1971. He would have been correct but for two factors. First, the 6.3 km long beam pipe turned out to have obstructions, sufficient to prevent protons from making a complete circuit of the ring. The desperate machine-builders even tried employing a ferret, named Felicia, to help in pulling magnets on wires through the pipe in an attempt to clear it! But the most devastating problem concerned the magnets. These were installed in a tunnel completed during a frozen Illinois winter. In summer, the tunnel became dripping wet, and when powered up, nearly half the magnets destroyed themselves in showers of sparks.

But by March 1972, the machine was accelerating protons to 200 GeV. It still proved a difficult beast to handle. To begin with it produced a proton beam only 50 per cent of the time. Despite these drawbacks, many groups from the association of universities that runs Fermilab, and from outside the US, successfully performed experiments in an exciting new energy region. And in May 1976, the machine reached 500 GeV—Wilson's dream had been fulfilled.

Meanwhile, Europe had also joined in the bid for higher energies. After some bitter wrangling among the member states, CERN had decided to build a machine that would take protons to 400 GeV. The Super Proton Synchrotron (SPS), as it became known, took six years to build, and it delivered its first beams almost as Fermilab's accelerator was finally reaching 500 GeV. Though lagging behind the American machine at first, the SPS was later to become the centre of world attention, thanks to a remarkable modification, described in Chapter 8.

The SPS and the Tevatron have become the foremost proton synchrotrons of the mid 1980s. But what of electron machines? Near Stanford in California, not far south from Berkeley, lies the machine that is at present the world's most powerful electron accelerator. This is not a circular synchrotron; it is a 3 km long linear accelerator, the longest linac in the world.

Why a linear accelerator? Electron synchrotrons work perfectly well apart from one fundamental problem: high-energy electrons radiate away energy when they travel on a circular path. The radiation—known as synchrotron radiation—is greater the tighter the radius of the orbit and the higher the energy of the particle. Protons also emit synchrotron radiation, but because they are 2000 times as massive as electrons, they can reach much higher energies before the amount of energy lost becomes significant. But even at only a few GeV, electrons circulating in a synchrotron radiate a great deal of energy. And this lost energy must be paid for by pumping in more energy through the radio waves in the accelerating cavities.

It was for these reasons that physicists at Stanford decided to build a huge linear machine, at the Stanford Linear Accelerator Center, or SLAC. The origins of SLAC go back as far as 1934, when William Hansen of Stanford University began to consider how to build a linear electron accelerator. This machine would be similar to Wideröe's pioneering device, but it would need a powerful source of high-frequency radio waves to accelerate the high-velocity lightweight electrons. (Recall that the frequency must match the time for the particles to travel from one accelerating gap to the next.)

Hansen was soon joined by Russell and Sigurd Varian, who had been working in their own private laboratory on a means for generating and detecting radio waves of centimetre wavelength. Sigurd had been an airline pilot and he was keen to develop better navigational aids for aircraft; Russell, a graduate of Stanford, had been working in radio and television research and had the right kind of expertise to invent a suitable device.

Fig. 6.37 *Robert Wilson (b.1914), director of Fermilab from 1967 to 1978, toasts his team of machine builders when the new synchrotron first accelerated protons to a world-beating 200 GeV in March 1972.*

Fig. 6.38 *William Hansen (1909–1949) with a section of his first linear electron accelerator, which operated at Stanford University in 1947. It was eventually 3.6 m long and could accelerate electrons to an energy of 6 MeV.*

Fig. 6.39 *The 3 km long linear accelerator at the Stanford Linear Accelerator Center (SLAC). The electrons start off from an electron 'gun', where they are released from a heated filament, at the end of the machine near the bottom right of the picture. The electrons in effect surf ride along radio waves set up in a chain of 100 000 cylindrical copper 'cavities', about 12 cm in diameter. The machine is aligned to 0.5 mm along its complete length. By the time they reach the far end of the accelerator, beyond the freeway, the electrons have an energy of 30 GeV, and are ready to be delivered to experiments.*

The arrival of the brothers at Stanford, unpaid except for $100 for materials, proved a turning point. While the Varians had no ostensible interest in accelerating electrons, together with Hansen they developed the 'klystron'—a powerful source of radio waves which has become standard equipment in electron and proton accelerators, as well as for other applications including the transmission of television signals. Russell Varian worked out the design of the klystron in 1937; and with the combination of Hansen's experience and Sigurd's skill, the first device was operating in August the same year.

After the Second World War, Hansen's thoughts turned once again to building a linear electron accelerator, in which the particles were fed energy by the new powerful source of radio waves—the klystron. He developed a series of machines, each more powerful than the previous one, and by 1953 Stanford could boast an electron accelerator 63.6 m long, which reached an energy of 600 MeV.

It was at about this time that the idea of 'the Monster' began to grow in the minds

of several researchers at Stanford, including Wolfgang Panofsky. At a meeting at Panofsky's home on 10 April 1956, the first unofficial ideas about the Monster were recorded: length, 3 km (or more precisely, 2 miles); energy, 15 GeV or more. By 1959, President Eisenhower had recommended federal funding for the machine as a national facility. Thus SLAC was born, and in January 1967 'the Monster' reached its design energy of 20 GeV.

The machine at SLAC has since been improved to reach a maximum energy of 30 GeV, comparable with the energy of the protons from the PS or AGS. But some of its most intriguing results came in the first few years of operation. It was then that the first clear evidence emerged that protons and neutrons are built from still smaller objects. SLAC's high-energy electrons, fired from one end of the accelerator, proved the ideal tool to probe protons.

In 40 years, from the first accelerators of Lawrence and the other pioneers, ideas of the basic constituents of matter had changed enormously. In 1932—the year the neutron was discovered—the picture had been of four basic particles: proton, neutron, electron, and the then hypothetical neutrino. Assuming Dirac was right—and the experiments at Berkeley in the mid 1950s showed that he was—then each of these 'elementary' particles should have its complementary antiparticle, thus bringing the grand total to eight. By 1973, the picture had at first become increasingly complex— with the discovery of many particles, especially at the Bevatron—and had then swung back again to simplicity. In 1973, the basic constituents of matter seemed to be the electron, the muon, two neutrinos (see pp. 143–6), and three varieties of quark— and of course their antiparticles. But as Chapter 8 describes, the course of physics was to change again the following year, in 1974.

Fig. 6.40 End Station A at SLAC is one of the experimental areas where the electron beam finally emerges from the linear accelerator. The beam enters through the pipe coming in from the left and collides with the 'target' (surrounded by concrete blocks). The electrons scatter from particles in the target's nuclei through a variety of angles, as if in a game of subatomic billiards; or they may produce other new particles from their high-energy collisions. Three 'spectrometers' record the results. One is the tall grey cylinder behind the target area; the second is the structure incorporating the large yellow 'container', with the third partially hidden behind it. The spectrometers contain banks of different detectors to track the scattered particles, and magnets to measure their momentum. They can be rotated to different positions around the target along the rails that are visible. The scale of these instruments, which provided the first direct evidence for quarks, is given by the man standing at the foot of the yellow 'container'.

CHAPTER 7

THE PARTICLE EXPLOSION

THE YEAR 1952 was a milestone in particle physics. It saw the invention of a new type of detector—the bubble chamber—which was to dominate discoveries for the best part of 30 years; and it witnessed the first of a new breed of accelerator—the synchrotron—designed with the express purpose of creating man-made versions of the particles found in cosmic rays. It was the beginning of a new era for particle physics. The subject, which had first been a branch of nuclear physics, then a branch of cosmic ray research, had now graduated to become a fertile field of discovery in its own right. By the early 1960s, particle physicists were almost falling over each other in their efforts to find new particles.

Experiments at accelerators allowed the physicists to fill in gaps in the pattern of particles that was beginning to emerge. The first particle to be discovered at an accelerator—the neutral pion—completed the pion family of three. Similarly, the neutral xi, when at last discovered in a bubble chamber, provided a partner for the negative xi, which had been found in cosmic rays. With increasing amounts of energy at their disposal, experimenters confirmed Dirac's theory of antimatter, finding antiparticles for each of the known particles. The antiproton, the antineutron, the antilambda, and so on, followed in quick succession.

Though the bubble chamber provides pictures that in some cases can be read almost as easily as a book, or which are readily adaptable to become works of art (Fig. 7.1), it is not always the best detector to use. In many cases, particularly with the rarer, shorter-lived particles, electronic techniques based on particle 'counters' have proved invaluable. In the following pages, we see how well bubble chamber and counter experiments complemented each other during the 1950s and 60s, rather as cloud chamber and emulsion techniques had done previously. Particles such as the pi-zero and the antiproton first succumbed to counters. And counter techniques were vital not only in demonstrating the existence of neutrinos, but also in showing that there are two different types of neutrino, one associated with the electron and the other with the muon; a pattern was also emerging among the leptons.

Having already revealed an amazing wealth of particles, in the early 1960s the two techniques together brought the startling evidence that the protons and neutrons are not the last word regarding the structure of matter. First came signs of the 'resonances'—very short-lived states that carry all the hallmarks of being complex vibrating structures. Their discovery came with the ability to analyse automatically hundreds of thousands of bubble chamber pictures. It was in fitting the resonances into the pattern of particles that Gell-Mann and Ne'eman produced evidence for a new symmetry of nature. The discovery of the omega-minus in 1964 vindicated these ideas and led to the suggestion that protons, pions, strange particles, and so on are all built from smaller entities—the quarks.

Then, at the end of the 1960s, electronic counter experiments at the Stanford Linear Accelerator Center began to show how high-energy electrons could penetrate the proton and pinpoint its granular nature. Sixty years after Rutherford had prised his way into the atom with alpha particles, the physicists at Stanford echoed his experiments but at a new, deeper level.

The effects that betray the presence of the quarks within the proton and other related particles are subtle in the extreme and stretch our concept of 'seeing' further than with any of the particles encountered so far in this book. Yet the effects are quite as real as with the other particles; it is just that we must look harder to perceive nature's deepest secrets.

Fig. 7.1 *The bubble chamber came to dominate images of subatomic particles in the 1960s, providing beautiful swirling pictures, with their own aesthetic appeal, in addition to their scientific merits. This bubble chamber photograph has been made still more striking by the photographic colouring techniques of Patrice Loïez at CERN.*

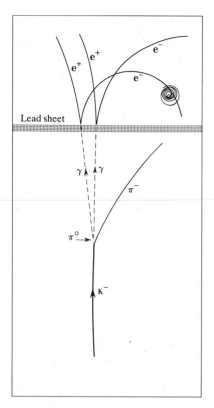

Fig. 7.2 *A negative kaon (K⁻) decays in a bubble chamber at the Lawrence Berkeley Laboratory, producing a negative pion (π⁻) and a neutral pion (π⁰). The neutral pion immediately decays into two gamma rays (γ) whose paths are marked by the dotted lines in the diagram. The gamma rays strike a lead sheet in the chamber and each turn into an electron (e⁻) and a positron (e⁺). The bubble chamber's magnetic field curls the negative particles to the right, the positive ones to the left. The tight spiral towards the end of the track of the lower electron is another electron knocked out of an atom in the bubble chamber liquid; other extraneous tracks have been removed.*

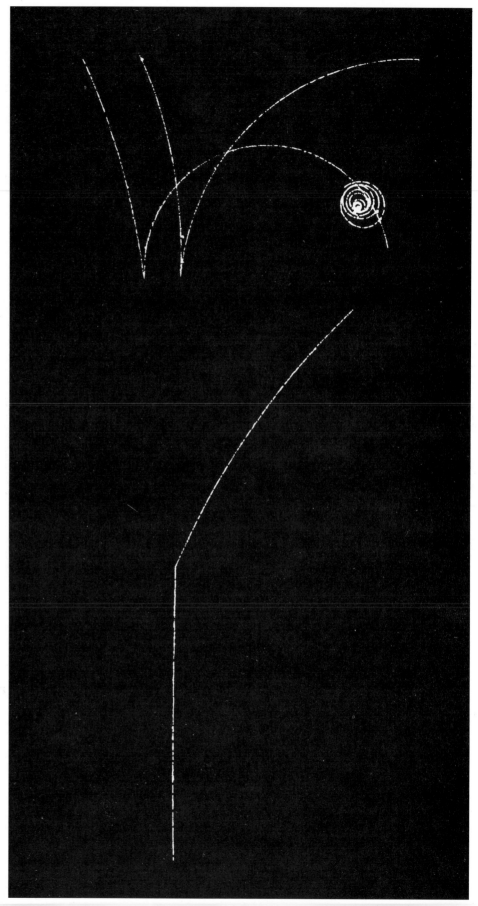

THE NEUTRAL PION

THE NEUTRAL pion, or pi-zero, was the first unstable subatomic particle to be discovered with the aid of an accelerator. It is the neutral partner to the positive and negative pions described on pp. 90–1, which were first observed in the interactions of cosmic rays. The pi-zero is produced just as readily as its charged siblings in cosmic ray collisions, but its lack of electric charge means that it behaves differently and is much more difficult to detect.

Charged pions live for 10^{-8} s before decaying into other particles, and as a result they leave relatively long tracks in cloud and bubble chambers. The pi-zero, on the other hand, lives for a mere 10^{-16} s, a hundred million times more briefly. And because it is electrically neutral, it does not leave any track in a cloud or bubble chamber. This invisibility is compounded by the fact that 99 per cent of the time the pi-zero decays into two very energetic photons, or gamma rays, which are also neutral. So how does one detect a short-lived particle that leaves no track and which decays into two particles that also leave no tracks?

The answer is to use some special tricks to make the elusive particle reveal itself. All detectors of pi-zeros incorporate a dense material such as lead, which will coerce the invisible gamma ray photons into producing electron–positron pairs. This effect is seen in Fig. 7.2, a photograph taken in a bubble chamber at the Lawrence Berkeley Laboratory. A negative kaon entering at the bottom produces a visible negative pion and an invisible pi-zero. The pi-zero immediately decays into two gamma ray photons which shoot, equally invisibly, towards the top of the picture. But where they meet a sheet of lead inserted into the bubble chamber, they turn into two visible electron–positron pairs.

Lead has proved to be an ideal material for 'converting' gamma ray photons into electron–positron pairs. Today, many experiments incorporate stacks of lead-doped glass blocks—just like the 'lead crystal' of fine glassware—to detect pi-zeros. The lead encourages the photons to convert to electrons and positrons, which in turn radiate Čerenkov light in the glass. A phototube on the end of a block detects the light and hence registers the arrival of a pi-zero. Lead will also induce electrons and positrons to radiate photons, which can in turn produce more pairs, and so on, creating a 'shower' of charged, visible particles. This is seen in Fig. 7.3, where a cosmic ray pi-zero entering a cloud chamber produces avalanches of electrons and positrons in successive lead sheets.

It was the quantity of gamma rays in the cosmic radiation that first led theorists to propose the existence of the pi-zero. In 1948, Robert Oppenheimer at Berkeley and two of his students, H. W. Lewis and S. A. Wouthuysen, published a paper suggesting that cosmic gamma rays originate from the decay of neutral pions, but no one could be certain.

In 1949, R. Bjorkland and colleagues searched for the particle at Berkeley's new 184 inch synchrocyclotron, and they employed electronic methods. When the protons accelerated in the cyclotron struck metal targets inside the machine, they produced a copious supply of pions. If neutral pions existed, they should have been produced too but should have immediately decayed into two gamma ray photons. Two strategically-placed holes in the concrete wall shielding the cyclotron allowed any gamma rays to escape and collide with foils of tantalum which, like lead, induced the production of electron–positron pairs.

The electrons and positrons emerging from the tantalum foil were deflected in opposite directions by a magnetic field and detected in coincidence by devices known as proportional counters. The degree to which the magnetic field deflected the particles, together with the amount of ionization they produced in the gas-filled tubes of the proportional counters, enabled the experimenters to determine the particles' energies. This in turn revealed the energy of the gamma rays that had produced the electrons and positrons. The results fitted well with the energies expected if the gamma rays did indeed come from neutral pions decaying in flight. No other explanation could account for the measured distribution of energies; the first evidence for the neutral pion had been found.

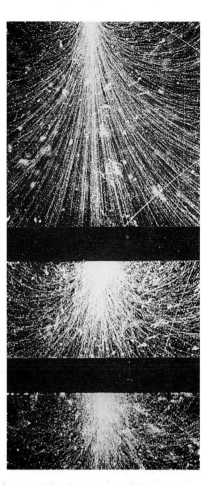

Fig. 7.3 *Pi-zeros produced in high-energy cosmic ray interactions in the upper atmosphere decay swiftly to gamma rays and can generate 'showers' of electrons and positrons. Here such a shower has been produced in a cloud chamber subject to a magnetic field. A high-energy gamma ray initiates a cascade of electron–positron pairs at the top of the chamber, which curve in opposite directions in the magnetic field. The electrons and positrons produce further gamma rays, through annihilation and radiation, and when these interact with two successive sheets of lead the process repeats itself, regenerating the shower each time.*

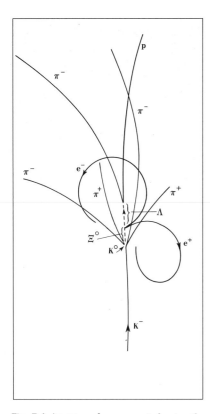

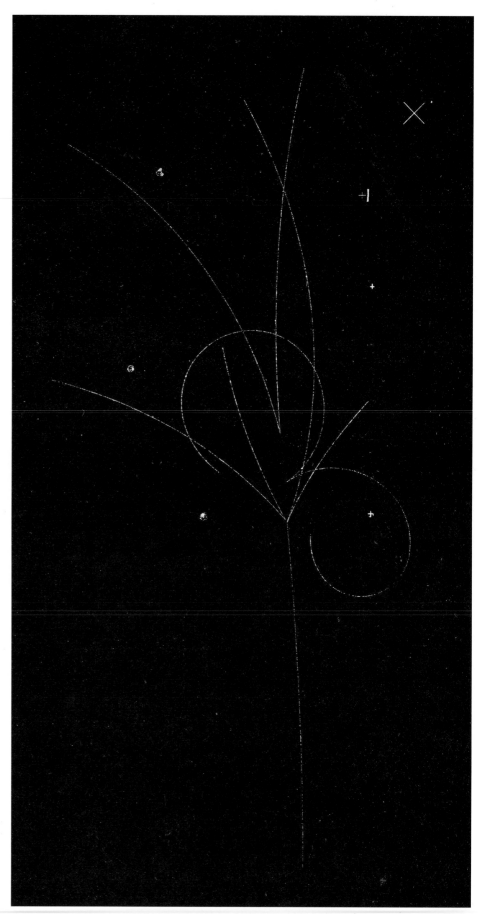

Fig. 7.4 *An extremely rare event showing the production and decay of a xi-zero. A negative kaon* (K^-) *strikes a proton in a bubble chamber at Berkeley, producing a neutral kaon* (K^0), *a xi-zero* (Ξ^0), *a negative pion* (π^-) *and a positive pion* (π^+). *The neutral kaon decays almost immediately into its own pair of negative and positive pions. The xi-zero travels up the picture (dotted line in the diagram) in almost exactly the same direction as the incoming negative kaon before decaying into a neutral lambda* (Λ) *and a pi-zero* (π^0). *The lambda, only marginally deflected, continues up the picture (dotted line) and then decays in its turn into a proton* (p) *and a negative pion. The pi-zero lives so briefly that it is not even marked by a dotted line on the diagram; it decays immediately into a single gamma ray photon (not marked because it leaves the bubble chamber undetected) and an electron-positron pair* (e^-, e^+). *Notice how this pair point back to the exact spot where the xi-zero decayed; without the electron and positron this point would be unknown. (Extraneous tracks have been removed from this image.)*

THE NEUTRAL CASCADE

THE NEUTRAL 'cascade' particle, or xi-zero, is, like the pi-zero, difficult to detect because it decays into two neutral particles—one of which is the pi-zero! Figure 7.4 shows a particularly good example of a xi-zero observed in a bubble chamber at the Lawrence Berkeley Laboratory.

A negative kaon enters at the bottom and produces a total of four particles: a xi-zero, a K-zero, a pi-plus and a pi-minus. The K-zero decays almost immediately into two pions, revealing itself only by the tiny gap between its birthplace and the point where it decays. The xi-zero, on the other hand, travels invisibly for some distance before decaying into a lambda-zero and a pi-zero. Being neutral, these particles also leave no tracks, but they show themselves by their own decays. The lambda turns into a proton and a pi-minus, while the pi-zero experiences one of its rare (one in a hundred times) decays into an electron–positron pair and a single gamma ray photon. The electron and positron curve more in the bubble chamber's magnetic field than the other, more massive particles.

The xi-zero was not discovered until 1959, although its existence had been predicted earlier. It was required by the 'strangeness scheme' of Gell-Mann and Nishijima, first put forward in 1953, which provided a simple rule that ties together particles with very similar masses. In this scheme, the proton and neutron are charged and neutral partners with masses just under 1 GeV, but with no units of the property Gell-Mann called 'strangeness'. Similarly, the negative xi discovered in cosmic rays, with a mass a little over 1.3 GeV, was expected to have a neutral partner of more or less the same mass—the xi-zero. The two xi particles each carry two units of negative strangeness—we say they have a strangeness of -2. This strangeness is divested one unit at a time when they decay. So the xi-zero decays into a pi-zero with no strangeness and a neutral lambda with strangeness -1. The neutral lambda in turn loses its unit of strangeness when it decays into a proton and a pion, neither of which is strange.

The neutral xi remained undiscovered until, late in 1958, Luis Alvarez and his group at Berkeley began a concerted effort to find the elusive particle in the interactions of negative kaons in their 38 centimetre (15 inch) bubble chamber, in the months before the famous 72-inch began work. The negative kaon, or K-minus, has a strangeness of -1. So to create a xi-zero, with strangeness -2, while keeping the *total* strangeness the same, requires that a particle of strangeness $+1$ must be produced together with the xi-zero.

Out of thousands of pictures, Alvarez's team found one example (Fig. 7.5) in which the visible tracks provided enough information to suggest that a xi-zero had been produced. By measuring the angles and momenta of the appropriate tracks, they concluded first that the two pions at the left came from a K-zero; this was the vital particle with strangeness $+1$. They then showed that the 'vee' formed at the top by a proton and a pi-minus came from a neutral lambda. This 'vee', however, did not point directly back to the origin of the K-zero; instead another neutral particle must have been produced, and it was *this* particle's decay that the 'vee' of the lambda pointed towards. Calculations based on the masses, angles, and momenta of the particles involved showed that this neutral particle must have approximately the same mass as the xi-minus. It bore all the hallmarks of the xi-zero.

(Note that a neutral particle that creates a 'vee' does not necessarily 'point' along a line bisecting the 'vee'; it depends on the masses and momenta of the two particles in the 'vee'. In the case of the lambda's 'vee', shown in the diagram accompanying Fig. 7.5, the relatively massive proton travels onwards in almost exactly the same direction as the neutral lambda, whereas the relatively lightweight pion is deflected sharply to the right).

The discovery of the xi-zero was a *tour de force* that demonstrated the power of both the liquid hydrogen bubble chamber and the analysis methods developed in the early days at Berkeley. Balancing the angles and energies of the charged particles revealed where the unseen neutrals had passed. The discovery also dramatically confirmed the predictive power of the strangeness scheme, which could now be used as a firm basis for ideas of a more fundamental nature.

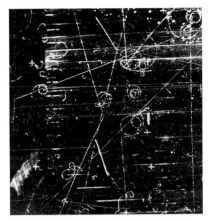

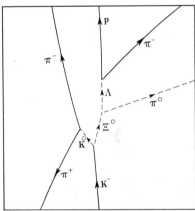

Fig. 7.5 *Although dramatically different in appearance, this xi-zero discovery picture, from the 38 cm (15 inch) hydrogen bubble chamber at Berkeley, almost reproduces the pattern of decays shown in Fig. 7.4. A negative kaon (K^-) with a faint and patchy track strikes a proton and produces a neutral kaon (K^0) and a neutral xi (Ξ^0). The neutral kaon decays into a pair of positive and negative pions (π^+, π^-). The xi-zero, as shown by the dotted lines in the diagram, decays into a lambda (Λ) and a pi-zero (π^0). The lambda, turns into a proton (p) and a pi-minus; the pi-zero presumably takes its usual decay path into a pair of gamma rays (γ), which are undetected.*

ANTIMATTER

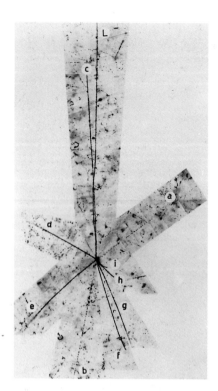

Fig. 7.6 *The first image of an antiproton annihilation 'star', found in emulsion exposed to antiprotons at the Bevatron in 1955. The antiproton enters the picture at the top (the track marked L) and travels about 430 micrometres before ending its life in an explosive act of mutual destruction with a proton. Nine charged particles emerge from the point of annihilation and move outwards, their tracks forming a characteristic star-like pattern. The tracks marked a and b are probably pions, the others probably protons.*

Fig. 7.7 *An antiproton (pale blue) strikes a proton in a bubble chamber at Berkeley. In the resulting annihilation, the energy released rematerializes as four positive pions (red) and four negative pions (green). In the bubble chamber's magnetic field, the negative pions and the negative antiproton curve in a clockwise direction, the positive particles anticlockwise. The two lower pions have less energy than the others, and they therefore curve more and leave thicker tracks. The one on the left travels only a short distance and stops when it is captured by a proton. The one on the right ends by decaying into a muon (yellow) and an invisible neutrino. Tracks not involved in the interaction, including the characteristic curlicues of low-energy electrons knocked from atoms, are coloured dark blue.*

OUR EARTH, the Solar System it inhabits, and indeed our whole galaxy and the millions of other galaxies in the Universe, all seem to consist of matter built from atoms in which electrons orbit nuclei that consist of protons and neutrons. But in 1932, in his experiments with cosmic rays, Carl Anderson discovered the 'anti-electron', or positron—a particle just like the electron except that it carries a positive rather than negative electric charge. The positron was in fact already required by equations in a theory written by Paul Dirac in 1928, and its discovery implied that other antiparticles should exist, for the equations applied equally well to protons, neutrons, and many other subatomic particles discovered since the early 1930s.

For each and every variety of matter there should exist a corresponding 'antimatter'—opposite in properties such as electric charge and strangeness, but with identical mass. Thus we can conceive of antiatoms in which positrons orbit antinuclei built from antiprotons and antineutrons. Dirac's theory also predicted that matter and antimatter are doomed never to coexist. When a particle meets its antiparticle, the two annihilate—a catastrophic process in which the mass of the two objects is converted instantly to energy. This energy can 'evaporate' as photons, or rematerialize as new particles and antiparticles that rush away from the point of their creation.

Today, we still have no firm evidence that large-scale clumps of antimatter, built from antiatoms, exist anywhere in our Universe. But physicists can readily make antiprotons, antineutrons, and other antiparticles in high-energy collisions at accelerators, and they can manipulate them to probe the mysteries of the subatomic world. Yet it was more than 20 years after Anderson's discovery of the positron that experiments proved the existence of the antiproton, and several more years before physicists could feel certain that for every particle of matter there exists an appropriate antiparticle.

The antiproton was the first of several antiparticles to be discovered at the Lawrence Berkeley Laboratory. Its well-defined attributes—the same mass as the proton, but with negative charge—made it a suitable subject for study with electronic counting techniques, as described on pp. 111-13. In this way, Emilio Segrè and his colleagues found the first signals of antiprotons in 1955. At the same time they sought visual confirmation of their discovery. Protons accelerated in the Bevatron produced many particles when they hit a target. The negatively-charged ones were filtered off and focused by magnetic fields and used to bombard stacks of emulsion. Occasionally a rare antiproton among the other negative particles should meet a proton in the emulsion, annihilate, and produce a distinctive starburst.

The exposed emulsions were studied both in Berkeley and by Eduardo Amaldi's group at Rome University. The Italians found the first proton–antiproton annihilation 'star' (Fig. 7.6), shortly after the discovery of the antiproton. Later, Segrè's team reinforced the discovery by finding a star in which the total energy of all the particles produced in the annihilation clearly added up to more than the energy of the incoming antiproton. Because energy cannot be created from nothing, this proved that the star did not result simply from the decay of the incoming particle; it had to result from the mutual annihilation of two particles—proton and antiproton.

Figure 7.7 shows an antiproton star captured later at Berkeley with the 72 inch liquid hydrogen bubble chamber. The image has been colour-coded to make it easier to identify the different particles. The incoming antiproton annihilates with a proton and produces eight pions—four positive (red) and four negative (green) with almost perfect symmetry.

The discovery of the antiproton opened the way for the search for its counterpart, the antineutron. When a proton and an antiproton have a near miss, they escape destruction but may neutralize each other's charge. The proton turns into a neutron and the antiproton turns into an antineutron. The antineutron is living in a hostile world of matter, and it is only a question of time before it annihilates with a neutron or a proton, producing a distinctive burst of energy.

Bruce Cork and colleagues at Berkeley decided to use this process of 'charge exchange' to hunt for antineutrons. They used a tank of liquid scintillator to detect

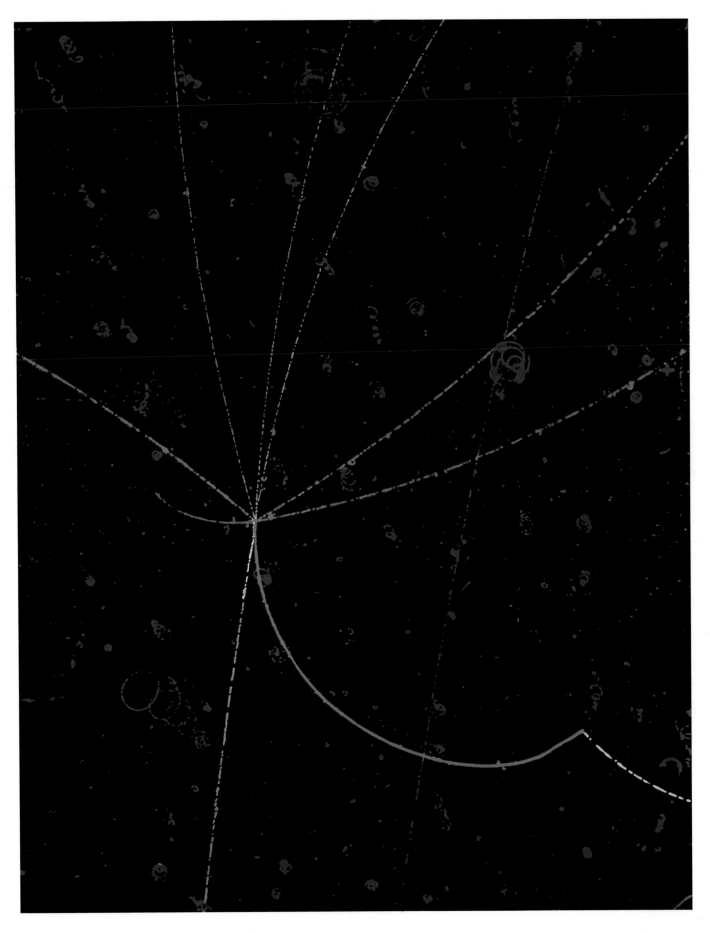

annihilations of antineutrons produced by charge exchange from antiprotons generated from the Bevatron. An annihilation in the scintillator created a burst of charged particles and this in turn produced a large, characteristic pulse of light, which was detected by photomultiplier tubes. In 1956, Cork's team found 114 antineutron annihilations in this way.

Berkeley continued to play an important role in tracking down antiparticles. The first antilambda was found in 1958 in emulsion exposed to a beam of negative pions from the Bevatron. Then, in 1959, the 72 inch bubble chamber came into operation and began to provide spectacular results. Figure 7.8 is the first picture of an antilambda produced in conjunction with a lambda. Lambdas are neutral and leave no tracks, but they reveal their existence by decaying into charged particles and forming visible 'vees'. The 'vee' from the lambda consists of a pi-minus and a proton; that from the antilambda consists of a pi-*plus* and an *antiproton*.

Another feature of this picture is that it records the passage of time as the particles travel around the chamber. The antiproton produced in the decay of the antilambda

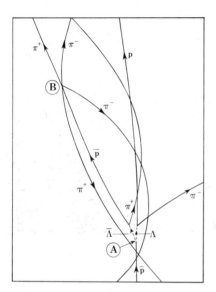

Fig. 7.8 *An antiproton (p̄) annihilates at A with a proton in the 180 cm (72 inch) hydrogen bubble chamber at Berkeley, in the first ever picture of a lambda (Λ) produced in conjunction with an antilambda (Λ̄). The lambda decays into a proton (p) and a pi-minus (π⁻), while the antilambda decays into the symmetrically opposite particles, an antiproton and a pi-plus (π⁺). The antiproton shoots off towards the top left before annihilating at B with another proton in the liquid hydrogen and giving rise to six new particles. Two of these—a neutron and a pi-zero—are neutral and therefore invisible. The others are two negative and two positive pions. This picture is particularly interesting because of its symmetry: the lambda and antilambda are almost exactly the same 'length'.*

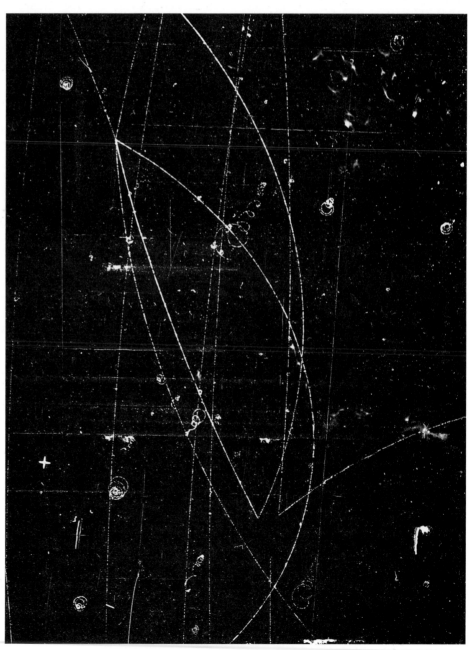

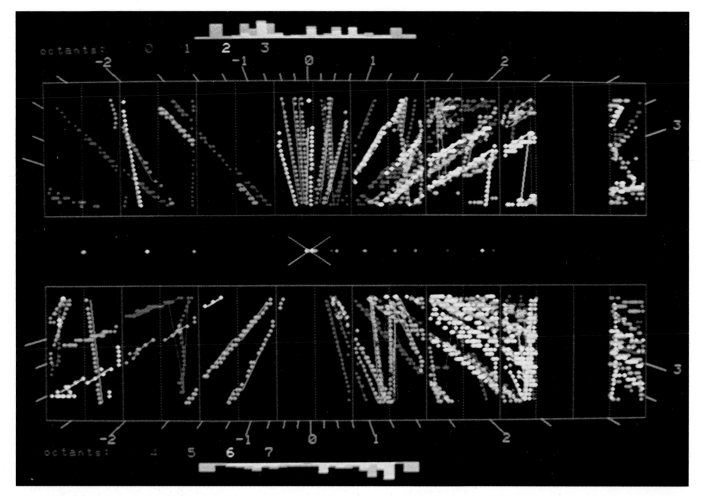

does not get far before its luck runs out and it annihilates with a proton in the bubble chamber's liquid. Two of the charged pions created in the annihilation are hurled back towards the bottom of the picture, crossing just beside the track of the original antiproton which entered the chamber billionths of a second earlier to initiate the whole sequence of events.

By the early 1960s, antimatter equivalents of most of the particles then known had been discovered, and today antiparticles—mainly positrons and antiprotons—are used routinely as tools of science, technology, and medicine. In high-energy physics, the largest and most successful modern accelerators are the colliders, in which beams of electrons and positrons, or protons and antiprotons, are accelerated in opposite directions inside the same beam pipe and brought to head-on collision at selected points.

These collisions of accelerated particles and antiparticles create enormous energies, and they enable physicists to probe the most fundamental structures of matter. Figure 7.9 is an electronic display of the collision between a proton and an antiproton, each accelerated to 800 GeV, in the Tevatron at Fermilab. The energy released is sufficient to create dozens of new particles, mainly pions, which spill out into the surrounding detector.

Despite our familiarity with antiparticles, the question of the existence of antimatter in bulk somewhere in the Universe remains a vexed issue. Surely matter and antimatter must have been produced in equal amounts in the energetic outpouring of the Big Bang. So where has all the antimatter gone? Why can we find no evidence for antigalaxies containing antistars, which might be orbited by antiplanets populated by antielephants and anticockroaches? The ability to explain this mystery is an important factor in the latest ideas about matter and the nature of the Universe—the so-called Grand Unified Theories or GUTs, described in Chapter 10.

Fig. 7.9 *The aftermath of one of the first proton-antiproton collisions at a record total energy of 1600 GeV (1.6 TeV) is captured in the Collider Detector Facility at Fermilab, in this image from October 1985. The electronic detector fits around the beam pipe of the superconducting ring in the Tevatron, like a giant jam roll, built from many components. The computer display shows a side view of the upper and lower halves of the detector's innermost tracking chamber, while the gap between them represents the beam pipe, where tracks are not detected. Protons and antiprotons travelling in the ring in opposite directions come in from the left and right respectively, and an annihilation has occurred at the point marked by the purple cross. Many particles created in the annihilation shoot out into the gas-filled tracking chamber, which contains wires that pick up the ionization due to charged particles. The chamber is divided into eight octants surrounding the beam pipe, four in each of the upper and lower halves, and the tracks are colour-coded (red, green, white, or blue) according to which of the four octants they traverse. (The four sections to the right appear empty because the detector was not quite complete when this event was recorded.)*

THE RESONANCES

MANY OF the particles that we have met so far live brief but visible lives. A particle moving near to the speed of light travels a few millimetres in a lifetime of 10^{-11} s, and its track can be seen clearly in enlargements of emulsion and bubble chamber photographs. But there are many particles whose lives are millions of times shorter. They do not produce visible tracks in detectors; instead, physicists infer their existence from the longer-lived particles into which they decay.

These extremely short-lived particles are generally known as 'resonances'. Their lifetimes are of the order of 10^{-23} s. A span of 10^{-23} s fits into a millionth of a second as does a millonth of a second into a thousand years! It is no wonder the resonances leave no visible tracks: even at the speed of light they travel barely further than their own diameter.

Something that dies before moving from its birth site, even if travelling at the speed of light: can such a bizarre thing be said to have existed at all? When the first resonance was discovered in the early 1950s, many physicists were initially reluctant to accept its reality for this very reason.

The first intimation of resonances came from the work of Enrico Fermi and his group at the University of Chicago. They wanted to study the way that pions interact with protons because they believed that this was the best strategy for understanding the nuclear forces. In 1951, a new synchrocyclotron started operation in Chicago and the following year Fermi and his team used it to produce pions at six different energies, which they then fired at the protons in a target of liquid hydrogen.

Many pions passed straight through the empty space in the hydrogen atoms, while others bounced off or were absorbed. In front of the target, two small pieces of scintillator, 2.5 cm across, recorded the number of incoming pions, while two larger pieces behind recorded how many of them had traversed the hydrogen. The researchers found that as they increased the energy of the pions, fewer particles reached

Fig. 7.10 *Early data on pion–proton scattering shows a typical 'resonance' peak in this sketch by Luke Yuan for the 4th Annual Rochester conference on High Energy Nuclear Physics in December 1954. Yuan, from the Brookhaven National Laboratory, and his colleague Sam Lindenbaum provided some of the first evidence to support the theory that pions can 'excite' short-lived resonances of the basic proton and neutron.*

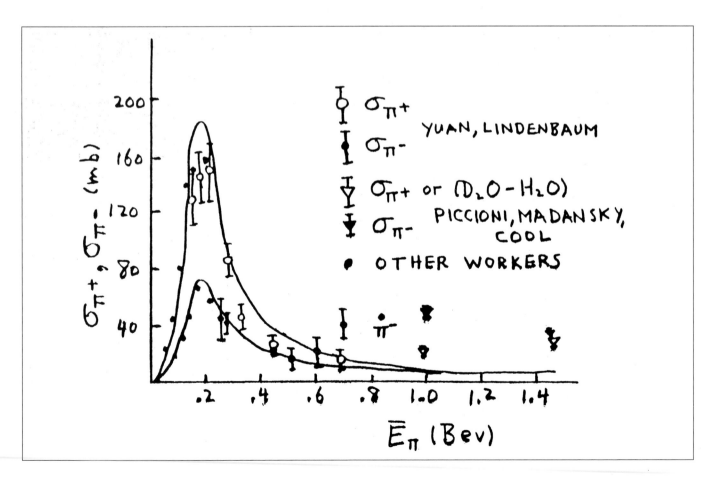

the scintillators behind the target. This effect occurred with both positively and negatively charged pions, although it was more noticeable with the positive ones. It seemed that they had perhaps caught the first glimpse of a phenomenon that at a larger scale is familiar to physicists and engineers—resonance.

It is resonance that enables an opera singer's voice to break a wine glass. The singer hits a note whose frequency exactly matches the natural frequency of the glass; the glass vibrates in response to the sound waves and shatters. Resonance also occurs at the atomic level, as when the sodium atoms in a street light absorb electrical energy and in turn radiate it as light. These resonant systems have the property of absorbing energy in the same characteristic way: if you plot the energy absorbed on a graph, it will rise smoothly to a peak as the frequency (or wavelength) changes, and then fall again.

In the experiment at Chicago, Fermi and his colleagues altered the energy of the pions, which quantum theory tells us is the same as altering the wavelength or frequency associated with them. The curve that they plotted as they fired increasingly energetic pions at the hydrogen target resembled a resonance curve; but the team could not confirm this idea because the Chicago cyclotron did not produce sufficiently energetic pions to show whether the curve reached a peak and then fell away. But in 1953, a team at the Brookhaven National Laboratory repeated the experiment with higher-energy pions produced by the newly-commissioned Cosmotron, and revealed a definite peak in the pions' absorption (Fig. 7.10). Somehow pions of a certain energy could excite protons into a resonant state. This state was so well-defined that it was given a name—the 'delta'.

How do we know that the lifetime of the delta resonance is as short as 10^{-23} s? The information comes from the width of the resonance 'spike' in the graph of the energy distribution. Quantum theory relates the lifetime and width of the spike in such a way that resonances with brief lives have relatively broad widths, while longer-lived states have narrower, sharper spikes. The widths of the delta and many other resonances are very large, implying that their lifetimes are indeed unimaginably brief.

For a long time the delta resonance was an isolated and unexplained phenomenon. Resonances had been seen before only in complex structures whose component parts can absorb energy in changing from one configuration to another. This is what happens when electrons absorb energy in a sodium atom: they are displaced for a brief period to a higher energy level. But in the 1950s the proton was regarded as a single, indivisible entity. No one dared take the discovery of the delta as a serious hint that the proton might itself consist of more elementary particles.

It was not until the early 1960s, nearly a decade after Fermi's work, that other examples of particle resonances were discovered. At the Lawrence Berkeley Laboratory, physicists had accumulated millions of bubble chamber photographs, which they set about analysing with the aid of computers. They could measure the energy of the various particles in each interaction, add the energies together in different ways, and plot the results as a graph showing how often a particular total energy occurred in each type of interaction. This kind of graph in effect records the energy distribution, or spectrum, of the particles produced in the interaction, and a resonance shows up as a spike or 'emission line' in the spectrum.

This is the reverse of what Fermi found. The delta was first seen through the absorption of pions by protons. In a similar way, the presence of various elements in the Sun was discovered because they absorb sunlight at particular wavelengths, producing dark 'absorption lines' in the solar spectrum (Fig. 7.11). But the solar spectrum also displays bright 'emission lines', where it emits much more light at particular wavelengths. The spectrum of the light from a sodium lamp (Fig. 7.12) also shows emission lines, and these correspond to specific movements of the electrons within the resonating sodium atoms.

Two young students, Stan Wojcicki and Bill Graziano, found evidence for the first such 'emission' resonance in bubble chamber film at Berkeley in 1960. They analysed pictures where a kaon had struck a proton to produce a lambda and two pions. They expected that the energy of each of the particles would be distributed smoothly, since particles emerging from an interaction usually share the available energy between

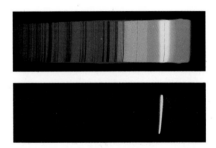

Fig. 7.11 (*top*) *The solar spectrum shows many black 'absorption lines' due to photons of sunlight of certain specific energies being absorbed by elements in the Sun's atmosphere. This absorption spectrum can be compared to the absorption spike which characterized Fermi's discovery of the delta resonance. The bright 'emission lines' also present in the solar spectrum are not visible in this photograph.*

Fig. 7.12 (*above*) *The emission spike of the Y-star resonance is comparable to the emission spectrum of sodium which, in the photograph, shows a single bright yellow line. In fact there are two emission lines, but they are so close together that the picture would have to be enlarged to a length of at least two metres before they became distinguishable.*

them in a democratic way, now one being more energetic, now another. Instead, and quite unexpectedly, their graph of the kinetic energies of the lambda and the pions displayed a distinctive spike. They concluded that this 'emission line' must be due to the extremely brief presence of a resonating lambda, which then gives rise to the longer-lived 'normal' lambda.

The Y-star (Y*), as it became known, was the first of a long list of resonances discovered at Berkeley. It turns out that not only protons and lambdas, but also pions and kaons have resonant states.

Figure 7.13 is a bubble-chamber photograph from Berkeley in which an antiproton has entered from the bottom and annihilated with a proton in the chamber's liquid to produce four visible charged pions and an invisible pi-zero. Analysis of the event led researchers to believe that, despite appearances, the pions do not all originate from the point, or 'vertex', of the annihilation. Instead, the two outer charged pions and the pi-zero probably came from the decay of a pion resonance known as the 'omega meson', denoted by the Greek character ω. Figure 7.14 shows how things might appear if we could magnify the event a million million times, with the annihilation vertex producing two charged pions and an ω, and the ω decaying a short distance further on into the other two charged pions and the pi-zero. (Despite its name, the omega meson is no relation to the omega-minus, which is described in the next portrait.)

We now know that the resonances exist because the proton, the pion, the kaon, and so on are built from smaller particles called 'quarks'. In much the same way that the constituent electrons rearrange themselves to form excited (resonating) states of atoms, so do the constituent quarks give rise to resonating states of the particles that are built from them.

Fig. 7.13 *Resonances last typically for a mere 10^{-23} s and therefore leave no discernible track in a bubble chamber. But by calculating back from the energies and angles of the particles that are detected, physicists can infer that a resonance has existed. In this picture from a bubble chamber at Berkeley, an antiproton, coming from below, annihilates with a proton to produce two negative pions, a neutral pion and two positive pions. The negative pions move off to the left, the positive pions to the right, while the pi-zero is undetected. The lower pi-plus decays to a muon, the short piece of track, and then to a positron, which curls out of the picture. The information in the picture is consistent with the lower-energy pions—the lower tracks on left and right—being the decay products of a resonance state known as the omega, ω.*

Fig. 7.14 *Here we see how the decay of the ω might look if we could magnify the area around the decay some million million times. The antiproton ($\bar{p}$) has annihilated with a proton in the bubble chamber to produce a pi-minus (π^-), a pi-plus (π^+), and an omega resonance (ω). After a short distance the neutral omega decays into a pi-minus, a pi-plus, and a pi-zero (π^0).*

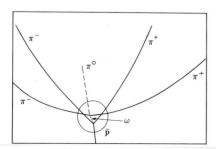

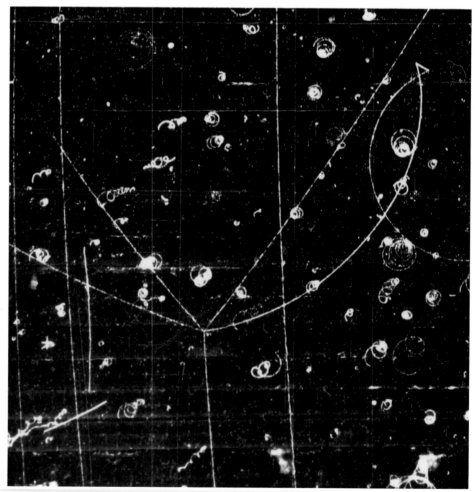

THE OMEGA-MINUS

FIGURE 7.15 shows one of the most famous pictures in particle research, a physicist's Mona Lisa. One of a set of 80 000 photographs from the 80 inch bubble chamber at the Brookhaven National Laboratory on Long Island, it was the first picture to show the production and decay of an omega-minus particle, and it caused tremendous excitement when it was announced in February 1964. Here was the final piece in a jigsaw that had been accumulating over the previous few years.

The proliferating quantity of subatomic particles, including the resonances, en-

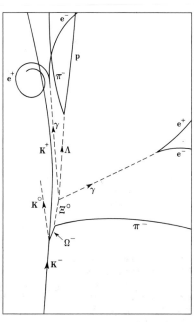

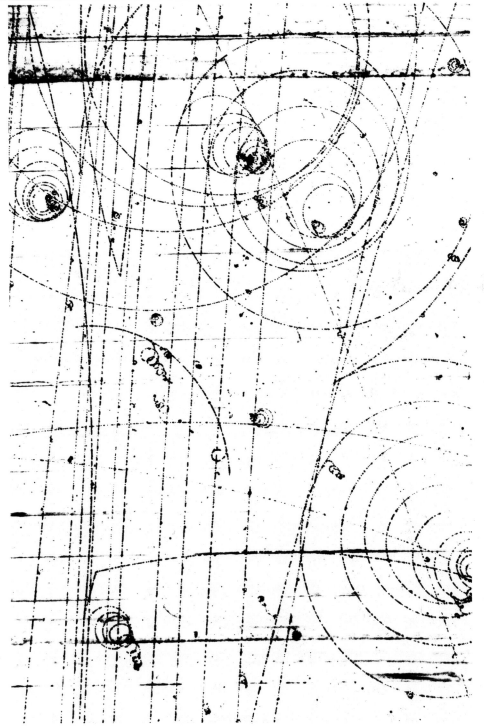

Fig. 7.15 *This historic picture from the 200 cm (80 inch) hydrogen bubble chamber at Brookhaven shows the first observation of the omega-minus. A negative kaon (K$^-$) collides with a proton to produce three particles: an omega-minus (Ω^-), a positive kaon (K$^+$), and an unseen neutral kaon (K^0), represented by a dotted line in the diagram. The omega-minus travels a short distance (2.5 cm) and then decays, emitting a pi-minus (π^-) that veers sharply to the right, and a neutral xi (Ξ^0) which itself decays into three more neutral particles—a lambda (Λ) and two gamma ray photons (γ). These neutrals, also marked by dotted lines in the diagram, finally reveal themselves by decaying into visible 'Vs': the gamma rays into electron-positron pairs (e$^-$, e$^+$), the lambda into a proton (p) and a pi-minus.*

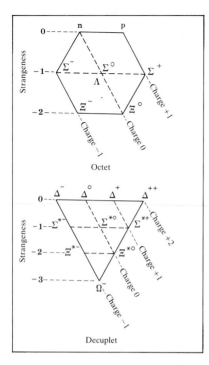

Fig. 7.16 *These diagrams show two 'families' of particles—an octet and a decuplet—described by the 'Eightfold Way', the classification scheme proposed by Gell-Mann and Ne'eman in 1962. The members of each family are arranged according to their qualities of strangeness (vertical axis) and electric charge (diagonal axis). The octet (top) consists of the familiar neutron (n) and proton (p), the trio of sigma particles (Σ), the lambda (Λ), and the two xi particles (Ξ). The central location in the octet is shared between the neutral sigma and the lambda (Λ), which is also neutral. (These two particles have the same values of charge and strangeness; in fact, they contain the same quarks, but in two slightly different arrangements.)*

The decuplet does not suffer from such complexities. Its ten members consist of four delta resonances (Δ), three sigma-star resonances (Σ^*), two xi-star resonances (Ξ^*), and the triply strange omega-minus (Ω^-)—the missing particle that Gell-Mann and Ne'eman predicted. (Note that one of the deltas has a double ration of positive charge.)

couraged the theorists to try to find some order in the confusion. In 1960–1, Murray Gell-Mann of Caltech and Yuval Ne'eman—a member of Israeli Defence who had been granted a leave of absence by Moshe Dayan to study physics in London—independently proposed a method for classifying all the particles then known. This method became known as the Eightfold Way, as suggested by Gell-Mann. What Mendeleyev's table of the elements had done for atoms and chemistry, the Eightfold Way did for particles and high-energy physics. In the Eightfold Way, the particles are classified into 'families' according to such characteristics as their electric charge and their strangeness. Figure 7.16 shows two such families, one with eight members (an 'octet') and one with ten members (a 'decuplet'). Each particle has a particular position in its family, according to the quantity of electric charge and strangeness the particle has.

A central clue to the deeper meaning of these patterns is that certain properties are conserved when particles interact or decay. A proton and an antiproton, for instance, have electric charge of $+1$ and -1 respectively, giving a net charge of zero. When they collide and annihilate, they can create four negative pions (-4) and four positive pions ($+4$) as in Fig. 7.7, or a neutral lambda and a neutral antilambda as in Fig. 7.8, since in both cases the net charge remains zero. But they cannot create three negative and four positive pions: that would give a net charge of $+1$ and violate the law of conservation of electric charge. Strangeness is another property that is conserved when strange particles are produced by the strong force, but it can be divested by fixed amounts when they decay.

These properties of charge and strangeness, together with a particle's intrinsic rate of spin, completely define that particle in the Eightfold Way. Each particle fits neatly into a position in one of the many families of particles.

In 1962, the Eightfold Way was still very new and poorly understood by all but a handful of theorists. Some critics thought its clever symmetries more accidental than fundamental, especially since it served only to classify known particles. In July 1962, Gell-Mann and Ne'eman attended an international conference at CERN and both were in the audience when a group from the University of Los Angeles announced the discovery of two new resonances, a negative and a neutral xi-star (Ξ^-* and Ξ^0*). Both Gell-Mann and Ne'eman realized that the two xi-stars would almost complete a new decuplet in the Eightfold Way.

This family is the inverted pyramid shown in Fig. 7.16. It contains four resonances with no strangeness (the deltas), three sigma-star resonances with one unit of strangeness, and the two newly-discovered xi-star resonances with two units of strangeness. Only the tenth, triply-strange member of the family was missing.

The next day's session, 10 July, included a review of the known strange particles. At the end of the talk the chairman called for comments from the floor. Both Ne'eman and Gell-Mann raised their hands. The chairman called for Gell-Mann, who was at the time the acknowledged leader in theoretical physics. Gell-Mann then strode to the blackboard and announced his prediction of a triply strange particle to complete the new decuplet.

Gell-Mann called it the omega-minus: minus because it should have negative charge, and omega—the last letter in the Greek alphabet—because it would complete the decuplet. Moreover, by extrapolating from the masses of the nine resonances already in the family, Gell-Mann could predict its mass. Heavier still than the two xi-stars, it should weigh in at 1680 MeV.

The challenge was on for the experimenters, and teams at Brookhaven and CERN set about scouring thousands of bubble chamber pictures. In February 1964, the Brookhaven team found the first 'gold-plated' example of an omega-minus—the picture shown in Fig. 7.15. Calculations gave a mass of around 1686 MeV. CERN found a similarly clear example a few weeks later.

At last, after years of mounting confusion, a viable classification system existed for many subatomic particles. The Eightfold Way clearly worked, but why it worked remained a mystery. Why did the particles fit so neatly into their various families? What principle underlay this subatomic ordering? The answer, as we shall see, is quarks.

THE NEUTRINO

THE NEUTRINO is one of the most pervasive forms of matter in the Universe, yet it is also one of the most elusive. It has no electric charge, little or no mass, and it can travel as easily through the Earth as a bullet through a bank of fog. As you read this sentence, billions of neutrinos are hurtling through your eyeballs as fast as light, but unseen. Theorists estimate that there are on average between 100 and 1000 neutrinos in every cubic centimetre of space.

An intense but unfelt 'wind' of neutrinos, emanating from the nuclear processes in the Sun, plays continually upon the Earth, and in addition there are lesser breezes of neutrinos and antineutrinos emitted from the collapse of stars and other catastrophic processes in our galaxy and beyond. Because the Earth is so transparent to neutrinos, as many of them rain up through our beds by night as rain down on our heads by day!

The neutrinos have no direct effect on us, but theorists have come to believe that they play a crucial role in the processes that formed and continue to shape our Universe. They are regarded today as one of the truly elementary particles of matter.

Although the neutrino interacts extremely rarely with other forms of matter, experimenters have devised ways of making the particle reveal itself. These methods rely on a straightforward 'brute force' technique: direct enough neutrinos at a large enough target and you are bound to detect a few interactions.

Figure 7.17 shows one of these rare interactions of a cosmic neutrino in an enormous tank of ultra-pure water located 600 m down in a salt mine under the bed of Lake Erie in Ohio. The tank was built to study the possible decay of protons, as described in Chapter 10, but occasionally one of the neutrinos traversing the Earth will collide head-on with an electron or a proton in the water, creating charged particles. In turn, these charged particles produce a cone of Čerenkov light, which is detected by photomultiplier tubes lining the tank's walls.

Fig. 7.17 A computer display of a tank of 10 000 tonnes of ultrapure water shows the effect of a burst of charged particles created in the collision of a cosmic neutrino with one of the protons in the tank's water. The detector, built by the Irvine–Michigan–Brookhaven collaboration, lies 600 m below ground, its walls lined with light-sensitive phototubes (see pp. 201–2). In this picture, we are looking down on the top of the tank (red lines). The neutrino, which has travelled right through the Earth, has entered the tank from below and interacted in the water. The resulting shower of charged particles has produced a cone of Čerenkov light which is detected by phototubes in a circular region on one of the side walls (bottom of image). Colour indicates the timing of the signals from the phototubes, ranging from red for the earliest to blue for the latest. The number of 'slashes' that make up each hit represent the number of photons detected by each phototube.

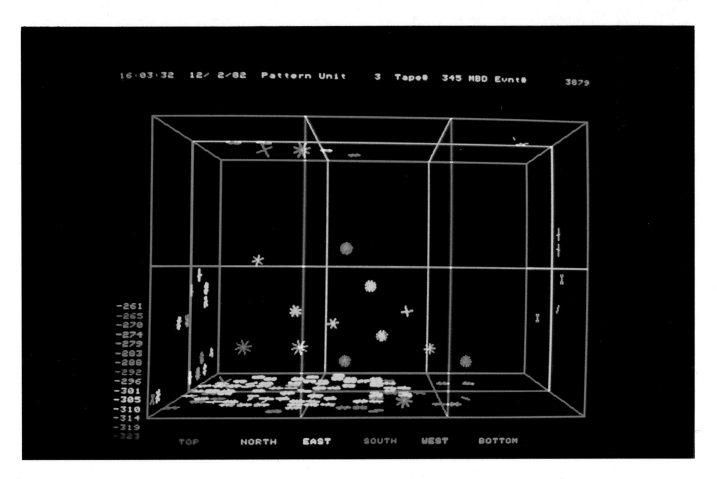

Today, particle physicists are able to produce beams of neutrinos to order, and to use them to explore the elementary constituents of matter. There are even neutrino and antineutrino 'telescopes', which detect the emission of these particles from the Sun or other galactic and extra-galactic objects. Though they are still elusive, neutrinos have been partially harnessed. Yet it is only 30 years since their existence was first proven.

In the late 1920s, physicists studying the radioactive beta decay of atomic nuclei were faced with a conundrum. In the beta decay process, a neutron was supposed to change into a proton, at the same time emitting an electron. And these two particles, the proton and the electron, should always share out the available energy in the same way. But experiments showed a quite different pattern: the electron, which was easier to measure than the sluggish proton, emerged with a range of energies. This apparently violated the fundamental law of conservation of energy.

The Austrian physicist Wolfgang Pauli took what was an extremely bold step at that time by suggesting that a hitherto unsuspected particle was responsible. The theory was simple: if three particles—the proton, the electron, and a mystery particle—shared out the energy of beta decay, then the energy released in the decay could be shared between them in any number of ways. And that would explain why the electrons produced when neutrons decay do not always have the same energy. Figure 7.18 shows the influence of this mystery particle in a cloud chamber picture taken in 1957. The electron and residual nucleus do not recoil back to back, as would be the case if they were the only particles involved.

Because of theoretical constraints, Pauli's proposed particle had to be quite bizarre compared with those already known. It had to be neutral, have little or no mass, but it had to spin about its axis like an electron or a proton. Enrico Fermi named it the neutrino—Italian for 'little neutral one'—and he gave it respectability by incorporating it into his theory of beta decay in 1933.

The neutrino remained a hypothesis until, early in the 1950s, two physicists at the Los Alamos National Laboratory in New Mexico were inspired by the idea of 'doing the hardest physics experiment they could think of'. Clyde Cowan and Fred Reines decided to show that neutrinos could do *something*, however rarely, and therefore had physical reality.

They thought at first that atomic bombs could provide a suitably copious supply of neutrinos, or rather antineutrinos, emitted in the decays of neutrons released in the explosions. In the end they realized that the proposed experiment should work equally well with antineutrinos produced in the more controlled conditions of a nuclear reactor.

Cowan and Reines chose to look for the process of 'inverse beta decay', in which a proton captures an antineutrino and converts to a neutron, at the same time emitting a positron. Work with a prototype detector (Fig. 7.19) encouraged them to build a full-scale apparatus at the Savannah River reactor in South Carolina. They called it Project Poltergeist because of their quarry's apparent undetectability.

To show that inverse beta decay had indeed occurred, Poltergeist was designed to detect two separate bursts of gamma rays emitted as a result of the process. The first burst of gamma rays would come when the positron annihilated with an electron. The second burst would be produced when the neutron was captured by a cadmium nucleus in 'target' tanks of cadmium chloride (Fig. 7.20). The timing of the two bursts was the crucial element. The positron would annihilate almost instantaneously, but the energetic neutron would need to be slowed by successive collisions before it could be captured by a cadmium nucleus. Cowan and Reines calculated that if inverse beta decay was really occurring, the interval between the two gamma ray bursts should be in the region of 5 microseconds (five millionths of a second). In the summer of 1956, Poltergeist triumphantly recorded gamma ray bursts separated by 5.5 microseconds, as Fig. 7.21 shows, and on 14 June Cowan and Reines sent Pauli a telegram to say that the neutrino he had invented almost 30 years earlier had finally been found.

Neutrons are not the only particles that can give rise to neutrinos and antineutrinos. Pions and muons, for instance, also do so. But the case of the muon presented

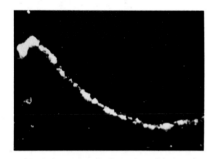

Fig. 7.18 *The influence of the invisible participant in nuclear beta decay is seen in this cloud chamber picture taken by S. Szalay and J. Csikay at the Nuclear Research Institute at Debrecen in Hungary in 1957. They fired nuclei of helium-6, which has two more neutrons than ordinary helium, into the chamber. One of the extra neutrons decays in less than a second, and what we see are the effects of this decay: the short thick track at the top left is the recoiling nucleus, and the lighter, curving track is the electron. The two tracks are not back-to-back, hinting at the invisible presence of the third decay product—the neutrino.*

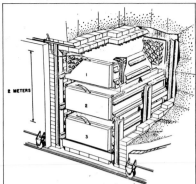

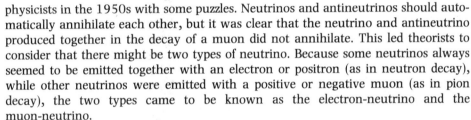

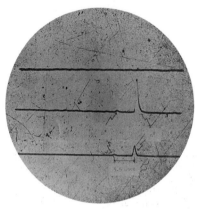

physicists in the 1950s with some puzzles. Neutrinos and antineutrinos should automatically annihilate each other, but it was clear that the neutrino and antineutrino produced together in the decay of a muon did not annihilate. This led theorists to consider that there might be two types of neutrino. Because some neutrinos always seemed to be emitted together with an electron or positron (as in neutron decay), while other neutrinos were emitted with a positive or negative muon (as in pion decay), the two types came to be known as the electron-neutrino and the muon-neutrino.

This idea of two neutrinos became established when the advent of the 30 GeV proton synchrotrons at CERN and Brookhaven made possible the creation of *beams* of neutrinos, as proposed in 1959 by Bruno Pontecorvo at Dubna in the Soviet Union and Melvin Schwartz at Columbia University in New York. The trick is to use a combination of electric and magnetic fields to select pions produced when the accelerator's protons strike a target. The pions are then allowed to decay, producing a beam of muons and muon neutrinos. After an appropriate distance—a few tens of metres at the energies of the CERN and Brookhaven accelerators—a massive wall of solid iron, many metres thick, filters out the muons and any other particles remaining in the beam. Only the extremely penetrating neutrinos pass through and enter a detector in which a tiny fraction of them will interact. A set-up of this kind was used in the famous two-neutrino experiment at Brookhaven in 1962, when Schwartz and several colleagues proved conclusively that neutrinos produced in association with muons always give rise to muons, never to electrons.

Despite the elusiveness of neutrinos, physicists now produce and manipulate them at will. At CERN and Fermilab, high-energy neutrino beams are used to probe the structure of protons and neutrons. They are also ideal for studying one of nature's fundamental forces—the weak force, which is responsible for the radioactive decay of neutrons and other particles, and for the nuclear interactions that fuel the Sun and other stars. In particular, neutrinos played a major role in establishing the so-called 'electroweak theory' which unifies the weak force and the more familiar force of electromagnetism.

In the 1860s, James Clerk Maxwell had brought together two of the physical forces then known—electricity and magnetism—in the unified theory of electromagnetism. A century later, in the 1960s, Sheldon Glashow and Steven Weinberg of Harvard University and Abdus Salam at Imperial College, London, independently put forward theoretical ideas that united electromagnetism and the weak force. But producing such an 'electroweak' theory is one thing; proving it is another.

Fig. 7.19 (*left*) *In 1956, Clyde Cowan (b. 1920), on the left, and Fred Reines (b. 1918), right, succeeded in demonstrating that antineutrinos can induce interactions. The photo shows part of the instrumentation of the prototype antineutrino detector (Project Poltergeist), which they built in 1953 at a nuclear reactor at the Hanford Engineering Works in Washington State.*

Fig. 7.20 (*top*) *Cowan and Reines went on to build a full-scale version of their antineutrino detector at the Savannah River reactor in South Carolina. This 10 tonne detector contained three tanks of liquid scintillator (1, 2, 3) in the form of a 'double-decker' sandwich. The filling between the decks consisted of two smaller tanks of water (A, B) in which cadmium chloride was dissolved. The idea was that an antineutrino would react with a proton in the water to produce a neutron and a positron. The positron would annihilate into gamma rays almost immediately; the neutron would slow down and be captured by a cadmium nucleus, giving off more gamma rays several microseconds later.*

Fig. 7.21 (*above*) *An oscilloscope displays three horizontal lines—traces showing the signal from each of the three tanks of scintillator in Fig. 7.20. An antineutrino interacted in tank B, between the lower two scintillator tanks. Each of the lower traces shows a small 'blip' (arrowed) due to the burst of gamma rays from the positron annihilation. They are followed 5.5 microseconds later by a larger pulse due to the gamma rays emitted after the capture of the neutron by a cadmium nucleus.*

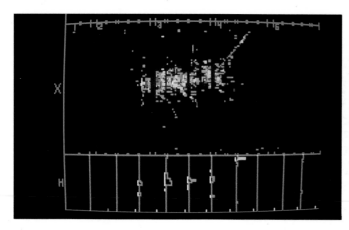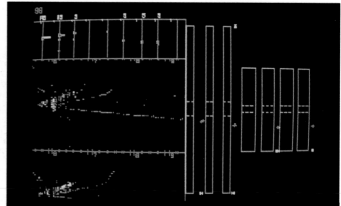

Since the 1920s, physicists have known that the electromagnetic force is 'transmitted' by the photon. According to their energy, photons manifest themselves as radio waves or gamma rays or the light we see with our eyes. When an electromagnetic interaction occurs between particles, it is said to be 'mediated' by a photon. The theory that embodies this concept works so well that physicists believe that other particle interactions, which do not involve electromagnetism, should also be mediated by 'carrier' particles. Each physical force should have its own 'carriers'.

In electroweak theory, Glashow, Weinberg, and Salam found that they required three 'carrier' particles for the weak force: two charged W particles (W^+, W^-) and a neutral Z particle (Z^0). These particles can be thought of as massive versions of the massless photon, and as a result they have sometimes been dubbed 'heavy light'. The existence of the W particles was implicit in Fermi's work in the 1930s, but the Z^0 was completely new, and evidence of its existence would be crucial in confirming the correctness of the electroweak hypothesis. This is where neutrinos came in. The theory predicted that the Z^0 would transmit a form of the weak force never seen before. This would cause neutrinos to bounce off electrons or quarks, simply knocking them into motion as in a game of billiards.

In 1972–3, physicists working on the Gargamelle bubble chamber at CERN pored over 290 000 photographs of interactions produced when beams of neutrinos or antineutrinos had entered the chamber. And they found just 166 examples of the new type of 'neutral current' interaction—so called because it is mediated by the neutral Z, rather than the charged W.

Figure 7.22 shows an example of such an interaction in an electronic neutrino detector at Fermilab. A high-energy muon neutrino enters unseen from the left and strikes an atomic nucleus, which in its turn produces a small shower of other particles; the neutrino has bounced off the electron. Figure 7.23 shows a complementary 'charged current' interaction, mediated by a W particle, in the same detector. Again, a muon-neutrino enters unseen from the left, strikes an atomic nucleus in the detector and gives rise to a small shower of particles together with the long, slightly curving track of a muon. Unlike the previous case, here the neutrino has turned into a muon.

Although charged current interactions are also rare—the Gargamelle search turned up only 576 examples among the 290 000 photos examined—they had been observed before 1973. It was the discovery of neutral current interactions, hinting at the existence of the vital Z particle, that made electroweak theory generally accepted—though it was another 10 years before the Ws and Z were observed directly and the theory fully confirmed (see pp. 192–5).

Today, particle physicists recognize the existence of three types of neutrino: the electron-neutrino, the muon-neutrino, and the tau-neutrino (see pp. 184–5). Discovered in 1975 at SLAC, the tau is a heavier version of the muon, just as the muon is a heavier version of the electron. If electrons and muons each have a neutrino associated with them, so should the tau. But evidence for the tau-neutrino is indirect at present and the experiment analogous to that of Schwartz and his colleagues has still to be done.

QUARKS

UP AND DOWN, charm and strange, top (or truth) and bottom (or beauty)—these are the rather whimsical names given by physicists to the six known quarks which, together with the six leptons, are now believed to be the fundamental constituents of matter. Headlines in the science pages of newspapers speak of 'The discovery of hidden charm' or 'Search for naked bottom', but these departures from the usually dry standard of scientific phraseology mask what is perhaps the most important development of modern physics.

In Chapter 2, we described how physicists established a new layer of reality when they discovered that the stuff everything is made of consists of atomic nuclei orbited by electrons. A few years later, a more fundamental layer was established when it was shown that the nuclei of atoms are not elementary but consist, in their turn, of protons and neutrons. Quarks are what protons and neutrons are made of, as are many other particles found in cosmic rays and created in accelerators. They are an even more fundamental layer of matter and one that the evidence to date suggests is truly elementary.

Quarks occur in clusters, either in pairs or triplets. The proton and neutron, for instance, are both clusters of three quarks; the proton is made of two up quarks and a down quark, and the neutron from two down quarks and an up. No one has yet seen a single or 'naked' quark, and most physicists believe quarks cannot exist as individual free particles. The strong force, which rules the quarks, appears to bind them so tightly together that they cannot be prised apart. When a proton is driven to high energy in an accelerator and smashed into a proton at rest in a target, we do not see individual quarks emerging. Instead, the quarks reform in new pairs and triplets, and additional quarks and antiquarks are created out of the energy of the collision, also forming pairs and triplets, to produce a spray of particles as shown in the bubble chamber photograph (Fig. 7.24) of a proton–proton collision recorded at Fermilab.

Fig. 7.24 *Smash two protons together energetically in an attempt to free the quarks they contain and all that happens is that more particles are created from the energy of the collision. This photograph taken in the 75 cm (30 inch) hydrogen bubble chamber at Fermilab shows a spray of 26 charged particles, mostly pions, produced when a 300 GeV proton enters from the left and strikes another proton in the bubble chamber's liquid. The sharply curving and spiralling tracks are lightweight electrons which are bent more easily than the other particles in the chamber's magnetic field. The distance across the image is about 60 cm.*

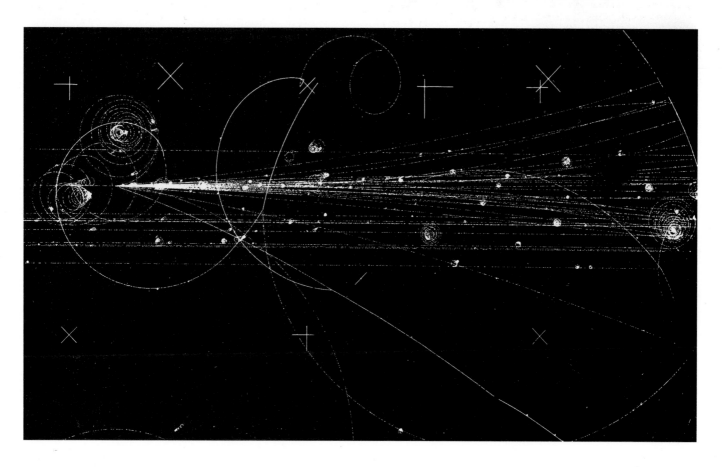

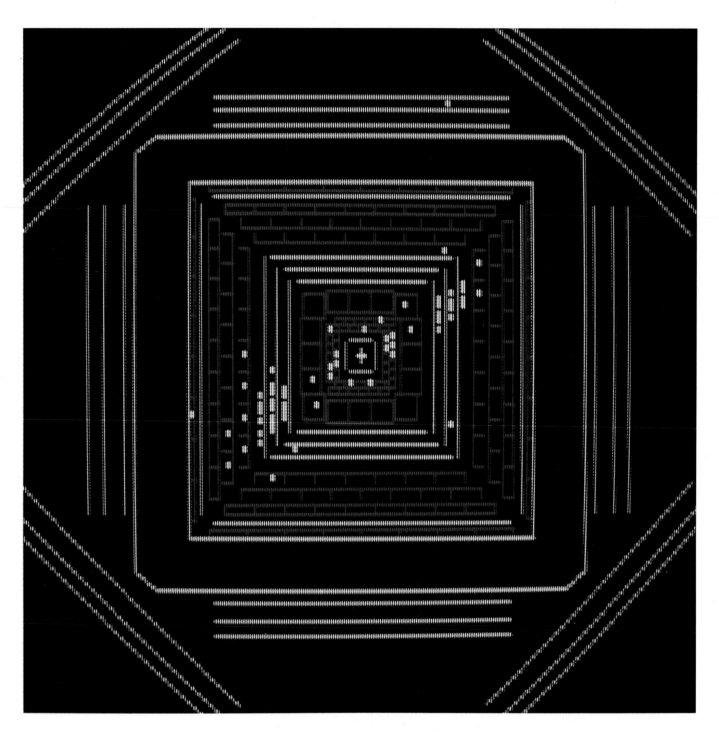

As we have seen in this chapter, there were hints in the 1950s and early 1960s that protons and pions and many other particles might not be as elementary as they seemed. One clue was the existence of the excited states of subatomic particles known as resonances. Physicists knew that atoms could be excited because they have an underlying substructure, and analogy suggested that the same should be true of excited protons and pions. Another clue was the neat manner in which the Eightfold Way classified particles into families. Surely these patterns reflected a deeper level of structure.

The man who 'invented' quarks and thereby helped to solve the conundrum was Murray Gell-Mann, the theorist from Caltech who had developed the Eightfold Way. In 1964, Gell-Mann proposed the existence of three quarks—up, down, and strange—which were all he needed at that time to explain all the particles other than leptons.

Why 'quarks'? The story goes that Gell-Mann liked the sound of the word and only later discovered it in a line in James Joyce's *Finnegans Wake*—'three quarks for Muster Mark'.

Gell-Mann was not the only person thinking along these lines. The same idea was put forward, also in 1964, by George Zweig, a fellow theorist from Caltech who was visiting CERN at the time. But the hypothesis of quarks was slow to catch on. There was no firm evidence for them. And they were also decidedly bizarre objects by the standards of the day. In particular, they would carry 1/3 or 2/3 of the basic unit of electric charge, which was unheard of. All other particles carried charge in full units—0, 1, 2 . . . .Thirds were beyond the pale.

The first solid evidence for quarks came towards the end of the 1960s from studies at the Stanford Linear Accelerator Center (SLAC) with the three enormous spectrometers located in the giant End Station A experimental hall (Fig. 6.40, see p. 127). The experiments were similar in principle to the work of Geiger and Marsden in 1911 in which they bombarded atoms with alpha particles and discovered that atomic nuclei exist. In the modern analogue, electrons accelerated to high energy in SLAC's linear accelerator were fired at protons.

The very high-energy electrons, which at that time could be produced only in SLAC's 3 km long linear accelerator, were a vital part of the experiment. This is because an electron in motion behaves like a wave, and the wavelength depends on the electron's energy. High-energy electrons travel in short waves, enabling them to penetrate something even as small as a proton; lower-energy electrons, by contrast, travel in longer waves and deflect off the proton as a whole.

If the proton was just a singular, elementary particle, the lightweight electrons fired at it should bounce off with almost the same energy as they arrived with; very little energy would go into making the massive proton recoil. But if the proton was a composite particle, consisting of quarks, the result would be very different. The quarks inside a proton would not be still, but would be in constant motion; in fact, the proton could be described as a vibrant cluster of quarks. So the electron might encounter a very energetic quark or merely a quiescent one.

In the experiments at SLAC, electrons would bounce from the quarks and enter a spectrometer. If the quark was moving faster than average, then the electron would enter the spectrometer with a higher energy than average; if the quark were slower, then the scattered electron's energy would be relatively low. In this way, the energies of the electrons coming into the spectrometers would provide a direct measure of the energies of any quarks lurking within the proton.

The answer from End Station A was clear: the energies of the returning electrons varied. Then the spectrometers were repositioned so as to capture electrons bouncing at large angles instead of small angles. The variation in the energy distributions and the rate of arrival of the electrons as the angle changed showed that the proton consisted of three quarks.

At about the same time, similar experiments with neutrinos, instead of electrons, took place at CERN. The detailed comparison of results from CERN and SLAC proved conclusively that the proton consists of quarks.

In the 1970s, accelerators that collide electrons with positrons at high energies began to show the existence of quarks in a different way. When electron and positron meet and annihilate, they convert into pure energy. This energy soon re-materializes, perhaps forming a new electron and positron, or a muon and an antimuon, or a quark and an antiquark. The quark and the antiquark fly off in opposite directions, but they cannot exist as individual entities and must form the clusters that we observe as particles such as pions, kaons, and protons. What happens is that they in effect catalyse the conversion of energy to mass, and generate additional quarks and antiquarks from the energy of the collision.

This all occurs within an instant—some 10^{-23} s—so what we see emerge from the electron–positron collision is not a quark and an antiquark but two streams or 'jets' of particles (mainly pions) formed around them. Figure 7.25 shows two such back-to-back jets produced in the Mark-J detector at the DESY laboratory in the suburbs of Hamburg. This may be the nearest we are able to come to 'seeing' a quark.

Fig. 7.25 (*opposite*) *A typical, back-to-back 'two-jet' event—the signature of a quark and an antiquark—appears in this computer display from the Mark-J detector at the PETRA electron–positron collider at DESY. The display shows a cross-section of the detector, with the beam pipe where the collision occurs marked by the white cross at the centre. Immediately around it are a variety of charged-particle detectors, both inside and outside the iron of the magnet, which is indicated by the two yellow squares. White dots depict the 'hits' recorded in the various elements of the detector. The electron and positron have come in at right angles to the image, from in front of and behind the page, to annihilate at the centre of the detector, producing these two sideways sprays of particles. The magnet iron is about 1.5 m from the centre.*

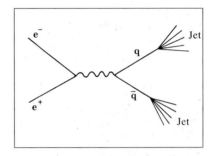

Fig. 7.26 *This Feynman diagram corresponds to the process that has occurred in Fig. 7.25. It shows how the collision of an electron (e^-) and a positron (e^+) gives rise to a photon (the wiggly line) that rematerializes as a quark (q) and an antiquark ($\bar{q}$), each of which produces a jet of particles.*

CHAPTER 8
COLLIDERS AND IMAGE CHAMBERS

SIXTY METRES below ground, among Swiss fields north of Geneva, close to the border with France, lies a huge particle detector. It is as big as a two-storey house, weighs as much as five Jumbo jets, and its name is 'UA1', for 'underground area 1'. The detector and its surroundings look like something from the imagination of H. G. Wells, and it produces its own fantastic visions—images such a Fig. 8.2, which shows the tracks of dozens of charged particles spilling out from the collision of a proton with its physical antithesis, an antiproton. Whereas a bubble chamber records visible tracks on film, UA1 and detectors like it register invisible tracks as electronic signals. A computer can then transform these signals into remarkable glowing images.

A visitor to UA1 first sees the detector sitting at the bottom of its huge cylindrical pit, looking rather like an aluminium-clad haulage container. But the aluminium forms only the outer layer of UA1. The main part of the apparatus is like a nest of Russian dolls, which wrap around the beam pipe of CERN's Super Proton Synchrotron. Protons and antiprotons collide head on in the beam pipe, at the heart of the detector, and the particles created fly outwards in all directions through UA1's many layers. It is only on pulling the nest apart that this complexity becomes apparent.

Each layer of UA1 yields its own brand of information about the particles produced in the proton–antiproton collisions. The innermost layer, appropriately known as the Central Detector, records the tracks of charged particles. The device is filled with a mixture of ethane and argon gas and is strung across with thousands of fine wires. The wires pick up tiny electrical signals from the ionized trails that the charged particles leave in their wake. The wire chamber sits within a magnetic field, which bends the tracks of the particles according to their electric charge and momentum, as in a bubble chamber.

Surrounding the Central Detector are further layers, each especially suited to revealing specific kinds of particle. There are two kinds of 'calorimeter', so-called because they measure the energy of the particles, just as other scientists use ordinary calorimeters to measure heat energy. The Electromagnetic Calorimeter, built from sandwiches of plastic scintillator and lead, highlights the passage of electrons and photons. These particles produce characteristic 'showers' of electrons in the lead, which are

Fig. 8.1 *A view down the access shaft shows the UA1 detector at the bottom of its pit, 60 m below ground. The aluminium-clad boxes visible along the top and sides of the detector contain muon chambers; they form the outer layer of UA1. In this position UA1 is not in fact in the tunnel of the Super Proton Synchrotron (SPS), where the proton–antiproton collisions occur, but in its 'garage' where it can be worked on while the SPS is running for other experiments. When the detector is operating it is rolled into the tunnel, to the bottom right of this picture. Yellow supports hold hanks of cables, connected to the various parts of UA1, which must be long enough to follow the apparatus into the tunnel.*

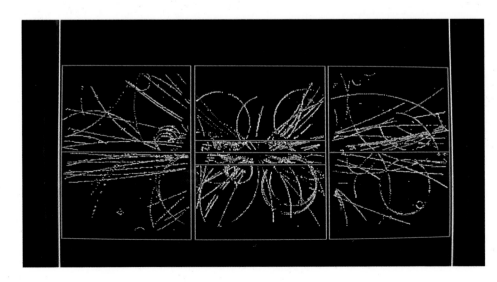

Fig. 8.2 *A computer display of charged particle tracks recorded in the central tracking chamber of the UA1 detector at CERN. A high-energy proton and antiproton have come in from the sides and collided head on at the centre of the chamber, which is a 2 m long cylinder in six sections, outlined here in blue. The gas-filled chamber is strung across with thousands of wires which pick up electrical pulses from the ionization that the charged particles leave behind. Each dot in this picture represents a 'hit' recorded on a wire. The colours show the 'depth' of the tracks: red and yellow are nearer to you, blue and green further away. The experimenters at UA1 can view this kind of display 'on line', in other words while the experiment is in full swing.*

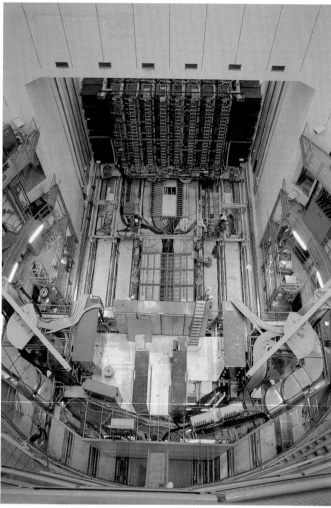

Fig. 8.3 (left) *Installing the Central Detector in UA1. The outside of the chamber is visible, its curved surface covered with electronic circuits to amplify the tiny signals from the wires within the cylinder. Cables, which pass along the vertical structure above the chamber, carry the signals away for processing. Pipes to supply the chamber with gas are also visible. The dark vertical 'walls' to the left and right are part of the ends of the Hadron Calorimeter, which consists of sheets of iron (part of the magnet) interleaved with scintillator. Plastic light pipes, wrapped in black, can be seen leading from the edges of the scintillator to phototubes mounted on the outer face of the calorimeter structure.*

Fig. 8.4 (right) *This view down the pit shows the UA1 detector, without its outer layer of muon chambers, rolling into position in the collider tunnel, beyond the concrete wall. The runners along which the huge structure moves are visible at the bottom of the pit.*

monitored by the layers of scintillator. The flashes of light produced each time a charged particle crosses the scintillator travel down long bars of plastic, known as light guides, towards phototubes, which convert the light into electrical signals.

In a similar manner, the next layer is tailor-made to reveal protons, pions, and other 'hadrons'—that is, particles built from quarks. This is the Hadron Calorimeter, and it consists of sandwiches of iron and scintillator. In this case, the iron serves a double purpose for it also forms part of the electromagnet that is used to bend the tracks of charged particles.

The outermost layer of UA1 detects muons, the only particles, apart from the elusive neutrinos, that can penetrate the large amount of lead and iron that has gone before. The muon detectors lie in the flat oblong aluminium boxes that cover UA1 and give it the appearance of a huge container.

UA1 is big. It contains some 2000 tonnes of iron and over 100 tonnes of lead. The Hadron Calorimeter alone incorporates about 6300 square metres of plastic scintillator in 7000 sheets each 1 cm thick. And there are 7000 optical fibres to transport pulses of laser light to the scintillator, to check its response as well as the behaviour of the light guides and phototubes.

A single proton–antiproton collision generates hundreds of electrical signals as the particles created fly out through the various layers of UA1. These signals pass via subsidiary microprocessors and computers to a main computer that records on magnetic tape all the information the signals contain. Each event yields about 70 000 computer 'words' and the main computer takes about a quarter of a second to write it all on tape. However, the counter-rotating bunches of protons and antiprotons in the Super Proton Synchrotron meet every 7.6 microseconds (7.6 millionths of a second). So while the main computer is writing the complete record of one collision—or 'event'—a further 3000 potentially interesting proton–antiproton collisions have occurred.

This is where the 'trigger system' of subsidiary microprocessors and computers comes in. They are programmed to perform, within four millionths of a second, a series of rapid evaluations to see if the signals from each collision fall into those categories the physicists have defined beforehand as 'interesting'. Of the 12 000 collisions that occur each second in the beam pipe within UA1, the trigger system generally filters out all but one or two. As a result, the physicists do not waste time recording vast quantities of information that is irrelevant to what they are seeking. On the other hand, they see only what they have defined in advance as interesting and are liable to miss those events that are unexpected. This is why the programming of the trigger system is such a vital element in the detector. It is also a flexible element, since the trigger system can be reprogrammed to look for different types of events.

As well as recording those events chosen by the trigger system, the main computer also generates colourful 'on-line' displays of the collisions it is recording, so that the researchers working with the apparatus have an immediate feel for the quality of the data. Once on magnetic tape, the data on the collisions pass to a standard number-crunching computer, which recreates the tracks of the particles and calculates how much energy has been deposited in the calorimeters. Just one collision occupies an IBM 370/108 computer for 20 seconds. Finally, the physicists can examine particularly interesting collisions on a Megatek graphics system. This can rotate the display of the tracks, zoom in to interesting places, and cut out extraneous information, just as a researcher chooses. It has proved vital in interpreting events that the computer may find ambiguous, and in checking that the computer's selections are indeed correct.

The whole £20 million colossus was put together over a period of roughly four years, from the first serious designs to the first observation of proton–antiproton collisions in July 1981. It was a truly cosmopolitan effort, involving teams of scientists and engineers not only from CERN, but also from many European countries and even the US. Initially, 52 physicists from places as far apart as Aachen, Annecy, Birmingham, College de France (Paris), Queen Mary College (London), California (Riverside), the Rutherford Appleton Laboratory (Oxfordshire), and Saclay (Paris), as well as CERN, put their names to the proposal to build UA1. By the time the researchers published the detector's first results, the team numbered around 135 and groups from Helsinki, Rome, and Vienna had joined, in addition to a number of 'visitors'.

The various parts of the detector had to come from far and wide. Sections of the Hadron Calorimeter were assembled in the UK before being shipped to CERN; the muon detectors were built in Aachen; and sections of the Electromagnetic Calorimeter were put together by teams from Annecy, Saclay, and Vienna. It was crucial to the whole project that all this work was properly coordinated. Naturally there were hiccoughs, but regular meetings among the teams, especially between the core of people domiciled at CERN, helped to ensure that eventually everything fitted together as planned. And commanding the whole business was a man of remarkable drive and determination, Carlo Rubbia.

Rubbia is an Italian who has sufficient energy both intellectually and physically to be a leading researcher at CERN as well as a professor across the Atlantic, at Harvard. As we shall see later in this chapter, Rubbia was instrumental in bringing about proton–antiproton collisions at CERN, and he is the leader of the UA1 team. His achievements were rewarded with the discovery in 1983 of the long-sought W and Z particles—the carriers of the weak force responsible for nuclear disintegrations—and with the Nobel prize in 1984.

The UA1 detector epitomizes the particle detectors of the 1980s—it is an 'electronic bubble chamber'. The apparatus covers the space around the collision region as completely as possible. Even though the apparatus cannot record the tracks of neutrinos, in certain circumstances UA1's computer system can deduce the energy carried by invisible neutrinos and can recreate their tracks in the computer displays of an

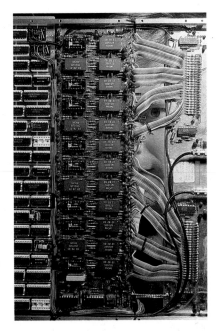

Fig. 8.5 *One of the printed circuit boards, covered with silicon chips, which forms part of UA1's 'trigger processor'. Thousands of proton–antiproton interactions occur within UA1 each second, but only a few of these produce interesting physics. The job of the trigger processor is to take signals from critical parts of the detector and 'decide' whether to go ahead with recording all the data from the thousands of individual wires and sheets of scintillator that UA1 contains. To set up the processor's circuitry for making such decisions, the physicists work out which patterns of signals characterize interesting interactions.*

Fig. 8.6 *Carlo Rubbia (b.1934), photographed at CERN in 1984. The computer display shows the decay of a Z particle in UA1.*

event. The only particles to escape detection completely are those that fly away along the pipe carrying the colliding beams.

But the most impressive part of UA1, when it comes to providing images, is undoubtedly the Central Detector, which records the tracks of charged particles in almost as much detail as a bubble chamber does. This detector, which Rubbia calls an 'image chamber', is based on the principle of the *drift chamber*, a device invented at CERN in the late 1960s. Drift chambers, and their relatives the multiwire proportional chambers, figure in almost every particle physics experiment at today's accelerators. In many instances they provide detailed images of particle tracks, just as in UA1. These devices, more than any other, have made the 'electronic bubble chamber' possible.

ELECTRONIC BUBBLE CHAMBERS

DURING THE 1970s, experiments in particle physics were gradually revolutionized by the inventiveness of a Frenchman at CERN—Georges Charpak. His work has led to particle detectors that combine speed with precision. In the 1960s, wire spark chambers proved valuable because they can operate much faster than a bubble chamber can, although they cannot provide the same amount of detailed information. Charpak's chambers challenged both these earlier devices; they run far faster than spark chambers, while at the same time they approach the precision of the bubble chamber.

When a charged particle travels through a gas, it leaves behind a trail of ionized atoms. A whole range of particle detectors, from the cloud chamber to the wire spark chamber, depend on sensing in some way this trail of ionization. In 1968, Charpak's group of researchers discovered new ways to put the ionization to work in revealing

Fig. 8.7 *The wires that make up the drift chamber of the old Mark II detector are seen during its construction at SLAC. It is typical of modern electronic tracking chambers, which contain thousands of wires to pick up the electrons released by ionizing particles.*

the tracks of particles. The team developed two basic types of detector—the multiwire proportional chamber and the drift chamber—both of which could work much faster and more precisely than the wire spark chambers. These new wire chambers now play a vital role in dealing with the copious numbers of particles created by the intense beams at modern accelerators.

The multiwire proportional chamber is superficially rather similar to a wire spark chamber; it is a sandwich of three planes of parallel wires (rather than two planes, as in a spark chamber) fitted into a gas-filled structure. The difference between the devices lies in the way they operate. With a spark chamber, you apply a high voltage (10–20 kV) for a brief period across the closely-separated planes of wires soon after a charged particle has passed through. The high voltage induces a spark to leap across the gap, but only where the gas has been ionized—that is, along the path taken earlier by the particle.

A multiwire chamber, on the other hand, behaves more like the single-wire counter that Rutherford and Geiger used (see p. 32). In this case you apply the voltage (some 3–5 kV) continuously, so that the central plane of wires is at a positive electrical potential relative to the two outer planes. Immediately a charged particle passes through the gas it triggers an avalanche of ionization electrons. This avalanche grows rapidly in the intense electric field around the wire in the central plane that is nearest to the original particle's path. It is vital that these central wires are fine—20 micrometres or so in diameter—so that the field near to them is very strong; this means that most of the avalanche develops close in to a single wire, and the central layer of wires acts like a row of independent wire counters.

Charpak and his colleagues discovered that they could readily pinpoint the wire nearest to the ionized trail through the distinctive signal it produces. So with a series of chambers they could follow a particle's path. They also found that a chamber with wires only 1–2 mm apart within the central plane produces a signal within a few hundredths of a microsecond after a particle has passed by. Thus a multiwire chamber can handle as many as a million particles per second passing each wire—a thousand-fold improvement on the wire spark chamber.

Multiwire chambers now form part of almost every particle physics experiment, and their use has spread into astronomy and medicine, where they are particularly valuable for forming images. They come in many shapes and sizes—from chambers a few square centimetres in area for measuring the size of particle beams, to arrays of several square metres, containing thousands of wires. And at colliding-beam

Fig. 8.8 *Georges Charpak (b.1925), at CERN in 1984. He is holding one of the first drift chambers, which he built in the late 1960s.*

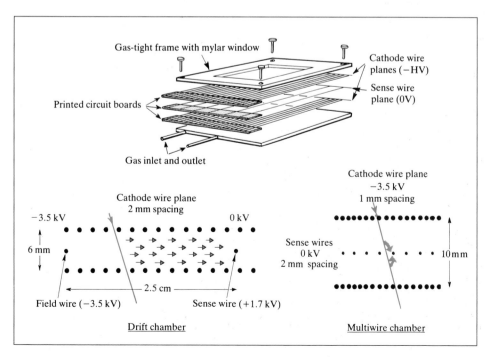

Gas-tight frame with mylar window

Cathode wire planes (−HV)

Sense wire plane (0V)

Printed circuit boards

Gas inlet and outlet

Cathode wire plane
2 mm spacing

−3.5 kV   0 kV

6 mm

2.5 cm

Field wire (−3.5 kV)   Sense wire (+1.7 kV)

Drift chamber

Cathode wire plane
−3.5 kV
1 mm spacing

Sense wires
0 kV
2 mm spacing   10 mm

Multiwire chamber

Fig. 8.9 *A multiwire chamber (top and right) is a sandwich of three layers of parallel wires held in a gas-filled framework with thin plastic windows. The two outer layers are held at a high negative voltage, while the central plane is at zero voltage. A charged particle passing through the chamber (red) ionizes the gas along its path, and the electrons released avalanche towards the nearest wire in the central plane (which is positive relative to the outer planes). The position of the wire that picks up the electrons gives a point along the particle's track.*

A drift chamber (left) has a similar construction, but with the wires in the central plane spaced further apart. Varying voltages applied to the cathode wires produce a field in which ionization electrons 'drift' at a constant velocity towards the nearest sense wire. This drift time is measured by an electronic 'stopwatch' started by a signal from a scintillation counter parallel to the chamber. The measured time is directly related to the distance between the track of the particle and the wire that produces a signal.

Fig. 8.10 *This view of one end of the Time Projection Chamber, built by SLAC and the Lawrence Berkeley Laboratory, shows the web of cables needed to transfer signals from the detector to the data acquisition system. Parallel wires in six equal sectors stretch across the ends of the gas-filled chamber (see Fig. 8.11), and pick up ionization electrons released along the tracks of charged particles. The six sectors are visible here, spanning a region from 20 to 100 cm from the central axis of the apparatus. The vacuum pipe carrying the counter-rotating beams of electrons and positrons normally occupies the hole at the centre.*

machines, where the apparatus must fit around the beam pipe, the multiwire chambers are often in the form of a cylindrical sandwich of concentric layers of wires surrounding the beam pipe.

A multiwire chamber easily out-performs a bubble chamber in terms of the rate at which it can accept particles; but it records the tracks of particles with far less precision, at best to within a little less than a millimetre. However, the drift chamber—the other device that Charpak's group developed in the late 1960s—overcomes this disadvantage to a large extent. Just as the multiwire chamber provided a new order of magnitude in the speed at which experiments can record tracks, so the drift chamber brought about a new degree of precision in measuring the positions of the tracks.

The drift chamber also consists of parallel wires strung across a volume of gas in an electric field. But this time the idea is to measure how long it takes for electrons released along the ionized trail of a charged particle to 'drift' towards a central 'sense' wire, before they initiate an avalanche close to the wire. If the electrons travel at a constant velocity, then the time taken gives a good measure of the distance of the track from the sense wire. Indeed, with this technique it has proved possible to locate particle tracks to an accuracy of some 50 micrometres (millionths of a metre).

Normally, electrons released in a gas will slow down as they lose energy in collisions with the gas molecules. Alternatively, a region of high electric field accelerates electrons, so that they gain energy. The skill with a drift chamber is to design an electric field such that the energy the electrons gain from the field matches the energy they lose in collisions, so that overall they travel at a constant, known velocity. Scintillation counters, which produce signals very rapidly, usually start an electronic 'stopwatch' when a particle passes through the chamber. When the drifting electrons reach a sense wire, the signal from the wire then stops the watch, and the time can be read out by a computer.

Drift chambers, like multiwire chambers, come in many shapes and sizes, and are common in many experiments. They not only give better precision, but also have the advantage of containing relatively few wires and therefore require less additional electronics. The wires in a drift chamber can be spaced at intervals of several centimetres, compared with the 1–2 mm required in multiwire chambers, because it is the drift *time* that provides the information on a particle track's position. However, multiwire chambers can operate much faster than drift chambers. They are valuable in regions of high particle intensities, and in providing signals rapidly for triggering other detectors.

A development of the drift chamber, which reduces the number of wires even further, figures in a detector with a name that sounds like something out of science fiction—the time projection chamber. This device was invented by David Nygren from the Lawrence Berkeley Laboratory. He is a member of a team working at SLAC, in California, on a machine called PEP, which collides electrons with positrons.

Nygren's idea was to have a cylinder of gas with a single electrode at high negative voltage cutting across the middle. Electrons released along the tracks of ionizing particles drift towards the ends of the cylinder, which are positive relative to the centre. The time of arrival of the electrons at the end planes gives a measure of how far *along* the cylinder the electrons originated; electrons from nearer the centre take longer to reach the end. And the positions of the electrons as they arrive at the end plate yields an image of a two-dimensional slice through the tracks. The measurements of the positions together with the arrival times provide enough information for a computer to reconstruct a three-dimensional image of the tracks.

The really ambitious part of Nygren's concept was to do all this with a cylinder 2 m in diameter and 2 m long, allowing the ionization electrons to drift across distances of up to 1 m. It took the best part of 10 years for the Time Projection Chamber (TPC) to come to life; it eventually started operating in 1983, as the key part of a huge detector surrounding PEP's colliding beams. Now the underlying principle of the TPC is well established and the late 1980s will see variations of the TPC in a number of experiments around the world.

SYNCHROCLASH

ELECTRONIC DETECTORS have produced their most spectacular results in an environment that is inaccessible to bubble chambers—at colliding-beam machines, where particles meet head-on within the beam pipe. These machines produce more violent collisions than accelerators that fire particles at a fixed target. In a collider, the target is neither a stationary lump of metal nor the liquid contents of a bubble chamber, but a second particle beam travelling equally rapidly in the opposite direction.

Why collide beams? Car crashes provide a useful analogy. Most drivers are familiar with the 'knock-on' effect: you are stationary at a traffic light when you receive a bump from behind and find yourself unavoidably propelled into the car in front. The driver behind may be at fault, but the momentum of the car behind has nevertheless been shared between you both, according to a fundamental law of nature, so you travel forwards at roughly half the other car's velocity.

Such 'knock-on' effects are equally frustrating to physicists studying the collisions of particle beams from accelerators. The high-energy particles plough into protons and neutrons within a fixed target and, as in the vehicle pile-up, the debris are propelled onwards. From the physicist's point of view this is undesirable because the hard-won energy of the beam particles is being transferred simply into energy of motion—kinetic energy—of particles in the target.

But suppose two similar cars are driving in opposite directions along the same road and at the same speed. If the cars crash head-on, the wreckage will fly off in opposite

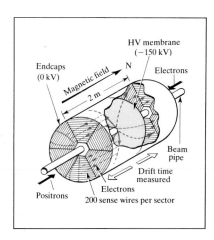

Fig. 8.11 *The Time Projection Chamber is a 2 m by 2 m cylinder filled with gas. It is in effect divided into two sections each 1 m long, which are separated by a membrane held at a very high negative voltage. Electrons released along the tracks of charged particles drift away from this membrane towards the ends of the chamber, which are at zero voltage. Wires are strung across the ends in an arrangement of six equal sectors. Each end plane records a two-dimensional view of tracks in its half of the chamber: one dimension comes from knowing the wire hit, and the other comes from 'pads' that divide up the wires along their length. A third dimension comes from timing the arrival of the electrons at the end plane. Earlier signals are due to electrons from a portion of track near the ends of the chamber; later signals are from electrons that have travelled from regions nearer the centre of the chamber.*

directions. All the energy of the two vehicles is redistributed among the debris; none is taken up in setting stationary objects in motion.

The same holds true with particle beams. If we can bring two particles to collide head-on, the problem of wasted energy is overcome, and the total energy of the two particles can be spent on the interaction between them. Indeed, with particles travelling close to the speed of light, as they do at particle accelerators, the benefits of head-on collisions are greatly increased. This is because relativity makes things even worse for the physicist with a fixed target than for the stationary driver. As particles approach the speed of light they become heavier and—because momentum is the product of mass and velocity—they have much more momentum to pass on to a fixed target. So the higher the energy, the more energy is wasted in moving the target, and the greater the benefits to be had from colliding beams, as we shall see.

Such arguments have not been lost on accelerator builders, who at one time coined the name 'synchroclash' for a machine that could collide particle beams. But though the basic principle was clear as early as the 1940s, it took 20 years for particle colliders to take shape, and another 15 years for them to become the dominant form of particle accelerator, as they are today.

In 1943 Rolf Wideröe—whose doctoral thesis had inspired Lawrence to invent the cyclotron—applied for a German patent on a scheme to store and then collide particles travelling in opposite directions around the same orbit. The patent does not mention the value of the higher effective energy in head-on collisions. Wideröe was advised that he would never be awarded a patent for such an obvious idea—so 'obvious' that it was another 15 years before other physicists independently put colliding beams into practice! However, the patent does describe different schemes to collide counter-rotating beams either of oppositely charged particles or of particles of the same charge. In the former case, a single magnetic storage ring would hold the particles; in the latter, they would be guided by electrostatic fields.

The key word here is 'storage'. If you fire two standard particle beams at each other, collisions of particles would be few and far between—imagine waiting for two pellets from a pair of shotguns to hit each other. Wideröe's idea of a 'storage ring' would improve the odds by accumulating successive bursts of particles from an accelerator, so creating much denser beams. Moreover, because relatively few particles actually interact when the two beams meet, the beams could circulate and meet many times and so provide still more collisions for a given number of orbiting particles.

The idea of colliding beams of particles arose again in the late 1950s when a number of people in the US independently suggested yet another way to collide particles of the same charge. Why not build *two* storage rings, using magnets to direct the beams around them, just as in ordinary accelerators? These rings could intersect at some point between the magnets; the stored beams, travelling in opposite directions, could then meet head on.

A fruitful partnership began between Gerard O'Neill from Princeton University (later to become famous for his work on space colonies), and Wolfgang ('Pief') Panofsky, of Stanford University. O'Neill was one of the pioneers of the idea of separate storage rings, and he decided to begin with *electrons*, which would be easier to work with than protons. Together with Panofsky, O'Neill gathered a small team of physicists who proposed building a pair of electron storage rings at Stanford, where there already existed a 1 GeV linear electron accelerator. One member of the team was a young man called Burton Richter, of whom we shall hear more later.

Construction of the storage rings began at Stanford in 1959. The two rings were built side-by-side, joined at one common point. Each ring was to store a circulating beam of 0.5 GeV electrons, which would give a total collision energy of 1 GeV. This may not sound very much, but to free this amount of energy in a collision with a stationary target, an electron beam would have to be accelerated to some 1000 GeV— 30 times the energy reached at SLAC by the world's largest linear electron accelerator!

By 1965, O'Neill and his collaborators had overcome all the problems and they were able to record the first physics results from colliding particle beams. They used banks of spark chambers set up on either side of the collision zone. Here the spark chamber proved the ideal tool; in particular, it could be triggered to respond only

Fig. 8.12 *Gerard O'Neill (b.1927).*

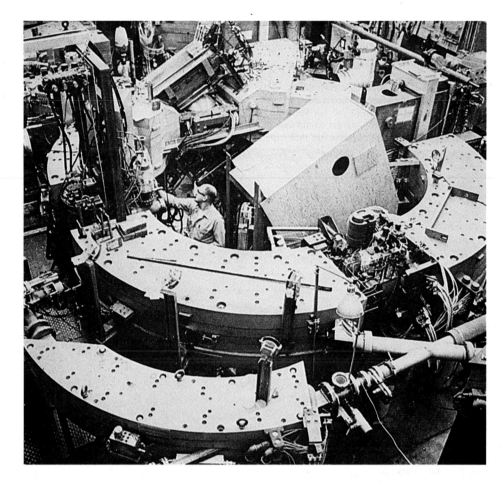

when two electrons actually interacted, rather than at random intervals when much of the time the particles would simply pass each other by.

The Princeton-Stanford collaboration was by no means the only team working on storage rings. In 1959 a group of Italian physicists, under the leadership of Bruno Touschek, began work at the Frascati Laboratory near Rome on a small machine to collide electrons with their antimatter equivalents, positrons. Positrons have the same mass as electrons, but opposite electrical charge. This means that a magnetic field that bends electrons to the right will bend positrons to the left. But suppose the positrons and electrons are moving in *opposite* directions, then the magnetic field will bend the two kinds of particle around the same curve. In other words, electrons and positrons travelling in opposite directions through the magnetic field will follow exactly the same path, providing they have equal energies. The magnets that guide electrons one way round a ring of magnets—clockwise, say—can guide positrons round the other way—anticlockwise.

The machine at Frascati was called ADA, for *Annelo d'Accumulazione* (accumulation rings), and was designed to store beams of 0.25 GeV energy. By the end of 1961 it had stored electrons, but it was then transported to Orsay, near Paris, where a more intense electron beam was available. There, towards the end of 1963, ADA's first electron-positron collisions were recorded; a new breed of machine had come into life that was to change the course of experimental particle physics in the years to come. But ADA was never used to collect high-energy physics data; rather it remained a device for learning about storage rings.

Several electron-positron colliders followed in ADA's footsteps, at Frascati and Orsay, and further afield at Novosibirsk in the USSR. In the US, the Cambridge Electron Accelerator, run by Harvard and MIT, was converted to a storage ring in the mid-1960s. But it was in the 1970s, at the Stanford Linear Accelerator Center (SLAC), that an electron-positron machine was to make an enormous impact on our knowledge of the fundamental particles.

Fig. 8.15 *The electron–positron collider SPEAR, built on a parking lot at SLAC in the early 1970s. This aerial view shows the completed SPEAR ring. The building on the left is End Station A (see Fig. 6.40, p. 127). The 3 km linear accelerator that feeds SPEAR with electrons and positrons lies off the top of the picture.*

Fig. 8.16 *Burton Richter (b.1931), shortly before he became the present director of SLAC in September 1984.*

THE RING ON THE PARKING LOT

SLAC IS famous for its 3 km long linear electron accelerator, which in the late 1980s is still the largest machine of its kind in the world. But even as the linear accelerator was taking shape in the early 1960s, Burton Richter (who had moved to SLAC) and David Ritson from Stanford University had put forward a proposal to build an electron–positron collider called SPEAR—for 'Stanford Positron Electron Asymmetric Rings'.

Richter, who had spent several years working with O'Neill on the Princeton–Stanford electron storage rings, could see that electron–positron collisions would provide an ideal way to study the behaviour of particles formed from the pure energy released in the annihilation of matter with antimatter. Though the first formal proposal for SPEAR was made in 1964, it was not until 1970 that the Atomic Energy Commission (which at the time provided money for particle physics) gave SLAC permission to build a simplified version of the machine, with only one oval ring, together with a large multipurpose detector. What is more, the money had to be found from the laboratory's normal annual budget.

Undaunted, and fired by enthusiasm, Richter and his team pushed ahead with SPEAR, building it on a parking lot at SLAC, close to the end of the linear accelerator. Dipole magnets to guide the beams on their curved path, and quadrupole magnets to keep them tightly focused were mounted together on 'girders' of reinforced concrete. Eighteen girders in all formed the oval ring, 63 to 80 m across, which was to store particles with energies between 1.3 and 2.4 GeV per beam. This machine, probably more than any other, gave American physicists a reputation for persevering despite financial difficulties. SPEAR was soon complete and early in 1972, only 20 months after approval, the first beams were colliding. It had cost only $5.3 million.

How does a machine like SPEAR work? First, the linear accelerator—the electron 'linac'—feeds the ring with successive bunches of electrons, which circulate clockwise. (Note that SPEAR operates at much lower energies than the 30 GeV that the linac can reach, so the accelerator does not operate at full capacity while it is filling the storage ring.) In the ring, the electrons merge into a single narrow bunch only a few centimetres long, and less than a millimetre across. After a few minutes, this bunch contains some 10^{11} electrons—sufficient to produce a reasonable number of interactions once the electrons and positrons are colliding.

But SPEAR also needs positrons. These are created by accelerating electrons about a third of the way along the linac before firing them at a copper target. Positrons in the resulting debris are picked off for acceleration along the remainder of the linac, in which the electric fields are flipped so as to accelerate the positively-charged positrons rather than the usual negative electrons. The positrons enter SPEAR in such a way that they circulate anticlockwise, in a bunch similar to the bunch of electrons. But the number of positrons entering SPEAR after each burst from the linac is about 10 times less than the number of electrons, so it takes 10–20 minutes to build up a sufficient number of positrons. Once enough particles are in the machine, they can circulate for 2 to 3 hours before the bunches need to be replenished with fresh particles from the linac.

The two short counter-rotating bunches of particles in SPEAR pass through each other twice per orbit when they meet head on at points on each side of the ring. These crossings occur in the straight 'sides' of the oval ring, so detectors there can record the products of the electron–positron annihilations. In one of these positions—the 'West pit'—Richter, Martin Perl, and other physicists from SLAC, together with Willy Chinowsky, Gerson Goldhaber, George Trilling, and colleagues from the Lawrence Berkeley Laboratory, installed the large detector they had been building while SPEAR was being constructed. This novel device—the Mark I—was destined to make great discoveries, and to be the blueprint for many a detector at later machines.

The Mark I covered 65 per cent of the space around the collision zone. It was the nearest approach to an electronic bubble chamber that had yet been used, though it still fell short of the bubble chamber's full coverage around the beam. The detector centred on a huge coil of wire, some 3 m long and 3 m in diameter, wrapped around the colliding particle beams. When electricity passed through the coil it provided a magnetic field to bend the tracks of charged particles. Sixteen layers of wire spark chambers—100 000 wires in all—packed the space within the coil. Together they

Fig. 8.17 *The Mark I detector at SPEAR was built by a team from SLAC and the Lawrence Berkeley Laboratory. In the mid 1970s it became famous for many discoveries, notably the J/psi particle and its relatives, and the tau lepton. The tracks of particles were recorded by wire spark chambers wrapped in concentric cylinders around the beam pipe, out to the ring where physicist Carl Friedberg has his right foot. Beyond this are two rings of protruding tubes, which are housings for photomultipliers that view various scintillation counters. The coil of the solenoidal electromagnet lies between the two layers of tubes; the magnet's iron forms the octagonal structure. To the left are rectangular magnets to guide the counter-rotating beams, which collide at the heart of the detector.*

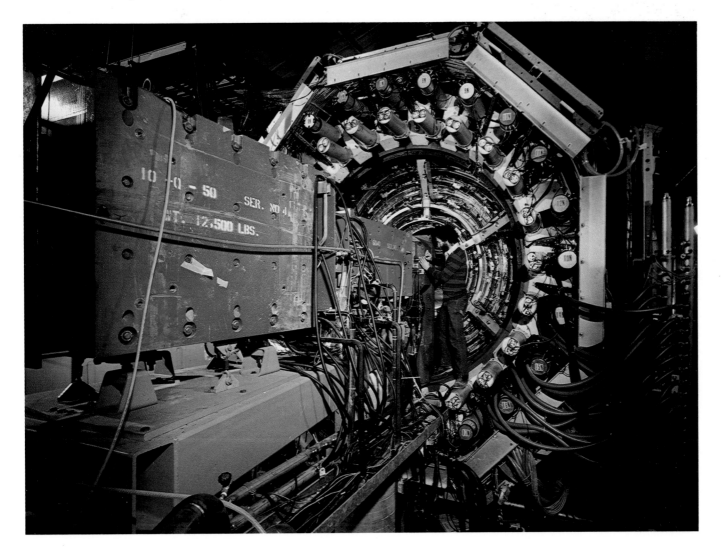

Fig. 8.18 (left) Sam Ting (b. 1936) with members of his team in their control room at Brookhaven. A plot of the data with the peak revealing the J/psi lies on the table at the front.

Fig. 8.19 (right) Members of the team that found the J/psi at SLAC peruse the log book. On the left is Martin Perl, with Burton Richter in the centre, and Gerson Goldhaber on the right. The display in the background shows tracks from the J/psi in the Mark I detector.

provided an accurate map of the tracks of charged particles as they flew away from the electron–positron collisions. Other types of detector, both inside and outside the coil, helped to reveal the identity of the particles, so that physicists could later differentiate between electrons, muons, pions, and so on.

A computer recorded all the information from the spark chambers and the other detectors on magnetic tape. The computer could also perform some simple analysis 'on line', and provided a TV display showing reconstructions of the tracks of charged particles passing through the spark chambers. This on-line display played a key role in the excitement that built up in the control room of the Mark I detector over the weekend of 9–10 November 1974. This was when SPEAR's collisions provided clear evidence for a brand new particle. It was a discovery that in some ways recalled that of the 'vees' by Rochester and Butler, for the particle heralded a new era in particle physics. Two years later, in 1976, Richter shared the Nobel prize with Sam Ting from MIT, whose group had discovered the same particle at Brookhaven.

NEW QUARKS, NEW LEPTONS

THE NEW particle has become known as the J/psi (see pp. 178–9), and we now know that it is built from a charm quark bound with its antiquark. It was the first example of a particle containing this kind of quark, and brought the total number of quarks known at the time to four. The charm quark seemed to bring with it a new natural symmetry, for now physicists knew of four kinds of quark—up, down, strange, and charm—and four kinds of leptons—the electron, the muon, the electron-neutrino, and the muon-neutrino.

During the months following the initial discovery, the Mark I detector at SPEAR collected a wealth of data on the J/psi and its excited states, in which the same type of quark and antiquark have enhanced energies. Moreover, physicists at the Deutsches Elektronen Synchrotron laboratory (DESY) in Hamburg were able to join those at SLAC in pursuit of the J/psi's family. A machine called DORIS, for Double Ring Storage Facility, started up at DESY during 1974. DORIS was built with two rings, one on top of the other, with the aim of being able at times to collide two electron beams, rather than electrons and positrons. The machine collided electrons with positrons at a total energy of up to 7 GeV. This was a little below the energy of SPEAR, which, after modifications during the summer of 1974, could reach a total of 8 GeV.

The Mark I, meanwhile, had still more surprises in store. In 1974, Martin Perl and colleagues began to see hints of a new particle akin to the electron and muon, but

Fig. 8.20 Martin Perl (b.1927), inside the disassembled Mark II detector at SLAC in 1984.

Fig. 8.21 *DORIS, the first electron–positron collider to be built at DESY, the accelerator laboratory in Hamburg. The machine was originally built with two rings of magnets, one on top of the other, so that it could store and collide two beams of electrons, if desired, in preference to electrons and positrons. In 1977, however, DORIS was modified and the two rings amalgamated into one, with the same radius, but now with unusually tall magnets (with blue tops). One of DORIS's important functions nowadays is to supply 'synchrotron radiation' to experiments. High-energy electrons emit radiation as they move on curved paths. This radiation, a waste in terms of accelerating particles, provides a useful source of X-rays and UV radiation for scientists studying the structure of atoms, molecules, and materials. Here we see the pipes (heading towards top left) through which the synchrotron radiation passes out of the accelerator ring to experiments beyond the wall.*

much heavier. At first they were confused by their results because the mass of the particle—nearly double that of the proton—was similar to that expected for charmed particles, that is, particles containing a single charm quark. It was rather like Anderson's discovery of the muon in 1935—a particle with the mass expected for Yukawa's meson. But this time the physicists were not to be mistaken and by 1975 they had convinced themselves that they were indeed seeing the creation and decay of a new lepton. They named it the 'tau', this being the first letter of the Greek word for 'third'. Once again, experiments on the DORIS rings at DESY provided valuable corroborative evidence. The tau soon joined the list of accepted particles, and seemed almost certain to be associated with its own neutrino—the tau-neutrino.

The appearance of the tau broke the neat symmetry of fundamental particles—four quarks, four leptons—that had only recently been established with the finding of charm. The possibility arose that nature might harbour still more quarks, to bring the total number of fundamental particles to a round dozen—six quarks and six leptons.

The new quarks were expected to be heavier than the charmed quark, and therefore to form particles yet more massive than those of the J/psi's family. Storage rings such as SPEAR and DORIS had proved the ideal hunting grounds for new particles, but their maximum energies precluded the discovery of particles heavier than around 8 GeV. New, larger electron–positron machines were on the horizon—PEP, being built at SLAC, and PETRA, under construction at DESY, were both designed to reach a total energy of 30 GeV and more. But during all the excitement at SPEAR, a new huge proton synchrotron had started up at the Fermi National Accelerator Laboratory.

It was here in 1977 that Leon Lederman (now director of Fermilab) and his team from Columbia University, the State University of New York at Stony Brook, and Fermilab, discovered a new particle some three times heavier than the J/psi and over nine times heavier than the proton. This particle became known as the 'upsilon' and turned out to be the first observed manifestation of a fifth kind of quark—the bottom quark (see pp. 186–8). The upsilon resembles the J/psi and is in fact built from a bottom quark bound with a bottom antiquark.

With a mass of 9.4 GeV, the upsilon was too heavy to be found either at SPEAR or at DORIS in its original form with two rings, one on top of the other. But by modifying the two rings to form a single ring that could carry much more intense beams of particles, the machine physicists at DESY were able reach the energy region of the upsilon. In May 1978, physicists at DORIS saw the first signs of the upsilon, and soon began to study the various ways in which the bottom quark and antiquark orbit each other in the system known as 'bottomonium'. DORIS has since been joined by another machine in studies of the upsilon and related phenomena. In June 1979, a new electron–positron collider began work at Cornell University in New York. The Cornell Electron Storage Ring, or CESR (pronounced Caesar), has a maximum energy of 8 GeV per beam—16 GeV total—and is well-suited to producing particles containing the heavy bottom quarks.

Fig. 8.22 *The Cornell Electron Storage Ring (CESR), at Cornell University in Ithaca, New York, occupies the same tunnel as the synchrotron that feeds it with particles. The 12 GeV synchrotron (the magnet ring on the left in this picture) accelerated its first electrons in 1967. However, in the mid 1970s, the proposal to build CESR was made, and by the end of 1977 the synchrotron was successfully accelerating positrons for injection into a prototype section of the storage ring. The first electron–positron collisions in CESR, the ring on the right, occurred in June 1979, and since then Cornell has implemented a thorough study of the heavy particles that contain the bottom quark. The storage ring was designed to work at a maximum energy of 8 GeV per beam, but normally operates at a little over 5 GeV per beam—the optimum collision energy for studying bottom particles.*

HOLOGRAPHY AND CHIPS

THE NEW particles of the 1970s live much longer than the resonances first observed in the previous decade. But they do not survive as long as the strange particles, which can be 'seen' through measurable tracks (or gaps in the case of neutral particles) in cloud and bubble chambers. Charmed particles live typically for only 10^{-13} s, decaying a thousand times more rapidly than their strange counterparts. The lifetime of the tau lepton is also about 10^{-13} s.

In experiments at synchrotrons with fixed targets, as opposed to colliders, the main difficulty with these short-lived particles is that, in their brief lives, they do not have time to move very far away from the general 'forward' direction of the beam that created them. A particle with a lifetime of 10^{-13} s, for instance, strays no more than 300 micrometres from this forward direction. This makes distinguishing the 'vees'— the points at which the particles decay—far more tricky, and sets the resolving power required for the detector.

In an average bubble chamber, for example, the bubbles are themselves typically 300 micrometres across when they are photographed; in a large chamber designed to capture neutrinos, the bubbles are even bigger, some 700 micrometres across. So the bubble chambers of the 1970s were at best able to record the decay of a charmed particle indirectly, through the pattern of tracks emerging from essentially the same point at which the decaying particle had been created. Figure 9.9 (see p. 183) shows the possible production and decay of a charmed relative of the sigma particle in the '7 foot' bubble chamber at Brookhaven. The birth and death of the particle are compressed into the same point; all that we see is a spray of particles materializing from the invisible track of an incident neutrino. The existence of the charmed sigma is inferred only from detailed calculations of the angles, energies, and so on of the emerging particles.

Why not simply take the photographs sooner, before the bubbles grow so large?

Fig. 8.23 *The body of the Rapid Cycling Bubble Chamber during its installation at CERN in 1980. The 250 litre chamber, designed and built at the Rutherford Appleton Laboratory in the UK, is designed to work at a rate of up to 30 expansions per second—some 30 times faster than conventional bubble chambers. It forms part of the European Hybrid Spectrometer, a large assembly that also includes many electronic detectors for measuring and identifying particles.*

The problem is that for the camera to resolve smaller bubbles—say, 30 micrometres across—its range of focus (depth of field) becomes much smaller, as any photographer knows, and it becomes impossible to focus on a complete track. But if in photographing small bubbles, only a small volume is in focus, why not use a small bubble chamber, which can be run through its cycle of compression, expansion, and recompression much more rapidly than a larger chamber? Such rapid cycling greatly improves the chances of capturing rare events like the decays of charmed particles.

This is precisely the route taken by an international team at CERN, who have built a tiny bubble chamber, only 20 cm in diameter and 4 cm deep. The researchers have used the chamber in conjunction with a sophisticated battery of electronic detectors. These detectors pinpoint the tracks of charged particles emerging from the chamber and reveal neutral pions, which leave no tracks in the chamber. The tiny bubble chamber provides a vital close-up view of the short lives of charmed particles.

At SLAC, a team of physicists from Japan, Israel, and the UK, as well as the US, have taken a different approach. They have used a 1 m bubble chamber equipped with a high-resolution camera to study charmed particles produced in liquid hydrogen by a beam of 20 GeV photons (see Fig. 9.7, p. 181). The camera takes the photographs when the bubbles are a mere 55 micrometres across, only 200 microseconds after the photon beam has passed through the chamber. To overcome the problem of the camera's small depth of field, detectors outside the bubble chamber provide a signal to trigger the flashlights only when tracks lead back to the 6 mm field of focus.

High-energy neutrino beams are useful tools for creating charmed and bottom particles, because they create these particles relatively frequently. But experiments with neutrinos in bubble chambers demand methods that are still more innovative. In this case, the volume of liquid *must* be large to give the neutrinos a good chance to interact. So is there another way of overcoming the problem of the camera's limited

Fig. 8.24 (left) The 'silicon strip' detector, or SSD, is a wafer of silicon divided into strips, where each strip forms a diode—an electronic device that passes electric current in only one direction. The detector acts rather like a wire chamber, collecting the charge released by an ionizing particle on the diode strips. The wafer is the bright square at the centre of the picture. The strips are so closely spaced that connections to them must be carefully fanned out on the surrounding board before they can be attached to the usual components to amplify and transmit the signals.

Fig. 8.25 (right) The charge-coupled device, or CCD, can be described as an electronic form of photographic emulsion. It is a single chip that is divided up into a two-dimensional array of some 250 000 picture elements, or 'pixels'. Each pixel responds to individual photons of light or to the passing of a charged particle, and so the whole device can form a two-dimensional image. CCDs are widely used in astronomy and for other specialist forms of imaging. In particle physics experiments, they are useful because they can resolve closely-spaced tracks passing through the detector more or less at right angles to the array of pixels.

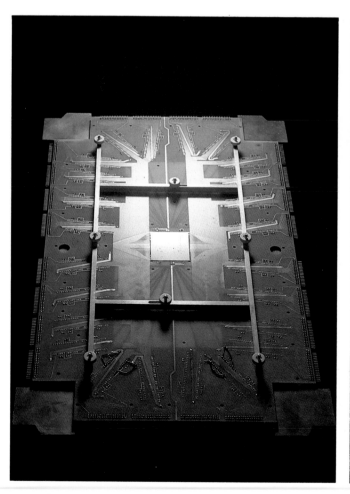

depth of field at high magnifications? One solution being tested in the big bubble chamber—the '15-foot' (4.5 m)—at Fermilab, is to use holography rather than photography to record the tracks.

A hologram is an interference pattern formed by laser light scattered from an object; the pattern contains more information than a conventional photograph and allows three-dimensional images to be reconstructed. A single hologram could in principle record the whole volume of a bubble chamber; however, its main advantage is that several metres of the chamber can be in focus on the one hologram.

Other researchers have turned to the miracle substance of the modern world— silicon—to record the new short-lived particles. The so-called silicon-strip detector collects ionization on *diodes*. These are simple electronic components that conduct electric current in one direction but not in the other. They can be constructed on a wafer of silicon in the form of narrow parallel strips, which conduct only when extra electrons have been freed by an ionizing particle. One team at CERN has resolved tracks in the decays of charmed particles only 0.1 mm apart, by using strips 0.02 mm wide.

The charge-coupled device, or CCD, provides another opportunity for particle physicists to exploit the silicon chip in their attempts to develop detectors with high resolution. A CCD consists of a two-dimensional array of 'picture elements', or 'pixels', each about 0.02 mm square. These devices are used in cameras where the levels of light are particularly low, for example in photographing faint astronomical objects. Electrons released by light, or by an ionizing particle, collect in the pixels, and can then be fed into electronic circuitry in a way that 'remembers' the locations of the relevant pixels. Though CCDs are difficult to use, requiring temperatures of −150°C or so, they have the advantage over strip detectors of providing two-dimensional information, and this makes the CCDs inherently more accurate.

CCDs, holographic bubble chambers, and other new types of detector are only just beginning to challenge the more traditional techniques. They will perhaps come into their own in the next few years, and help to establish accurately the lifetimes and other important properties of the 'new' particles.

THE ANTIPROTON ALTERNATIVE

THE DISCOVERY in 1977 of the bottom quark, concealed within the upsilon particle, strengthened the idea of a natural symmetry between nature's basic building bricks: six types of quark and six different leptons. However, the bottom quark was only the fifth kind of quark. A sixth quark—the top quark—was still missing. The search for the top quark became one of the main priorities of particle physics in the late 1970s. In addition, physicists were keen to observe the W and Z particles, assumed to transmit the weak nuclear force in much the same way as photons transmit the electromagnetic force.

By the beginning of the 1980s, a new generation of electron–positron machines had begun to reach for higher energies. In 1974, proposals had been put forward to build larger colliders both at DESY and at SLAC. The West German government seized the opportunity to help out an ailing construction industry, and plans for PETRA— Positron-Electron Tandem Ring Accelerator—were approved in the following year. Before the end of 1978, the new collider, which just fits into the DESY site, was complete and ready to produce the first collisions.

PETRA got off to a slow start and did not come close to its design energy of 19 GeV per beam until 1980. But when it started up, first at 6.5 GeV and then at 8.5 GeV per beam, the total collision energy was the highest that had ever been achieved with electrons and positrons. Moreover, there was no competition from the Positron-Electron Project (PEP) at SLAC, which was designed to reach a total energy of 36 GeV. A series of difficulties, financial and technical, delayed completion of this machine until 1980.

One difficulty with accelerating electrons in circular machines like PETRA and PEP is that they radiate energy as they swing round bends. This is not a problem with

Fig. 8.26 (top) An aerial view of DESY shows how the underground ring for PETRA just fits into the laboratory's site in a Hamburg suburb. Roads and tracks mark most of the ring's path, which passes from behind the chimney near the centre top of the picture, round by the houses at the right, close to the sports field near the bottom, and back up across the fields at the left.

Fig. 8.27 (left) Inside PETRA's tunnel some of the 224 bending magnets, each 5.4 m long, are visible. The shorter, square magnets are for focusing the beams. The tunnel is 2.3 km in circumference — hence the bicycle !

Fig. 8.28 (right) The klystrons that generate up to 4.8 MW of radio-wave power to accelerate the beams in PETRA.

proton synchrotrons, because the heavy protons do not radiate as readily as light-weight electrons do. But in electron synchrotrons the effect soon becomes trouble-some—double the energy of the electrons, and the amount of 'synchrotron radiation', as it is known, rises sixteen times! To accelerate its electron and positron beams, PETRA uses a powerful supply of radio waves to make up the energy lost through synchrotron radiation. Each of the klystrons that generate the radio waves for PETRA can supply 500 kW of power, while the whole machine can use as much as 10 MW just in accelerating particles.

In 1979, PETRA came up trumps with some remarkable results concerning the strong nuclear force. The products of collisions at the machine's higher energies revealed evidence for the radiation of *gluons*, the carriers of the strong force presumed to flit between quarks within the particles they form. This was important evidence for the theory known as quantum chromodynamics (QCD), which is based on analogy with the quantum theory of charged particles—quantum electrodynamics (QED). The discovery showed that just as electrons radiate photons, so can quarks radiate gluons.

By the spring of 1984, after various improvements, PETRA reached a new world record for positron-electron colliders of a little over 23 GeV per beam. But still the top quark eluded the teams at PETRA; all they could say was that if the top quark did exist then its mass must be greater than around 23 GeV. And in that case a machine that could reach higher energies would be necessary to produce it.

As we have already seen, there are problems in accelerating electrons and positrons to high energies in circular machines because of the huge energy losses through synchrotron radiation. So why not build proton-proton colliders, or even proton-antiproton colliders?

In the late 1950s, Gerard O'Neill had opted for building rings to collide electrons because no one at the time was quite sure how protons could be stored. A decade or so later, in 1971, engineers at CERN had solved the problem and succeeded in producing the first head-on collisions between protons in a machine called the ISR, or Intersecting Storage Rings.

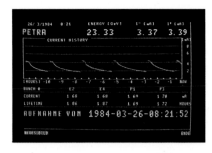

Fig. 8.29 *A computer display in PETRA's main control room shows the record-breaking energy of 23.3 GeV per beam, reached in the spring of 1984.*

Fig. 8.30 *A crossing between the two beam pipes of CERN's Intersecting Storage Rings— the world's first, and so far only, proton-proton collider. The machine collided protons that were stored in counter-rotating beams in two interlaced rings. It operated from 1971 to 1984.*

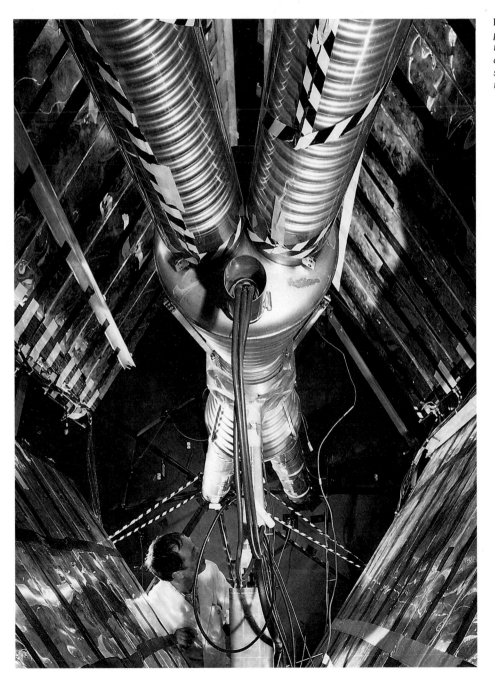

Fig. 8.31 *Simon Van der Meer (b.1925).*

The ISR consisted of two interlaced rings of magnets, with two beam pipes that crossed at eight places. It was fed by 26 GeV protons from the Proton Synchrotron, and brought the two beams to collide after they had each been accelerated to 31.5 GeV. The total collision energy of 63 GeV was equivalent to the effect of a stationary target being struck by a beam of protons at a fantastic 1800 GeV! In one jump, the ISR had catapulted particle physics into a new energy region, and one that conventional accelerators had no chance of reaching; the highest energy planned for a synchrotron was 'a mere' 400 GeV.

The physicists from around the world who worked on the ISR in the early days were almost bowled over by the new experience, and it took several years to learn what were the best designs for apparatus, and how the data might best be analysed. In the meantime, the people who built and ran the machine were also learning how to handle stored beams of protons. In particular, a Dutch engineer, Simon Van der Meer, started to look at a way of concentrating the beams in the ISR so that more particles would collide each time the beams crossed.

Van der Meer's idea was subtle and relied on some very clever manipulations of the particle beam. Put simply, you measure the positions of random samples of protons in the beam and then nudge these particles accordingly. By repeating this procedure many times, the whole beam is slowly concentrated closer to the optimum orbit. In practice, you sense the average position of a sample of protons in a 'slice' through a beam at one point in its orbit. You then use this information to send a signal across the ring to a 'kicker' that generates just the right amount of electric field to push these protons, on average, towards the ideal path. This requires some sophisticated electronics, to ensure that the signal gets across the ring before the swift-moving particles arrive.

The protons do not travel around in a fixed formation like marching soldiers, but move about relative to one another. As a result, the sensor picks out a slightly different bunch of protons on each orbit. Thus the kicker acts on different protons each time, and the net effect of doing this many times is to squeeze the beam in towards its ideal path. The method is called 'stochastic cooling'. 'Stochastic' because it operates on random samples of a beam; 'cooling' because squeezing the beam reduces its sideways motion, and smaller motions are usually related to lower temperatures.

The machine physicists at CERN showed that stochastic cooling would work on the proton beams of the ISR in 1975. But it was in another development at CERN that Van der Meer's idea was to have a much greater impact and win him a share of the Nobel prize for physics in 1984. The technique proved vital in allowing CERN to use its 400 GeV Super Proton Synchrotron as a proton–antiproton collider.

The ISR had shown colliding protons to be a valuable route to high energies. And in the mid-1970s, researchers in the US began to consider building a giant proton–proton collider that would bring together beams at energies of 200 or even 400 GeV each. Such an enterprise would be costly and would take time, in effect requiring the construction of two entire machines, each one equivalent to CERN's SPS or the 400 GeV machine at Fermilab. An ill-fated project, named ISABELLE, did in fact begin at the Brookhaven National Laboratory in 1978. But two years earlier, three physicists had proposed a simpler scheme to reach the energies of ISABELLE.

Carlo Rubbia from CERN, together with Americans David Cline and Peter McIntyre, proposed putting antiprotons into either or both of the big synchrotrons at CERN and Fermilab. The machine could then be made to operate like the electron–positron colliders, with the antiprotons being bent by the same magnets and accelerated by the same electric fields while travelling in the opposite direction to the protons. It was a beautifully simple idea. The problem was to get enough antiprotons into the machine for any collisions to occur. And that is where Van der Meer's work on stochastic cooling came in.

Antiprotons are produced in large numbers when a beam of high-energy protons strikes a metal target. The antiprotons emerge from the target with a large variety of velocities and over a wide range of angles, and so they cannot pass directly into a synchrotron, which operates on well-defined beams of particles all travelling at the same velocity. To tame the antiprotons before injecting them into the Super Proton

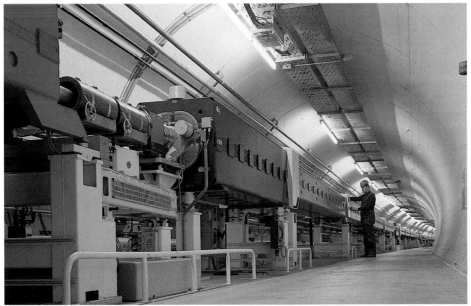

Fig. 8.32 *The antiprotons for CERN's proton-antiproton collider are created in collisions of protons with a metal target and then passed to this machine, the Antiproton Accumulator. Here successive bursts of antiprotons are added together, and 'cooled'—a process whereby the beam is made sufficiently well-behaved for injection into the Super Proton Synchrotron. The magnets (blue) are unusually fat because they have to accommodate a beam pipe wide enough to carry the unruly antiprotons as they emerge at a variety of angles from the production target.*

Fig. 8.33 *The tunnel of the Super Proton Synchrotron, in the weeks before it accelerated its first protons in 1976. Since 1981 the machine has also operated as a proton–antiproton collider. The same bending magnets (red) and focusing magnets (blue) can guide protons and antiprotons, provided the two types of particle, which have opposite electric charge, are travelling in opposite directions.*

Synchrotron, CERN decided to build a small machine called the Antiproton Accumulator. This takes antiprotons from a target and uses the principle of stochastic cooling to concentrate them into a well-behaved beam.

In 1978, CERN gave the official go-ahead for the proton–antiproton project and the building of the Antiproton Accumulator. Three years later, in August 1981, the accumulator delivered the first antiprotons to the Super Proton Synchrotron and ecstatic physicists at CERN detected the first collisions of matter with antimatter at a record energy of 270 GeV per beam—equivalent to a beam of 150 000 GeV striking a stationary target.

Operating the proton–antiproton collider calls for some remarkable juggling with particle beams. The 26 GeV protons that create the antiprotons come from CERN's smaller synchrotron, the PS, every two seconds. The antiprotons enter the Antiproton Accumulator one burst at a time and are cooled there for nearly two seconds before the next burst is due. The accumulator is rather like two rings in one, with a wide beam pipe that is divided in places by metre-long metal 'shutters'. On one side of the shutters—toward the outer edge of the ring—antiprotons fresh from the PS are cooled; on the other side is the 'stack' of antiprotons accumulated from successive bursts. Just before the next burst of particles is due, the shutters rise, and the antiprotons are moved across the pipe to join the stack. The shutters then fall and the accumulator is ready to accept the new burst of antiprotons from the target.

Once in the stack, the antiprotons are cooled further, more and more antiprotons joining them every two seconds. After about 40 hours there are some three hundred thousand million (3×10^{11}) antiprotons in the dense core of the stack and at last there are enough to be useful in the Super Proton Synchrotron. But the antiprotons do not go there directly. The Antiproton Accumulator works on particles with an energy of 3.5 GeV, the most common energy for antiprotons emerging from the target. This is too low an energy for injection into the big synchrotron. Instead, the antiprotons must first pass back into the smaller PS—which accelerated the protons that created the antiprotons up to 40 hours earlier! The PS then accelerates the antiprotons to 26 GeV. Only then are they ready to enter the SPS, where 26 GeV protons are already circulating. In all, three bunches of antiprotons are pulled from the accumulator in this way; likewise there are three bunches of protons.

At long last the protons and antiprotons are all set for the big kick that sends them to energies of hundreds of giga electronvolts. In 1981, the first collisions occurred at a total energy of 540 GeV. However, modifications have allowed the normal collision energy to rise to a total of 630 GeV, and in special circumstances it can reach 900 GeV—equivalent to a beam of protons of 400 000 GeV striking a fixed target.

540 GeV . . . AND BEYOND

SOON AFTER the first proton–antiproton collisions in the Super Proton Synchrotron, several experiments were ready to explore the new energy region that had opened up. The hunt was on for the top quark and the W and Z particles. But there was also plenty of excitement in simply observing the character of the collisions, and making basic measurements of the particles created in conditions similar to those formed high in the atmosphere by the arrival of energetic cosmic rays. Some of the first information came from a detector known as UA5, which recorded the first images of the collisions both at 540 GeV and, four years later in 1985, at 900 GeV. However, most of the time the proton–antiproton collisions come under the scrutiny of two large detectors, in underground caverns at separate locations round the ring. These detectors are

Fig. 8.34 *The aftermath of a proton–antiproton annihilation at a total energy of 900 GeV. Normally CERN's protons and antiprotons collide at a total energy of 630 GeV, but in special circumstances the beams can collide at 900 GeV. Here the tracks of charged particles produced in one of the first collisions at this high energy have been captured in the UA5 streamer chamber—a gas-filled device in which luminous streamers form along ionized trails under the influence of an electric field. This image was recorded by a TV camera and then enhanced by computer. The light intensity has been colour-coded so that the faintest areas are at the red end of the spectrum, and the brightest purple.*

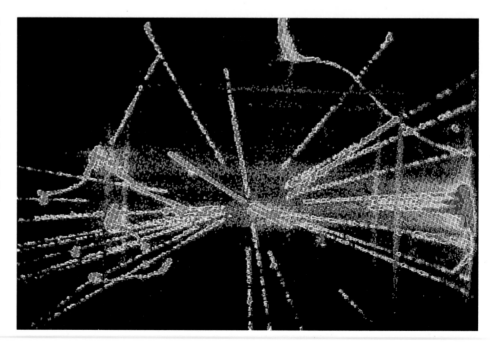

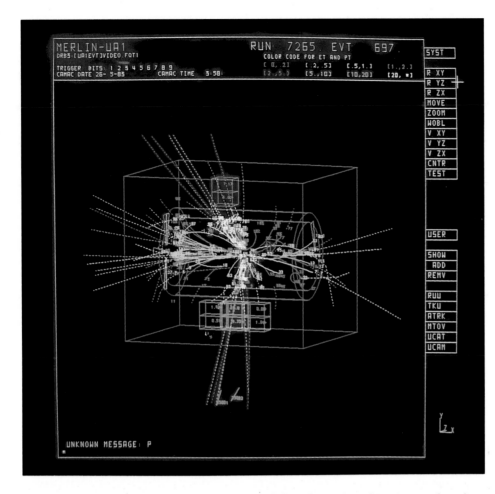

MERLIN-UA1   RUN 7265. EVT   697.
DRB3:[UA1EVT]VIDEO.FOT1   COLOR CODE FOR ET AND PT
TRIGGER BITS: 1 2 3 4 5 6 7 8 9   [0,.2]   [.2,.5]   [.5,1.]   [1.,2.]
CRMAC DATE 26- 5-85   CRMAC TIME 3:58: [2,.5.]   [5.,10]   [10,20]   [20,*]

UNKNOWN MESSAGE: P

Fig. 8.35 *In this computer display of a proton–antiproton collision in the UA1 detector, two distinct back-to-back high-energy jets are easily seen. The tracks are colour-coded according to momentum: lower momenta are represented by the red and yellow end of the spectrum, higher momenta are blue and purple. The proton and antiproton, containing quarks and antiquarks respectively, have come in from the left and right. The low-momentum tracks going out sideways are due to particles which have materialized from quarks and antiquarks that made glancing collisions. The two jets of high-momentum particles shooting out to the top and bottom of the picture are from a quark and antiquark that have met head-on and rebounded violently at 90° to their original directions.*

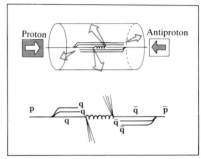

Fig. 8.36 *The diagram shows the event in Fig. 8.36 at the level of the quarks (q) and antiquarks (q̄). A quark and an antiquark within the colliding proton (p) and antiproton (p̄) can interact via the exchange of a gluon (wiggly line), and then shoot off sideways. They materialize in the detector as jets of high-momentum particles (blue arrows); the remaining quarks and antiquarks that formed the original proton and antiproton give rise to jets of low-momentum particles (red arrows) that move off horizontally. The entire process takes place inside the beam pipe, in a tiny 'femtouniverse' only 10^{-15} m across.*

called UA1 (which we met at the beginning of this chapter) and UA2; together they have yielded many of the riches of this new energy region.

One of the first phenomena that UA1 and UA2 observed were 'jets' of particles emerging from the collision site. In the jargon of particle physics, a jet is an individual shower of particles, generated by a single quark, antiquark, or gluon. As Chapter 7 described, quarks appear to be confined permanently within particles such as protons and pions. Even when a quark and an antiquark are produced afresh in electron-positron annihilations, they do not emerge as free particles themselves, but in some way use their energy to materialize more quarks and antiquarks to form the clusters we detect as protons, pions, and the like. If the quark and antiquark are moving rapidly they produce individual showers or 'jets' of particles, as a team working on the SPEAR electron-positron collider first observed. This production of jets is the nearest we come to seeing an individual quark—the jet bears the memory of the direction in which the parent quark was initially moving.

Observing jets is not restricted to electron-positron machines. Jets can also occur when two particles built from quarks—hadrons—collide, provided the quarks clash violently and bounce off at large angles relative to the main thrust of the debris. The chance of separating quarks from their fellows increases with the energy of the collision, so CERN's proton-antiproton collider offered an opportunity to see clear examples of jets in a new environment.

Figure 8.35 shows an example of jets produced at the heart of the UA1 detector. The image shows the complicated jumble of particles produced in the collision, but even here two jets of blue tracks stand out from the red and yellow tracks. The computer has colour-coded the tracks according to their momentum, measured by the curvature in the magnetic field. The coding follows the colours of the rainbow, from red for low momentum to blue for high momentum. The low-momentum tracks come from quarks and antiquarks that made glancing collisions, rather than a violent head-on collision of the kind that gives birth to the energetic jets.

A bonus with proton–antiproton collisions is that the colliding particles contain gluons, which flit around binding the quarks (and antiquarks) together. The collisions can cause these gluons to materialize as jets. Indeed, the operating conditions of CERN's proton–antiproton collider make it rather easier to detect gluon jets; electron–positron colliders, on the other hand, produce mainly quark and antiquark jets.

As well as jets, UA1 and UA2 have also found the evidence that everyone was waiting for. Early in 1983, the teams from the two experiments announced the discovery of the W particle, followed a few months later by the discovery of the related, but rarer, Z. The top quark proved a more difficult beast to locate, but in 1984, UA1 presented the first evidence that could indicate the decays of particles containing top quarks. The proton–antiproton collider seems to have at last completed the balanced picture of nature's building blocks. Yet it is also providing tantalizing hints that it is on the brink of a new world of high-energy phenomena.

Jets show up so cleanly in the proton–antiproton collisions at CERN that the physicists can study fine details. And they are discovering some bizarre behaviour. There are, for example, rare events where a jet shoots off on one side, with nothing visible balancing its flight in the opposite direction. One or more neutral particles must be recoiling unseen in order to balance the thrust of the visible jet. But the physicists have difficulty in explaining these neutral particles in terms of conventional physics. It seems either that known particles could be behaving in a new way, or that these high energies are creating new forms of matter. However, there are only a handful of these puzzling events and it may be that they are due to conventional processes after all; nature may have tricked the physicists by showing them some very rare processes in the first burst of data.

The nature of these and other odd events at CERN will probably become clear only with new machines that will enter a domain of still higher energies. Several such machines are already being built. Fermilab in the US is following CERN's example by converting its large synchrotron—the Tevatron—into a proton–antiproton collider. The Tevatron can accelerate protons to 1000 GeV, or 1 TeV, and it should provide proton–antiproton collisions with a total energy of 2 TeV, over three times CERN's usual maximum energy.

Meanwhile, CERN is building a huge new electron–positron collider, 27 km in circumference. The size of the ring will allow the electrons and positrons to curve as gently as possible, reducing the emission of synchrotron radiation as the particles are accelerated. The machine, known as LEP for Large Electron–Positron collider, will reach 100 GeV total energy—double the energy of PETRA at the DESY laboratory in West Germany. This may not sound much compared with the 630 GeV of CERN's proton–antiproton collisions, but in electron–positron colliders the full energy goes into the annihilation. In a proton–antiproton machine, the energy in each particle is shared among the constituent quarks and gluons; the amount carried by a single quark is on average only a fifth of the quoted beam energy. So LEP's electron–positron collisions will explore a similar energy range to the current proton–antiproton experiments. LEP should start up in 1989, and there are plans to increase its energy later to 100 GeV per beam—200 GeV total. Another possibility is to build a proton–antiproton collider in LEP's 27 km tunnel. This would be able to reach energies of about 10 TeV per beam.

Over at DESY, the machine physicists are already experts with electrons and positrons, but now they are having to learn about accelerating protons. The laboratory is building an entirely new type of machine that will collide protons with electrons. Because electrons and protons are so different they each require their own specially designed ring. So HERA, for Hadron Electron Ring Accelerator, will consist of two separate rings, one above the other, in the same 6.3 km long tunnel. The rings will cross so that the beams collide at four points during each circuit. With electrons at 30 GeV and protons at 820 GeV, HERA's collisions will use electrons to probe protons at far higher energies than has been achieved at SLAC or with secondary electron beams at CERN and Fermilab.

American particle physicists are meanwhile determined not to be left behind. At

SLAC, Burton Richter is leading a challenge to LEP by building a new kind of colliding-beam machine. The plan is to use the existing 3 km long linear accelerator to feed electrons and positrons in opposite directions around two intersecting arcs. The bunches of particles will collide only once, unlike in a conventional collider. But by upgrading the linear accelerator so that its maximum energy rises from 30 to 50 GeV, Richter's 'linear collider' scheme will reach the same energy as LEP, but maybe a year sooner.

Many other Americans are working on a plan to build a Superconducting Super Collider, or SSC, which will accelerate protons and antiprotons to 20 TeV—20 000 GeV—per beam. The SSC would use the technology of superconducting magnets, which has already worked successfully in Fermilab's Tevatron. But even with the higher fields produced by superconducting magnets, the SSC will need to be about 100 km in circumference to reach energies of 20 TeV. As of Summer 1986, this project has still to be approved.

The late 1980s and early 1990s should be an exciting time of exploration at new energies. CERN's proton–antiproton collider has already provided intriguing glimpses of what may lie in store. However, higher energies alone will not provide all the answers to the questions particle physicists still ask. Why, for example, does the electric charge of the electron so precisely balance the proton's charge, when the simple electron and the complex proton seem to be such utterly different forms of matter? Why do we live in a Universe of four dimensions, three of space and one of time? Why is the Universe apparently made of matter to the exclusion of antimatter? The answers to these questions, and others, may well come from experiments quite different from those at particle accelerators, as Chapter 10 describes.

Fig. 8.37 *CERN's next big collider will follow the 27 km path shown by the circular dotted line on this aerial view of the countryside north of Geneva. The Large Electron–Positron collider (LEP) will initially accelerate electrons and positrons to 50 GeV, before bringing the two beams to meet head-on at a total collision energy of 100 GeV. The larger of the solid circles marks out the tunnel of the Super Proton Synchrotron, with the main site of CERN (site Meyrin) to its left. The SPS serves experiments in two areas, one on the main site, and one in France (site Prévessin). The dashed line follows the border between France (top) and Switzerland (bottom). Geneva airport is in the foreground and the Jura mountains are just visible beneath the clouds.*

CHAPTER 9
FROM CHARM TO TOP

MATTER IS built from quarks and leptons, held together by fundamental forces, which in turn are mediated by particles known collectively as gauge bosons. This statement summarizes what particle physicists call the 'standard model' and it represents the state of our understanding of the nature of matter in the mid-1980s. Such a straightforward statement would not have been possible in the early 1970s, and it provides some indication of how much progress was made in particle physics in the decade from 1974 to 1984.

The year 1974 marked the beginning of what is still often referred to as the 'new' physics. The discovery of the J/psi particle in November of that year took the number of quarks known to four, when for the previous ten years three quarks had seemed adequate. Then, within another three years, evidence had emerged for a fifth type of quark ('bottom') and for a third kind of electrically-charged lepton, the tau.

All this might have confused the picture had it not been for important advances in understanding the forces between particles, which occurred over the same period. The idea of unifying the weak and electromagnetic forces within one framework began to seem more and more to fit with the reality of nature. More than merely accomodating the fourth quark ('charm'), the 'electroweak' theory turned out to demand it. This put increased pressure on experimenters to search for the predicted carriers of the electroweak force—the W^+, W^-, and Z^0 particles.

At the same time, a new theory for the strong force, modelled on the same concept of force-carrying particles, made great headway. This was based on a new property known as 'colour', which is analogous to electric charge. In other words, the strong force is a 'colour force'. According to the theory, known as quantum chromodynamics, quarks carry colour and are bound together by particles called gluons which mediate the colour force. In addition, there was a growing conviction that quarks and leptons occurred in pairs, and that three pairs of each occurred in nature—suggesting six quarks and six leptons in all.

By 1984, experimenters had not only found convincing evidence for gluons but also for the elusive Ws and Z. Moreover, they had caught the first signs of a sixth type of quark—top—required by the standard model to complete the pattern of six quarks and six leptons.

From 1974 to 1984, experimenters continued to use bubble chambers and counter experiments just as in the previous two decades. But most of the major discoveries—and the images in this chapter—came from electronic detectors set up in a new configuration: in a 'barrel' surrounding the site of a head-on clash between two particle beams travelling in opposite directions.

The earliest notable successes with a colliding-beam machine, or 'collider', came at SLAC, where a small magnet ring called SPEAR was set up to collide electrons with positrons. The physicists there were rewarded not only with some of the first evidence for the J/psi and its charmed relatives, but also the tau. Later, a larger electron–positron collider at DESY, in Hamburg, claimed the discovery of the gluon. Then, in the early 1980s, CERN stole the scene with its proton–antiproton collider.

During the time from the first experiments at SPEAR to the successes at CERN, the electronic detectors used at colliders became increasingly sophisticated. It is not possible to use a bubble chamber at a collider because the detector must surround the beam pipe, but the modern electronic detectors can provide almost as much detail. Moreover, computers can translate the signals from the detectors into colourful images of the particle tracks, which can be manipulated in a variety of useful ways.

Fig. 9.1 A computer display of a proton–antiproton annihilation in the UA1 detector at CERN takes on an exotic appearance after photographic treatment by Patrice Loïez. The image contains the decay of a Z particle (see pp. 192–5) into an electron and a positron (tracks coloured yellow).

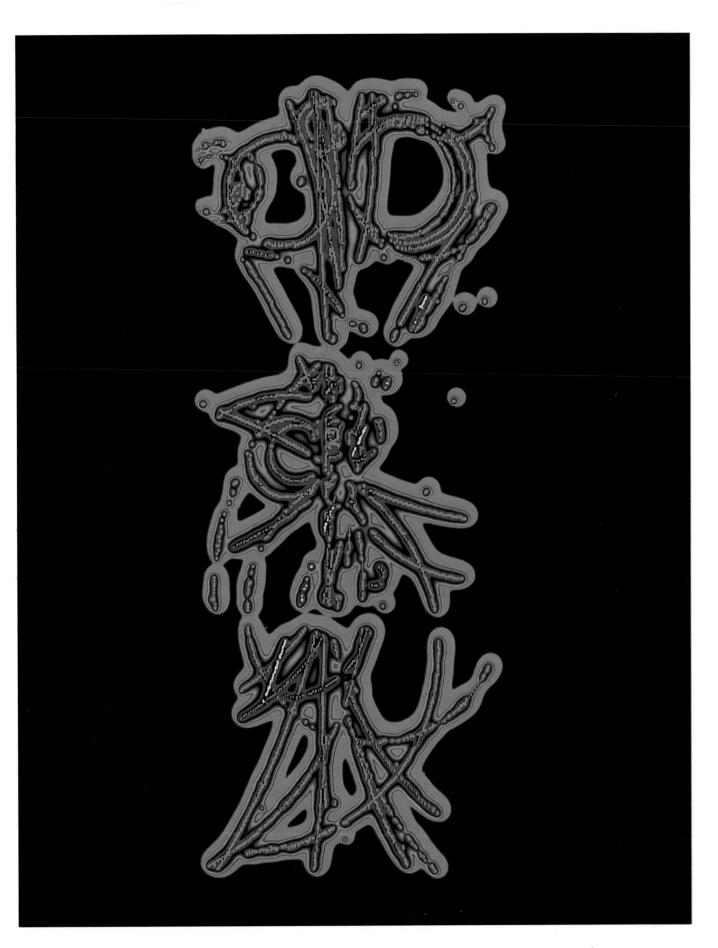

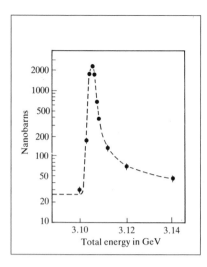

Fig. 9.2 *This dramatic hundredfold increase in the number of hadrons produced in electron–positron annihilations at a total energy of 3.1 GeV signalled the production and decay of the J/psi particle in the Mark I detector at SPEAR.*

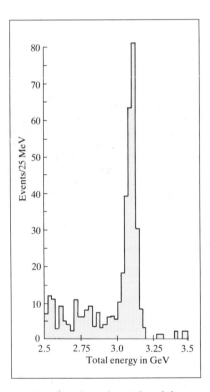

Fig. 9.3 *The spike in the number of electron–positron pairs produced in collisions of a proton beam with a beryllium target at Brookhaven also revealed the existence of the J/psi.*

CHARMONIUM

ON THE morning of Monday, 11 November 1974, members of the Program Advisory Committee at SLAC were assembling for one of their regular meetings. When one of the committee members, Sam Ting from the Brookhaven National Laboratory, met Burt Richter, a leading experimenter at SLAC, he said, 'Burt, I have some interesting physics to tell you about.' Richter responded immediately, 'Sam, I have some interesting physics to tell *you* about.' Neither realized that they had each discovered the same particle in entirely different experiments, nearly 5000 km apart. Richter's team had already called the particle 'psi'; Ting had opted for 'J', the Chinese character for Ting. To this day it is known as the J/psi, an unwieldy name for a particle that opened a new era in particle physics.

The J/psi revealed itself as a resonance. At SLAC's electron–positron storage ring, SPEAR, it produced a sharp spike in the number of charged particles emerging from the electron–positron annihilations (Fig. 9.2). In Ting's experiment at Brookhaven, it was responsible for a similar spike in the number of pairs of electrons and positrons produced in the collisions of high-energy protons with a beryllium target (Fig. 9.3). In both experiments the spike occurred at a total electron–positron energy of 3.1 GeV—the mass of the new particle, more than three times as heavy as the proton.

By 1974, resonances were nothing new; but the J/psi was remarkable because its spike was very narrow. In the quantum world of subatomic particles, the narrower a resonance, the longer its lifetime; and the width of the J/psi corresponded to a life of 10^{-20} s. This does not sound very long, but it was a thousand times longer than expected for a particle as heavy as the J/psi, which should have decayed very rapidly to lighter particles.

As if one new particle were not enough, at 3.20 a.m. on 21 November, 10 days after their first discovery, Richter's team at SPEAR found a second spike at a slightly higher energy, just below 3.7 GeV. Like the J/psi, the new particle—called the Ψ'' (pronounced 'psi-prime')—has a narrow width and therefore a relatively long lifetime. The discoveries of the Ψ'' and the J/psi stunned the physicists. It was as if anthropologists had stumbled on a tribe of people who lived to the age of 70 000 years. What could be prolonging the life of the new particles?

The most likely possibility was that the J/psi and the Ψ'' possessed some new property, which they could not easily divest themselves of, and which prohibited a rapid decay. In the months following these discoveries, an enormous number of papers were published offering explanations of the new particles, but one theory began to emerge head and shoulders above the others.

In 1970, theorists Sheldon Glashow, John Iliopoulos, and Luciano Maiani had been considering how to incorporate the behaviour of quarks into a single 'unified' theory of electromagnetic and weak forces. They discovered that the way was clear to realizing such a theory if a fourth type of quark existed, which they called 'charm'. The observation of the 'neutral currents' in neutrino experiments in 1973 (see p. 146) strengthened the idea of the unified theory; and with the discovery of the J/psi, the idea of a charmed quark came right to the forefront of theoretical wisdom. The properties of the J/psi and its heavier relative could easily be explained if they were each built from a charmed quark bound with its antiquark.

A charmed quark and a charmed antiquark make a particle that contains charm within it but which has no net charm overall; the charm carried by the quark and the 'anticharm' of the antiquark cancel out. A close analogy is with the 'exotic atom' positronium, which consists of an electron and its antiparticle, the positron. This contains electric charge within but has none overall. Moreover it is unstable, surviving only so long as the electron and positron do not touch and annihilate. In the case of a charmed quark and a charmed antiquark we have 'charmonium', which survives only so long as the quark and antiquark do not come too close.

The charmed quark and antiquark move around one another, as do the electron and proton in hydrogen or the electron and positron in positronium. And as in a hydrogen atom or in positronium, a variety of orbitals of different energies are possible. If the quark and antiquark orbit with high energy they form a relatively heavy particle, since mass is equivalent to energy. This heavier particle can emit energy as the quark and antiquark move to a state of lower energy, and so become a lighter particle. The emitted energy materializes as pions, muons, electrons, or photons. Once the quark and antiquark are in the lowest energy state possible they can no longer radiate particles. Instead, they mutually attract and annihilate one another, their energy rematerializing as lighter particles.

The J/psi is the lowest energy state of charmonium that is directly accessible in electron-positron interactions, and the Ψ'' is the second lowest. In Fig. 9.4, we see the results of the decay of a Ψ'' to a J/psi accompanied by the emission of two pions (one positive, one negative), which carry away the excess energy. This is followed almost immediately by the J/psi's decay, as the charmed quark and antiquark annihilate, their energy appearing as a positron and an electron. Appropriately, the particle tracks write out the Greek letter Ψ.

Charmonium is an inhabitant of a high-energy world populated by particles far heavier than those we have encountered so far. Its prolonged lifetime provided powerful confirmation of the idea that the strength of the strong force—the force between quarks—*diminishes* as energy increases. This leads to the exciting prospect that at ultra-high energies the enormous gap between the electromagnetic force and the strong force may disappear altogether. At such energies, the force between the quarks may be no stronger than the force between electrically-charged particles. This concept of the equivalence of the fundamental forces of nature at very high energies is central to the grand unified theories (GUTs) now at the forefront of particle physics.

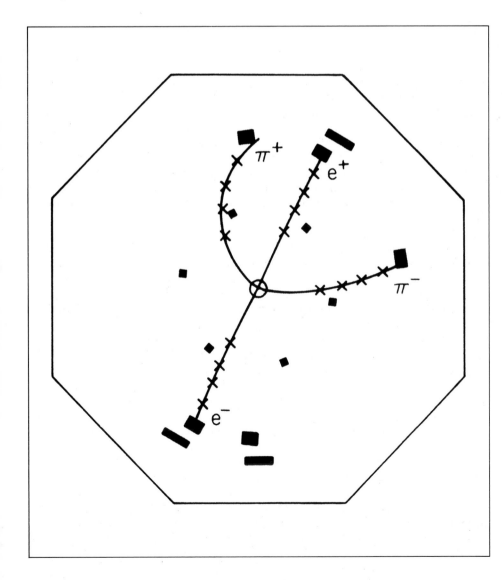

Fig. 9.4 *A psi-prime (Ψ'') particle writes its own Greek name in the Mark I detector as it decays to two pions (the curved tracks—π^+ and π^-) and a J/psi, which immediately decays to a positron (e^+) and an electron (e^-). The octagon outlines the basic shape of the detector, approximately 2 m from the centre. The crosses mark hits in four layers of concentric cylindrical spark chambers, and the dark bars indicate scintillation counters that have fired. The Ψ'' was formed in the annihilation of an electron and a positron, which came from in front of and behind the page to collide at the centre of the detector.*

CHARMED PARTICLES

UP, DOWN, and strange quarks combine to form either three-quark particles (baryons) such as the proton, neutron, and lambda, or quark–antiquark pairs (mesons) such as the pion and kaon. The charmed quark can also combine with any of these lighter quarks, obeying the same rules of attraction and repulsion, to produce charmed baryons and charmed mesons. Just as there is a world of strange matter, as physicists discovered in the early 1950s, so there must be a whole range of charmed matter, containing charmed quarks and antiquarks. After the discovery of the charmonium family, the search was on for this new world of charmed matter. But the charmed particles have shorter lives than their strange counterparts, and the search proved long and difficult.

Figure 9.5 illustrates the problems facing the charm hunters. It shows the aftermath of an electron–positron annihilation observed by the experiment called TASSO at the PETRA collider in Hamburg. Among the many tracks are three that come from the decay of a charmed meson, which are also shown in more detail. An electrically-charged D*(pronounced 'D-star'), consisting of a charmed quark and a down anti-quark, decays into a neutral D-zero, emitting a charged pion as it does so. This decay happens very quickly, and the pion (track 7) appears to come from the point of annihilation. The D-zero (a charmed quark and an up antiquark) takes more time—some 5×10^{-13} s—and travels a little less than a millimetre before it too decays, to a kaon (track 8) and a pion (track 6).

The point of annihilation (the 'vertex') and the subsequent production and decay of the D-zero all take place within PETRA's 13 cm diameter beam pipe, and are therefore invisible to TASSO's detectors. But the events in the immediate region of

Fig. 9.5 *The decay of a charmed particle is captured in the TASSO detector at DESY. The image shows only the hits recorded in the vertex detector—a cylindrical wire chamber that encircles the beam pipe and records charged particles for the first 7 cm of their paths beyond the pipe. (The blank circular region at the centre indicates the beam pipe.) A computer has fitted tracks to the hits, and extrapolated them back into the 'invisible' region within the beam pipe.*

Fig. 9.6 *When this region is enlarged, we can see that tracks 6 and 8 come from a 'V', a little less than a millimetre from the beam spot (indicated by the white bar). The 'V' is due to the decay of a charmed particle called the D-zero, which was itself produced, together with a pion (track 7), in the decay of a D* created in the electron–positron annihilation.*

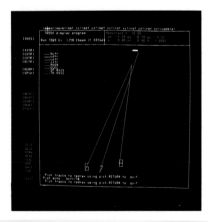

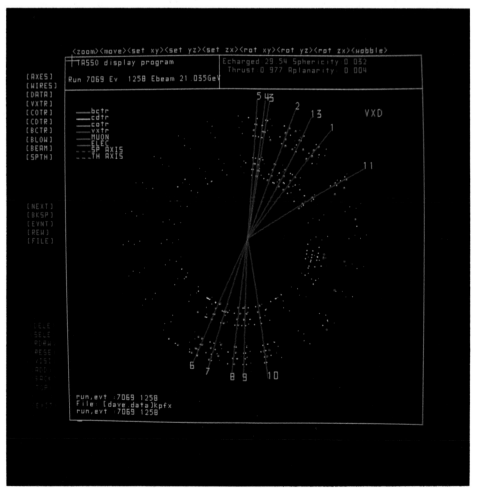

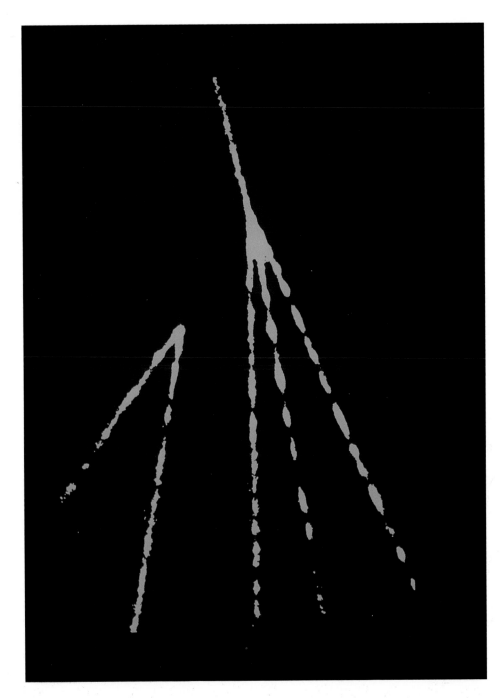

Fig. 9.7 *The 'footprint' of the associated production of charmed particles in the Hybrid Facility bubble chamber at SLAC. An invisible photon comes in at the top of this false-colour picture and collides with a proton in the liquid hydrogen in the bubble chamber, producing two charmed particles—one neutral and one charged. The neutral charmed particle, probably a D-zero, leaves no track but decays, forming the 'V' on the left of the picture. The charged charmed particle travels about 2 mm before it too decays, to three charged particles. In this bubble chamber, the bubbles are allowed to grow only to about 0.055 mm before they are photographed. This provides the high resolution needed to observe the tracks of the short-lived charmed particles (compare Fig. 9.9 overleaf).*

the vertex can be reconstructed, as in Fig. 9.6, by computer. In TASSO, a high-precision wire chamber maps the tracks of particles for the first 7 cm after they emerge from the beam pipe, and these tracks can then be accurately projected back to reveal what has happened around the vertex.

We saw in Chapter 5 that in particle collisions the strong force always creates strange particles in pairs—the phenomenon known as associated production. This is because the creation of a strange quark by the strong force is always balanced by the creation of a strange antiquark. Precisely the same rule applies to charmed particles: when the strong force produces a charmed quark, it must also produce a charmed antiquark. Such charm–anticharm pairs are readily formed in electron–positron annihilations. Sometimes the charmed quark and antiquark form charmonium; but if they are moving fast enough as they emerge they can escape from one another and associate instead with up, down, and strange quarks and antiquarks also created from the energy of the annihilation. In this way, the charmed quark and antiquark

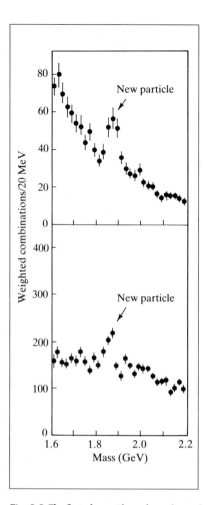

Fig. 9.8 *The first clear evidence for a charmed particle—the D-zero—came in these graphs of data from the Mark I detector. The top graph shows the number of times a positive (or negative) pion and a negative (or positive) kaon are produced together with a specific total energy. The bottom graph is similar but shows the production of a pion together with three kaons. Both graphs show a clear bump at a total energy of around 1.865 GeV, indicating that the measured clusters of particles come from the decay of a particle with a mass of 1.865 GeV— the D-zero.*

can form two charmed particles. These fly apart and can decay individually in a process analogous to the beta decay of a neutron, in which one type of quark transmutes into another variety. Thus the charm created by the strong force in the initial collisions leaks away slowly through the agency of the weak force, which is responsible for beta decay.

Theory indicates that the charmed quark prefers to decay to the strange quark, rather than the up or down varieties. So, strange particles such as kaons should bear witness to the brief existence and subsequent decay of a charmed particle. The first clear evidence for a charmed particle came in this way in 1975. Gerson Goldhaber from Berkeley had led a persistent attack on the data from electron–positron collisions at SPEAR. Eventually he was able to show that when he added up the energy and momenta of the various pions and kaons emerging as products of the annihilations, some combinations of pions and kaons were produced more frequently with specific energies. These energies corresponded to the decays of a neutral charmed meson with a mass of 1.865 GeV (Fig. 9.8). This was the first observation of the D-zero, and it confirmed that charmed quarks prefer to decay weakly to strange quarks.

The discovery of a charmed particle, through a spike in an energy distribution, is reminiscent of the way in which we observe the J/psi and other short-lived resonances. But the lifetimes of the charmed particles—some 10^{-13} s—are much longer than the J/psi's (10^{-20} s) and are close to the borderline of what could be detected directly in the bubble chambers of the mid-1970s. With the advent of high-resolution bubble chambers it is now possible to see charmed particles directly by their trails. This confirms that identifying particles by the energy-momentum plot is as real as seeing their trails directly.

An example is shown in Fig. 9.7. In this case a photon provides the energy to create a pair of charmed particles as it hits a proton in a bubble chamber at SLAC. Two charmed mesons are produced, one charged and one neutral. The charged meson leaves a visible track and then decays into three charged particles whose trails are also seen. The neutral charmed meson decays into two charged particles, one negative and one positive, whose tracks form a 'vee'. The beauty of this picture is that we can measure the distances that the charmed particles travel before they decay, which provides useful information on the lifetimes.

Over the past decade particle physicists have accumulated details of a set of charmed particles, both mesons and baryons, all of which are roughly twice as heavy as the proton. Figure 9.9 shows one of the first images consistent with the decay of a charmed sigma particle—a charmed baryon—taken in the '7 foot' bubble chamber at Brookhaven in 1975. In this case, we cannot discern the track of the charmed particle; we can infer its existence only from the information contained in the tracks of its decay products. An invisible neutrino enters the bubble chamber and interacts with a proton, producing a muon and a 'charmed sigma'—a particle resembling the sigma particle but with a charmed quark in place of the strange quark. The muon moves swiftly from the scene, leaving a long, almost straight track; the charmed sigma decays before it can leave a discernible track.

The discovery of charmed particles in the 1970s showed that nature exhibits a symmetry between quarks and leptons. The electron and its neutrino are matched by the up and down quarks; the muon and its neutrino are matched by the quarks bearing strangeness and charm. Our everyday world comprises matter whose nuclei contain up and down quarks. Moreover the formation of this matter, in nuclear reactions at the heart of stars and in the explosions of supernovae, relies upon radioactive transmutations between up and down quarks. The beta decay of a neutron to a proton, in which a down quark changes to an up quark while emitting an electron and an antineutrino, is a classic example of such a transmutation.

We can, however, imagine another, 'Mark 2' universe built from strange and charmed quarks, in which the radioactive transmutations mimic those between up and down quarks. Such strange and charmed matter probably existed fleetingly after the Big Bang, alongside up and down matter. Today we are left with only glimpses of strangeness and charm; the precise significance of these qualities remains to be understood.

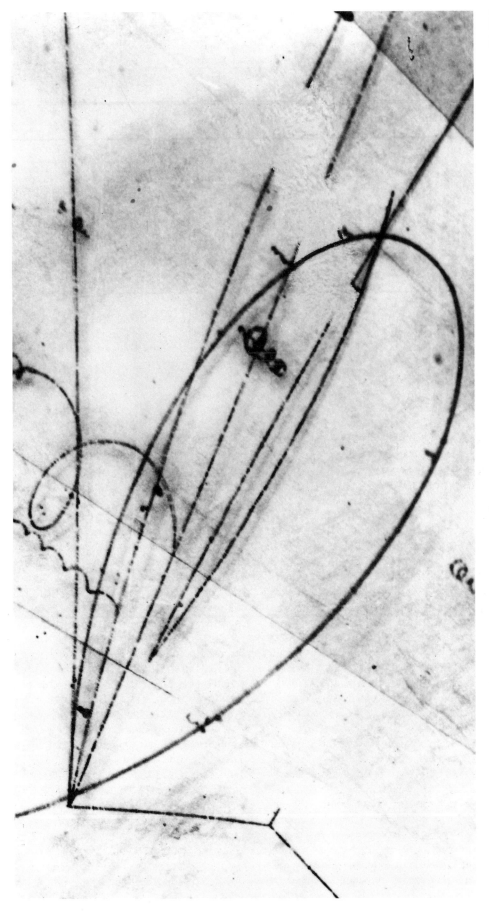

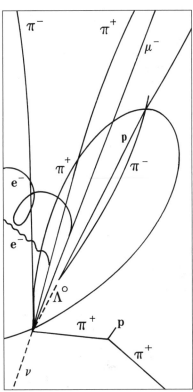

Fig. 9.9 *One of the first examples of an event consistent with the production and decay of a charmed baryon (three-quark particle), photographed in the '7 foot' (2.1 m) bubble chamber at Brookhaven in 1974. A neutrino (ν) enters the picture from below and collides with a proton in the chamber's liquid. The collision produces five charged particles—a negative muon (μ⁻), three positive pions (π⁺), and a negative pion (π⁻)—and a neutral lambda (Λ). (Note how the muon and one of the pions knock electrons (e⁻) out of the liquid, which spiral round in the chamber's magnetic field.) The lambda produces a characteristic 'V' when it decays to a proton (p) and a pi-minus (π⁻). The momenta and angles of the tracks together imply that the lambda and the four pions produced with it have come from the decay of a charmed sigma particle, with a mass of around 2.4 GeV. But the decay happened too quickly—within 10^{-12} s—for the original charmed particle to leave an observable track in this chamber (compare Fig. 9.7).*

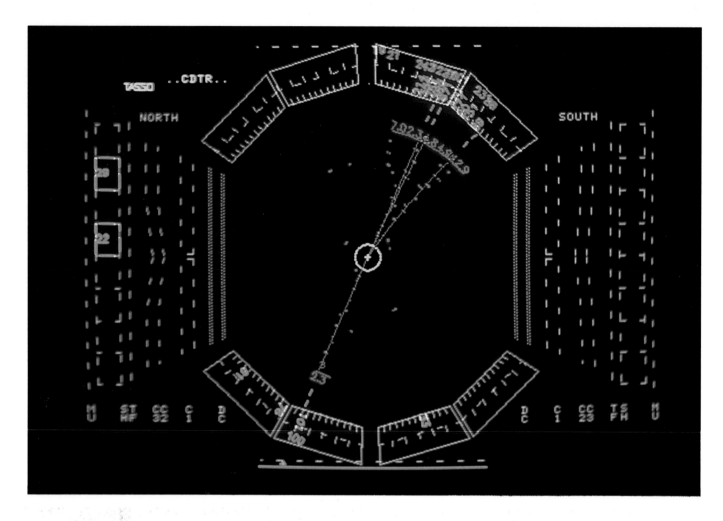

Fig. 9.10 *The distinctive '3 + 1' decay of a tau-plus and a tau-minus in the TASSO detector at DESY. An electron and positron have annihilated at the heart of the detector at the point marked by the yellow cross. A tau-plus and a tau-minus materialize from the annihilation and shoot off in opposite directions, but they soon decay, while still within the beam pipe (the yellow circle). The tau-plus decays to two invisible neutrinos and a positive muon, which travels towards the bottom left. The muon can be identified because it penetrates the argon-lead shower counters (the purple blocks) to score a hit in one of the muon chambers (the blue line indicates the extent of the chamber, and the cross shows the location of the hit). The tau-minus, on the other hand, has decayed to an invisible neutrino and three charged pions, which travel up towards the top right. The pions and the muon all leave tracks in the drift chamber which fills most of the space between the beam pipe and the shower counters. Hits in the drift chamber wires are marked by blue bars, and the tracks of the particles, as calculated by the computer, are shown in red. The drift chamber extends to about 1.3 m from the centre of the detector.*

THE TAU

PARTICLE PHYSICISTS were still congratulating themselves on having discovered the symmetry between four types of quark and four leptons when they were thrown into disarray. The same experiment that had caught some of the first glimpses of charm unearthed an unexpected guest at the feast: the tau. The tau is an electrically charged lepton, a much heavier version of the electron and muon. It weighs about twice as much as a proton, 20 times as much as the muon and a staggering 4000 times as much as the electron. The law of nature that determines this bizarre numerology is one of the major puzzles in particle physics today.

Like the electron and muon, the tau has negative charge and exists in an antimatter version with positive charge. It is not affected by the strong force, but it does take part in electromagnetic and weak interactions. And just as the electron and muon are partnered by their distinctive neutrinos, we believe that the tau is partnered by a third variety, the tau-neutrino, bringing the total number of leptons to six.

When an electron and a positron annihilate in a head-on collision, they can rematerialize as new forms of matter, provided the total energy is high enough to create the appropriate antiparticle along with a new particle. If the total energy is above about 3.6 GeV, a tau and an antitau can emerge back to back. Four times in every hundred the negative tau decays into an electron and two neutrinos while the positive antitau decays into a positive muon and two neutrinos. Alternatively, the tau and antitau can produce a negative muon and a positron together with unseen neutrinos. These are very distinctive reactions, because an electron and a positron annihilate, but an electron and a muon (or a positron and a muon) emerge from the annihilation. It was events such as these that gave the first hints of the tau to Martin Perl and his colleagues working on the Mark I detector at the SPEAR electron–

positron collider in 1974, though it was 1975 before the researchers were sure of what they were seeing.

Today, at electron–positron colliders that reach far higher energies than the 8 GeV of SPEAR, taus can be distinguished even when they decay in other ways. Unlike the muon, the tau can also decay to particles containing quarks, such as pions and kaons. This is because the tau is so heavy—heavy enough to produce the quark and antiquark of a pion or kaon as well as an energetic neutrino to carry the 'lepton-ness' of the original tau. In low-energy electron-positron annihilations, the pions and kaons resulting from tau decays are difficult to disentangle from other products of the annihilations. At higher energies, on the other hand, annihilations are typically characterized by jets of many particles, which contrast strongly with the relatively few particles created by tau decays.

A particularly distinctive signature is the so-called '3 + 1 decay', as shown in Fig. 9.10. Here a positive and a negative tau have been produced together. One of the taus—the negative one, say—decays into three charged particles, which leave tracks in one direction; these are balanced in the opposite direction by the track of a lone charged particle emerging from the other tau.

The lifetime of the tau is an important quantity. Theory predicts that if the tau is a heavier version of the electron and the muon, then the lifetimes of the muon and tau are precisely related. Accordingly, the tau's lifetime should be 3×10^{-13} s, which is indeed the value that experiments find.

This lifetime is an average; in fact, it is the 'half-life', which tells you for how long half of a large sample of taus will live. Individual particles may live much less or much more than this. The occasional tau that lives significantly longer can reveal its flight in precision detectors. Figure 9.11 shows an example where two taus have decayed in the TASSO detector at the PETRA collider in Hamburg. The taus decayed before reaching the first cylinder of detectors but the computer can extrapolate the tracks of their decay products back to the points of decay, not far from the point where the taus were born in an electron–positron annihilation.

Fig. 9.11 *A '3 + 3' decay of two tau particles within the beam pipe at the heart of the TASSO detector. An electron and a positron have annihilated and produced a tau-plus and a tau-minus, each of which decays almost immediately into three charged pions and a neutrino. The pions leave tracks in the vertex chamber surrounding the beam pipe (see Fig. 9.5). The computer draws lines through the hits in the chamber and reveals that the pion tracks emanate from two separate points, close to the beam spot. The small gap is the total distance travelled by the tau and antitau, which were moving out in opposite directions from a common origin, before they decayed. The dotted circle marks the outline of the beam pipe, which is 13 cm in diameter.*

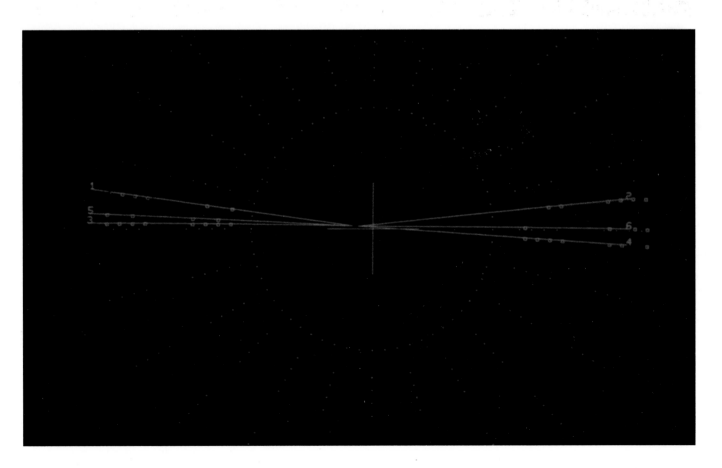

Fig. 9.12 *Leon Lederman (b.1922), director of Fermilab since 1979.*

BOTTOM PARTICLES

'CONGRATULATIONS LEON. You've done it this time!' is how one physicist greeted Leon Lederman after the discovery of the particle now known as the upsilon. Today Lederman is director of Fermilab, near Chicago, where on 30 June 1977 his team made its historic announcement. The upsilon particle provided the first indication of a fifth variety of quark. Although sometimes known as 'beauty'—after charm, beauty!—it is most often called, rather prosaically, 'bottom', in analogy with 'down'. The bottom quark binds with its antiquark to form 'bottomonium', and Lederman's upsilon, with a mass of 9.46 GeV, is the state with the lowest energy.

The analogies with the J/psi and other members of the charmonium family containing charm and anticharm are strong. With the bottom quark, nature has repeated the pattern of particles, but at still greater masses; the bottom quark is some three times as heavy as the charmed quark. In bottomonium, the heavy quark and antiquark orbit each other more closely than in charmonium, forming a particle 1/10 to 1/5 the size of a proton. Together, charmonium and bottomonium provide a fascinating 'laboratory' for physicists to study the behaviour of quarks in close proximity to each other, and to learn how the interquark force behaves as distances decrease.

Charmonium had been known for less than three years when Lederman made his discovery. Theorists were just beginning to believe that there is a symmetry between the quarks and leptons, each type of particle occurring two at a time. In the everyday world up and down quarks are matched by the electron and its neutrino; at higher energies the strange and charmed quarks are matched by the muon and its neutrino.

Then came the tau lepton. The similarity of the tau to earlier 'generations' of leptons naturally suggested that there is a third generation of quarks. So confident were many physicists in the symmetry between quarks and leptons that they took the discovery of the tau as a prophecy of further quarks whose discovery would restore the symmetry. Theorists predicted a bottom (or beauty) quark with charge −1/3 and a heavier top (or truth) quark with charge +2/3. Like the up–down and charm–strange pairs, this third generation of quarks is related by beta radioactivity; the top quark decays into the lighter bottom quark, emitting a positron to balance the difference in charge, and a neutrino. Lederman's discovery of the upsilon was a clear sign that the theorists were on the right track.

Lederman had performed a similar experiment to the one that had produced the J/psi for Ting. (Indeed, some years earlier in an experiment at Brookhaven, Lederman's group narrowly missed discovering the J/psi.) Ting's team detected electron–positron pairs produced when a high-energy proton beam smashed into a beryllium target. Lederman and his colleagues chose instead to detect pairs of negative and positive muons created in a similar way by Fermilab's high-energy protons. They used beryllium to absorb most of the charged particles from the collisions; muons, however, being more penetrating, would survive. Once they emerged from the beryllium, the muons passed through magnets, which directed positive and negative particles along two separate 'arms' of apparatus, each 2 m wide, 2 m high, and some 35 m long. In each arm, a series of detectors measured the energies and angles of the muons. From this the researchers could calculate the energy associated with the muon pairs. They were rewarded with the discovery of peaks at particular values, just as if the muons were the decay products of new particles. The team found first two, and later three, members of the bottomonium family

A year later, in 1978, the DORIS electron–positron collider at Hamburg began operations with enough energy to produce bottomonium, followed in 1979 by the higher-energy Cornell Electron Storage Ring or CESR. Today, details of several members of the family are well known.

In the more energetic excited states of bottomonium, the bottom quark and antiquark can give up energy by emitting photons and so return to the ground state— the upsilon. Such a decay has occurred in Fig. 9.13, captured in the CLEO detector at the collider at Cornell. An excited state of bottomonium emits a photon of energy 128 MeV and temporarily forms a less energetic state. The photon converts into an electron and a positron, which leave characteristic curly tracks in the detector, bending left

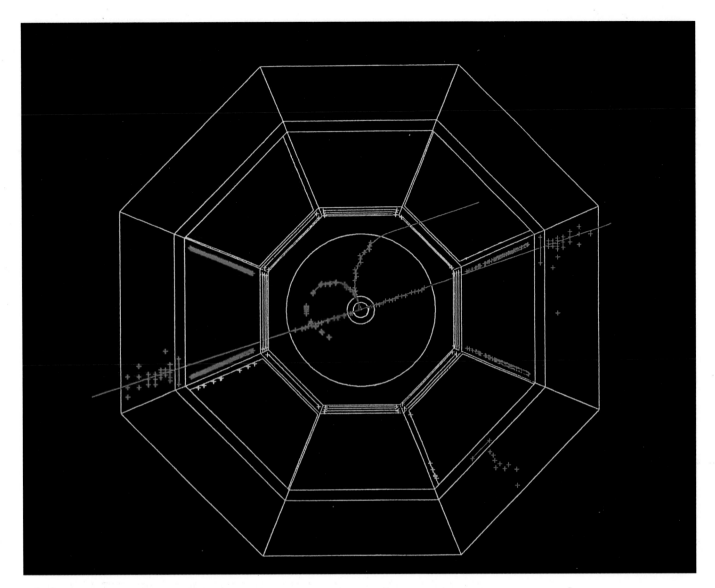

and right in the magnetic field. The new bottomonium state then emits a second photon, which deposits measurable energy, and the bottomonium system falls to the lowest possible level: the upsilon. Finally the upsilon decays into an electron and a positron—a relatively rare occurrence.

As in the case of charm, where there exist charmed particles as well as charmonium, so there exist bottom particles as well as bottomonium. Bottom particles consist of a single bottom quark or antiquark accompanied by quarks or antiquarks of the other varieties. The bottom quark has an electrical charge of $-1/3$, which is the same as the strange quark. Thus the varieties of bottom particles should mirror the strange particles, though the bottom particles are five to ten times heavier than their strange analogues. For example, the negative kaon, built from a strange quark and an up antiquark, weighs about 0.5 GeV; its bottom analogue, known as the B-minus, consists of a bottom quark and an up antiquark, and weighs in at about 5 GeV.

The heavy bottom quarks soon decay to lighter charmed quarks, which in their turn also decay rapidly, so the bottom particles quickly disintegrate into more familiar particles. The lifetime of the bottom particles is typically some 10^{-12} s. This is ten times longer than the charmed particles, yet studying the bottom particles has still proved to be a tedious process because the particles are difficult to identify.

The basic problem is the large mass of the bottom particles. Because they are much heavier than even the charmed particles, they can decay in many ways and to many particles. Experimenters look for clues such as the fast lepton released when a bottom

Fig. 9.13 *A computer display of an electron-positron annihilation in the CLEO detector at Cornell shows the decay of an excited upsilon—a particle in which a bottom quark is bound together with its antiquark. The electron and positron have annihilated within the beam pipe—the innermost circle at the centre. The excited upsilon decays into another excited state, of lower energy, emitting an invisible photon as it does so; this new excited state then also decays, this time to the ground-state upsilon (with lowest energy) by emitting a second photon. The first photon converts to an electron (green) and a positron (red) before it enters the drift chamber, indicated by the large circle. The second photon is detected at the bottom right only when it converts (blue crosses) in the lead of the shower counter—the outermost octagonal layer. The ground-state upsilon also decays to an electron and a positron. These are of higher energy than the first pair and shoot straight off as far as the outer layer, where they produce showers. The total width of the apparatus on this display is a little more than 6 m.*

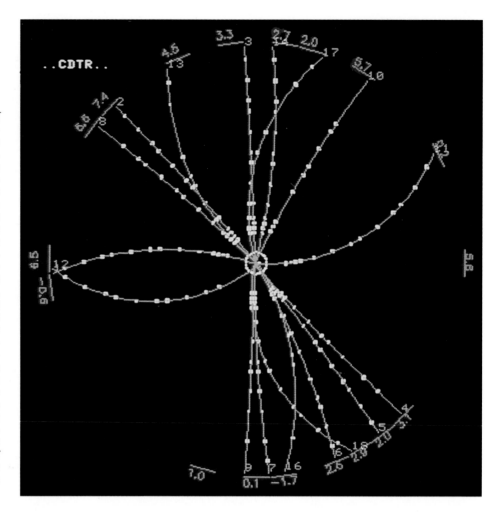

Fig. 9.14 *A pattern of tracks characteristic of the decay of a B meson, formed in the annihilation of an electron and a positron, appears in the main drift chamber of the TASSO detector. Yellow dots mark drift chamber wires that have been hit, and the red lines show the tracks calculated by computer. The blue bars and numbers refer to 'time-of-flight' counters. These are scintillation counters which send signals to stop an electronic 'stopwatch' that is started when the electrons and positrons approach along the beam pipe; this information helps in identifying particles, because faster particles give earlier signals, and shorter times. Three clues suggest that a B meson could have been produced. First, one of the tracks, going straight up almost at 12 o'clock (track 3), is identified as a muon because it scores a hit in a muon detector (not shown); a muon or an electron should be emitted when the bottom quark in the B meson decays to a lighter quark. Secondly, when the computer reconstructs the point from which some of the tracks emanate, it is slightly displaced from the annihilation point. This shows that something with a very short lifetime has travelled out from the annihilation and then decayed. Thirdly, the tracks are spread around the detector, in contrast to the distinct pattern of two jets that often occurs. (This is the same event as in Fig. 1.13 on p. 18, which gives a more complete display of the detector.)*

quark decays to a charmed quark, or the production of charmed particles. Figure 9.14 shows the progeny of charged particles that are likely to have come from the decay of a B-meson produced by an electron–positron annihilation in the TASSO detector at DESY.

Decays of bottom particles identified individually are rare. More generally, experimenters have learned about bottom particles by analysing a number of events likely to contain bottom particles and plotting graphs of angles and energies of the decay products. Clusters of events at specific masses reveal indirectly the existence of the bottom particles.

As of Summer 1986, only one experiment, at CERN, has succeeded in recording the tracks of bottom particles before they decayed, and so claimed to be the first to see 'naked bottom'! In the experiment, code-named WA75, a beam of pions was directed into a stack of emulsion. Information from electronic detectors located beyond the stack was used to pinpoint the area within the emulsion where each interaction occurred. By looking for appropriate clues in the emerging particles, the researchers were able to find one event in which a negative and a neutral B-meson had been produced together, and where the 'vees' associated with the decays of the two bottom particles were visible.

Bottom particles are not yielding their secrets easily, yet they hold valuable information about how the third generation of matter transforms to second-generation matter—particles containing charmed and strange quarks. The properties of the bottom particles, in particular their precise lifetimes, provide stringent tests of modern theories of the decays of quarks and also of the weak force, which is responsible for the decays. In this way measurements of bottom particles test the electroweak theory, which unites the weak force with the electromagnetic force, and help to pave the way to an understanding of the different generations of quarks and leptons.

GLUONS

THE MOST powerful force we know of in the Universe—the strong force—binds together the quarks from which protons and neutrons and all the other hadrons are made. The inter-quark force is so strong that it is apparently impossible to prise a single 'naked' quark out of a hadron. It is as if quarks are stuck together by a kind of superglue. Elucidating the nature of this glue was one of the major achievements of particle physics in the 1970s.

Quantum theory implies that all the fundamental forces of nature are transmitted by carrier particles, which physicists call gauge bosons. In the case of the electromagnetic force, the carrier particle is the photon. In the case of the strong force, the carriers are the *gluons*, and according to theory, there are eight varieties of them. The gluons are massless bundles of strong radiation just as the photon is a massless bundle of electromagnetic radiation. But whereas photons are free to travel indefinitely through space, gluons appear to be free only within the narrow confines of a 'femtouniverse'—a region some 10^{-15} m, or 1 femtometre, in radius. This is the typical size of a particle such as a proton or a pion. Gluons are confined within hadrons much as quarks are; and like the quarks, they can advertize their presence indirectly by generating jets of particles in energetic collisions.

One example is when an electron and a positron annihilate and create a quark and an antiquark. If the energy is high enough the quark and antiquark fly apart from each other and shower into two jets of hadrons, such as pions and kaons. This violent separation may shake loose one or more gluons. If the gluon comes off with enough energy, it will produce its own jet of particles distinct from those created by the quark and antiquark. In electron–positron collisions at PEP and PETRA, 'three-jet events', due to a quark, antiquark, and gluon, occur about a tenth as often as the two jets from a quark and antiquark; and four jets, produced when two fast gluons are emitted, occur approximately once for every hundred two-jet events.

Figure 9.15 shows a three-jet event, recorded in the HRS detector at the PEP collider at Stanford. Here the gluon is emitted at a large angle relative to the quark and antiquark, and the three jets produce a pattern like the 'star' of the Mercedes

Fig. 9.15 *A typical 'three-jet' event—an example of the signature that provided the first conclusive evidence for gluons. Such events were found first in experiments at the PETRA collider in Hamburg. This computer display of the aftermath of an electron–positron annihilation shows the tracks of charged particles in the central drift chamber of the High Resolution Spectrometer at the PEP collider at SLAC. The coloured dots mark hits in the 2 m diameter drift chamber, and a computer has drawn in tracks that fit the hits. The electron and positron have annihilated at the centre of the detector to produce a quark and an antiquark. As they race off sideways, either the quark or the antiquark has radiated a gluon. However, before the quark, antiquark, and gluon proceed beyond 10^{-15} m or so, they materialize into particles, producing the three jets.*

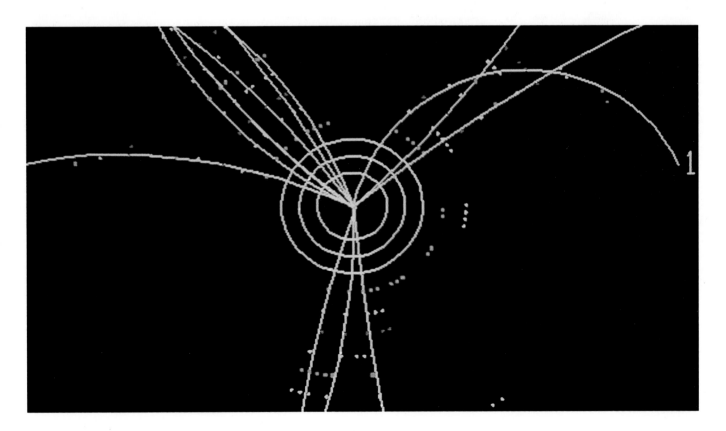

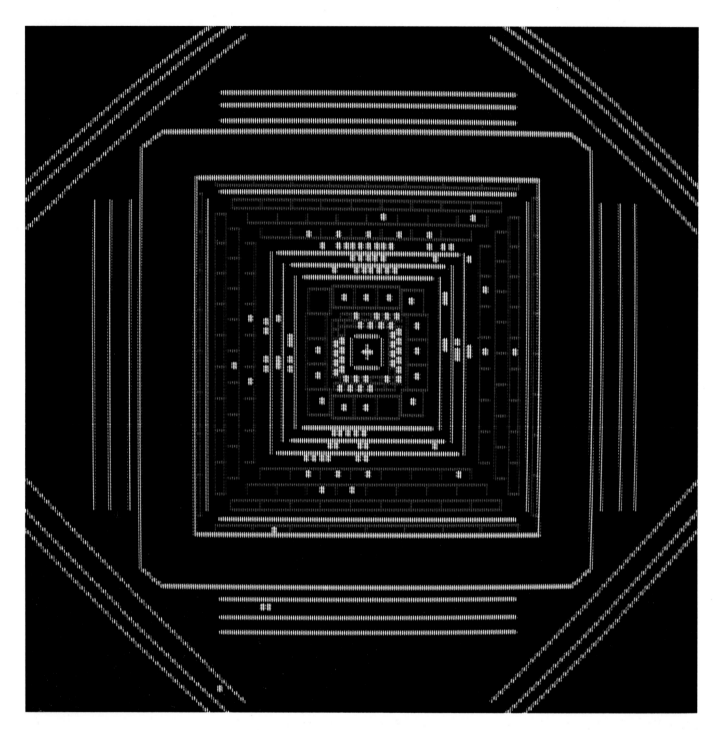

Fig. 9.16 *This computer display from the Mark-J detector at the PETRA electron–positron collider shows four jets of charged particles bursting out sideways from the central collision region. Such events are consistent with the radiation of two gluons by the quark and antiquark formed from the energy of the annihilating electron and positron. The yellow squares mark the iron of the magnet and are about 2 m from the centre of the detector.*

symbol. Figure 9.16 shows the rarer occurrence where two fast gluons are radiated to produce a total of four jets in the detector—this time the Mark J apparatus at the PETRA collider in Hamburg. Such clean examples of three- and four-jet events are relatively rare; often the gluon is emitted closer to the path of the quark (or antiquark) and the jets merge into each other. It is through a statistical analysis of directions of energy flow and the relative orientations of the jets that the physicists find their best evidence for the radiation of gluons. This evidence helps to confirm the theories of quark behaviour and the strong force that physicists have developed since the early 1960s.

Quarks carry electrical charge and feel the electromagnetic force; but they also carry a further form of charge known as *colour*. The successful theory that evolved from this idea is known as quantum chromodynamics (QCD) and is closely modelled on the theory of quantum electrodynamics (QED) developed in the 1940s to explain how the electromagnetic force applies to subatomic particles.

Colour is considered to be a new kind of charge, analogous to electric charge. Electric charge can be positive or negative and so can each type of colour charge. If we choose to call the colour charges carried by quarks 'positive', then antiquarks have negative colour charge. Identical electric charges repel each other, and opposite charges attract and neutralize each other. The same is true of colour charge, and this is why quarks attract antiquarks to form mesons—particles such as the pions, kaons, J/psi, and upsilon.

The great difference between electric and colour charge is that while the former comes in only one type, which can be positive or negative, colour charge effectively comes in three types, each of which can be positive or negative. The three types of colour charge are called red, blue, and yellow (or green), by analogy with the three primary colours. Not only do positive and negative colour charges attract each other, so do the different colours themselves: unlike colours attract, while like colours repel. Red attracts blue and yellow, for instance, but repels red.

The result is that three quarks of different colour can form the particles we call baryons—for example, the proton, the neutron, and the omega-minus. They may consist of any combination of the six available quark 'flavours'—such as the two ups and one down of the proton, or the three strange quarks of the omega-minus—but they always consist of three differently-coloured quarks. Indeed, only those particles exist in which the overall colour charge is neutral—or 'white'. In baryons, the three primary colours combine to form a white particle; in mesons, on the other hand, the positive colour of a quark neutralizes the negative colour of an antiquark to give a white particle. This is the reason why we do not observe particles that consist of two quarks, say, or two antiquarks.

If colour is hidden within baryons and mesons, how have we discovered its existence? The clue lies in particles such as the omega-minus, which at first sight appears to contain three identical quarks. According to a basic rule of quantum theory—Pauli's Exclusion Principle—a particle should not be able to contain more than one quark in a given quantum state. To overcome this paradox, American theorist Oscar Greenberg proposed in 1964 that quarks not only come in different 'flavours'—up, down, strange, and so on—but also in three different 'colours'. If the three strange quarks forming the omega-minus each have a different colour, then they are no longer identical and Pauli's principle is not violated.

Back in the 1940s and 1950s, theorists thought that pions were the transmitters of the strong force. But experiments later showed that pions and other hadrons are composite particles, built from quarks, and the theory of the strong force had to be revised completely. We now believe that it is the colour *within* the proton and neutron that attracts them to each other to build nuclei. This process may have similarities to the way that electrical charge *within* atoms manages to build up complex molecules. Just as electrons are exchanged between atoms bound within a molecule, so are quarks and antiquarks—in clusters we call 'pions'—exchanged between the protons and neutrons in a nucleus.

Gluons participate in this colour world as 'carrier' particles. They transmit the colour force between one quark and another, in the way that the photon transmits the electromagnetic force between electrical particles. But there is one crucial difference between the photon and the gluons. The photon is not electrically charged itself, it is neutral; and because it is neutral it does not interact with other photons. Gluons, on the other hand, *are* colour charged and they therefore interact strongly with each other as well as with the quarks.

The neutrality of photons enables them to transmit the electromagnetic force throughout space, the power of the force decreasing with distance. The gluons, by contrast, tend to pull on each other because they are colour charged. This appears to be why the quarks and gluons are confined to their femtouniverse. No one, however, has yet proven from first principles that quarks are always confined. It remains possible that they might be released at extremely high energies—energies beyond those presently available in accelerators.

THE W AND Z PARTICLES

IN 1979, three theorists shared the Nobel prize for physics. Sheldon Glashow, Abdus Salam, and Steven Weinberg had all made important contributions to the theory that unites the electromagnetic force with the weak force—the 'electroweak' theory. The award of the prize to these three came as no real surprise, for there was plenty of evidence to support the theory. Yet there was one sense in which the Nobel committee was gambling on the results of experiments that were still in the future.

The electroweak theory demands the existence of four gauge bosons, or force-carrying particles. One of these is the familiar photon, which mediates electromagnetic effects. The other three are responsible for weak interactions: they are the two W particles, one positively charged and one negatively charged, and the neutral Z. In 1979 there was no direct experimental evidence for the Ws and the Z. Particle physicists had to wait another four years before CERN triumphantly announced the discovery, first of the W particles, in January 1983, and then of the Z particle (Figs. 9.17 and 9.18), in the following May.

Physicists had suspected since the 1930s that particles carry the weak force in the way that the photon carries the electromagnetic force. These particles, called W for 'weak', trigger the beta decay of the neutron and other particles. But the Ws had to differ from the photon in one important way: they had to be heavy because the weak force is limited in its range—it is not felt beyond distances greater than 10^{-15} m.

This last assertion, about the mass of the W particles, is a consequence of the 'fine print' in nature's book of rules. In our everyday macroworld, energy is conserved. But in the quantum microworld of subatomic particles, the balance of energy need not be exactly maintained, provided any imbalance occurs over a very short time and hence over a very short distance—so short that we are oblivious to it with our macroscopic senses. (We know now that the strong force is also confined to a 'femtouniverse' of less than 10^{-15} m. In this case the range of the force is restricted not because the carriers—the gluons—are heavy, but because they are coloured and interact strongly between themselves.)

When a particle such as a neutron decays by the weak force, in a sense it withdraws on its energy account for just long enough to make a W particle. We know that the energy account is overdrawn for only a very short time, because the force is very short-ranged, so we can infer that the amount borrowed is large—in other words, the W particles are heavy. How much does a W particle weigh? The answer came from the electroweak theory, which once fully developed predicted a mass of 83 GeV, or nearly 90 times heavier than a neutron.

We can never capture an 'overdrawn' W particle, of the kind that triggers the decay of a neutron. This inability is not due to a lack of technology; rather it is forbidden by a principle of nature. The W in this case is like the Cheshire cat, and we can detect only its grin—the electron and the antineutrino that the W leaves behind. To succeed in producing a W in the laboratory requires an input of the full 83 GeV of energy from elsewhere, so that the energy account has enough credit to pay for the W's mass. In the 1970s, physicists began to think of ways of providing this much energy to create a W. They were encouraged by another prediction of electroweak theory—the existence of a neutral boson, the Z, a partner to the two charged Ws.

The theory predicts that the Z has a mass of 93 GeV, slightly more than the W. The Z mediates 'neutral' weak interactions, in which no electric charge is transferred between the participating particles. In the 1960s, when the electroweak theory first exposed the need for the Z, there was no evidence for such interactions. But all changed in 1973, when a bubble chamber in a neutrino beam at CERN revealed 'neutral current' events that could be explained in terms of the Z (see p. 146).

These observations strengthened confidence in the electroweak theory and helped Glashow, Salam, and Weinberg to win the Nobel prize. Moreover, a theory of the strong force that was also based on the notion of gauge bosons—this time the gluons described in the last section—was gathering support. An important corner in the jigsaw of particle physics seemed almost complete. But one vital piece was missing: no one had yet shaken W and Z particles loose in high-energy collisions.

The challenge of producing Ws and Zs became a kind of Holy Grail to many particle physicists. The particles would live for less than 10^{-24} s before the weak force destroyed them and they decayed to lighter particles. Thus they could be observed only through their decays. However, the electroweak theory indicated precisely into what particles the Ws and Z would decay and how frequently each kind of decay would occur.

So the stage was set in the late 1970s for the proposal to convert CERN's large proton synchrotron, the SPS, into a proton–antiproton collider that would reach energies high enough to yield a few W and Z particles—provided, of course, the electroweak theory was indeed right. International teams of physicists began to build two large detectors, UA1 and UA2, to explore the proton–antiproton collisions.

The Ws and Zs would be rare, but fortunately they should have distinctive decay patterns. The neutral Z can decay to two leptons of the same kind but opposite charge; decays to an electron and a positron or a positive and a negative muon would prove particularly noticeable. Similarly, the charged W particles can decay to a charged lepton and a neutrino. In this case, a single energetic electron or muon would signal a W's decay, while the apparatus would 'detect' the neutrino by showing that energy has been taken away by an unseen participant.

In January 1983, the UA1 and UA2 teams triumphantly announced success: they had seen W particles decay. News of the rarer Z particles—crucial to the electroweak theory—came a few months later. Electroweak theory was in good shape after all. Not only did the particles have the masses prescribed by the theory, they also appeared in the proton–antiproton collisions at the rate the theory predicted.

When we collide protons with antiprotons, we are in fact colliding a cluster of three quarks (and attendant gluons) with a similar cluster of three antiquarks. A W or Z particle may be formed in the annihilation of one of the quarks with one of the antiquarks. Thus the attendant maelstrom from the other quarks and antiquarks can mask the vital 'signatures' of the W and Z particles as they decay. But modern electronics and computers provide powerful techniques that allow the physicists to find the needles in the haystack, and make the Ws and Zs clear for all to see.

Figure 9.17 shows an example from the UA1 experiment. We do not see the colliding proton and antiproton, which have entered from the left and right, only the debris that they create. Most of the debris consists of pions and kaons, which bend in the magnetic field surrounding the collision zone. The computer has removed these low-energy tracks from the display, leaving only a few high-energy ones. Two of these

Fig. 9.17 (left) A Z particle decays into an electron and a positron in this computer display from the UA1 detector at CERN. A proton and an antiproton have come in along the axis of the cylindrical Central Detector (outlined in red) and collided head on. The Central Detector records the tracks of the charged particles produced, which may also register in other parts of the detector. In this case, the computer has removed all low-momentum tracks from the display, leaving only a few medium-momentum tracks moving off to the left (green) and two high-momentum tracks. Each of the high-momentum tracks fires a cell in the Electromagnetic Calorimeter (the cylindrical layer around the Central Detector; the cells hit are indicated by the white bars). This identifies these particles as electrons or positrons; a slight bending in the magnetic field reveals that the white track is due to a positron, the pink track to an electron. Measurements of the energy deposited in the Electromagnetic Calorimeter allow the physicists to calculate that the total energy of the electron and positron is close to 93 GeV, the predicted mass of the Z particle.

Fig. 9.18 (right) In this display from UA1, a Z particle formed in the annihilation of a proton with an antiproton has decayed into two muons. The image shows all the tracks measured in the Central Detector, including the low-momentum ones. The computer has matched two of these tracks to hits in the muon chambers—indicated by the blue slashes—which lie outside the iron of the magnet (the red rectangular outline). Muons are the only charged particles that can penetrate as far as this outer layer. Measurements of the momentum of the muons, from their slight curvature in the magnetic field, shows that together they add up to the expected mass of the Z particle.

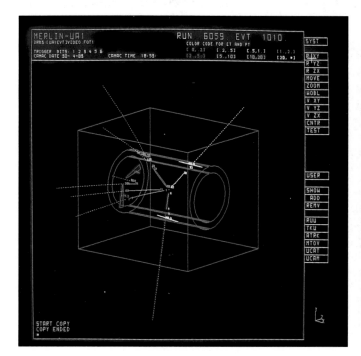

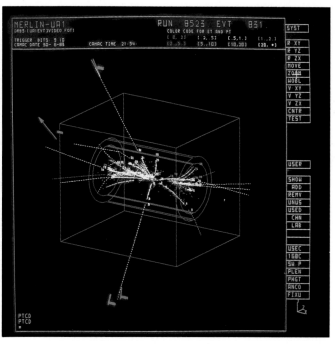

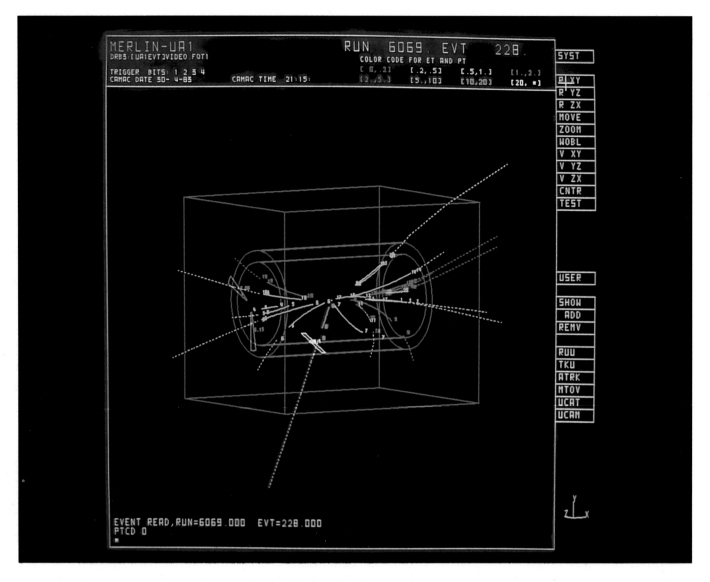

MERLIN-UA1
DRB3:[UA1EVT]VIDEO.FQT1
TRIGGER BITS: 1 2 3 4
CAMAC DATE 30- 4-85   CAMAC TIME 21:15:

RUN 6069. EVT   228.
COLOR CODE FOR ET AND PT
[0,.2]   [.2,.5]   [.5,1.]   [1.,2.]
[2.,5.]   [5.,10]   [10,20]   [20, *]

SYST
P XY
R YZ
R ZX
MOVE
ZOOM
WOBL
V XY
V YZ
V ZX
CNTR
TEST

USER

SHOW
ADD
REMV

RUU
TKU
ATRK
MTOV
UCAT
UCAM

EVENT READ,RUN=6069.000   EVT=228.000
PTCD 0

Fig. 9.19 *This display from UA1, and the one opposite, shows the decay of a W particle into an electron and a neutrino. Here the tracks of charged particles through the Central Detector are colour coded according to momentum. Most of the tracks are of low momentum (red and yellow). However, there is one high-momentum track (blue), and this connects to a cell in the Electromagnetic Calorimeter that has fired (white bars). This indicates that the track belongs to an electron. Notice how the low-momentum tracks flow generally to the right and left sides of the detector, counterbalancing each other. The high-momentum electron, on the other hand, shoots out through the front of the detector at bottom left and appears to have nothing travelling in the opposite direction to balance it. A neutrino must have passed that way. Calculating the missing energy indicates that the neutrino and electron together possess the right energy to have come from the decay of a W particle (mass 83 GeV).*

tracks provide the tell-tale sign of a decaying Z, for they end in cells in the electron detector. One track belongs to an electron, the other to a positron; together, the energies associated with the tracks add up to close to 93 GeV—the mass of the Z.

The Z's decay to a positive and a negative muon also stands out clearly from the surroundings, as Fig. 9.18 shows. Here two straight tracks connect with cells in the muon detectors forming the outer layer of the UA1 apparatus. In this case, the low-energy tracks have been left on the display, but the two muons are clear to see.

Like the Z, the W is accompanied by a maelstrom of low-energy particles when it is produced in a proton–antiproton annihilation. But as mentioned above, the W decays into only one charged lepton that can leave a track; the neutrino emitted simultaneously leaves the apparatus undetected—though not quite unnoticed. In Fig. 9.19 the single electron from the decay of a W particle is clearly visible even among the accompanying low-energy particles. It is the light blue track similar to those already seen in Z decays, but this time without a visible compensating partner. However, once we realize that the low-momentum red and yellow tracks contribute almost nothing to the balance of momentum compared with the single blue track, we can spot an imbalance. The electron is recoiling against an *invisible* partner.

The UA1 detector cannot reveal the neutrino's track directly, but it does provide sufficient information for a computer to draw in the neutrino's path. The apparatus has recorded the energy and directions of all the other particles, except for those that escape along the pipe containing the colliding proton and antiproton beams. The

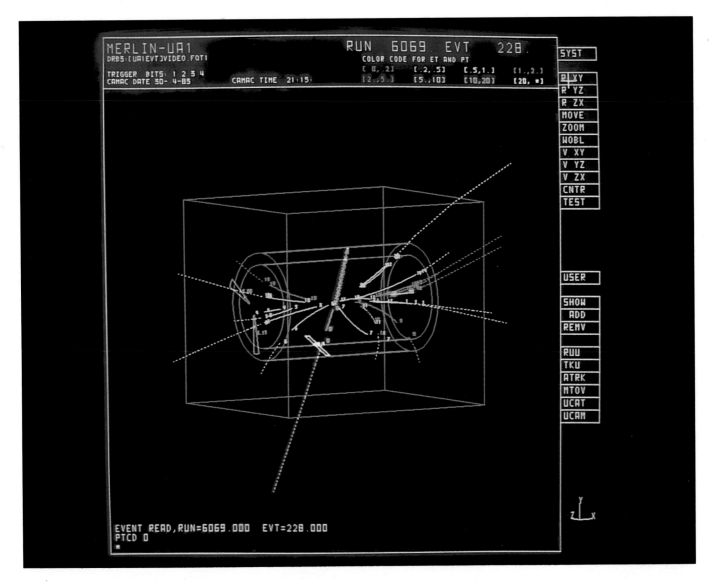

computer adds up the amounts of energy and momentum carried off in each direction, and if it finds a direction in which the amounts do not balance, then a neutrino must have passed that way, carrying the undetected energy. In Fig. 9.20, the computer has drawn in an arrowed track to indicate the neutrino's path; it has made the neutrino visible! Once it has 'seen' both the electron and the neutrino, the computer can calculate the mass of the W. It is 83 GeV, just as theory predicts.

With the observation of the W particle at CERN our story has almost come full circle, for the W particle is a key player in our modern understanding of radioactivity—an understanding that in fact links radioactivity with electricity and magnetism in the electroweak theory. This theory reveals that we perceive these phenomena as diverse only because we live in a low-energy Universe. At higher energies, the heavy Ws and Z can be produced as readily as photons, and weak interactions occur as readily as electromagnetic interactions. The experiments at CERN have given us our first glimpse of such a high-energy Universe.

Electroweak theory successfully ties together two of the four fundamental forces. Quantum chromodynamics, a mathematically similar theory based on gluons, describes strong interactions. Together, these theories indicate that nature does indeed use gauge particles to mediate the forces that mould the Universe. For the time being, the fourth force—gravity—lies outside such theories; there is as yet no direct experimental evidence for the 'graviton', the hypothesized gauge particle of gravity.

Fig. 9.20 *It is possible by balancing the energies of all the tracks to work out precisely the direction of the neutrino. This picture shows the same event as Fig.9.19, but this time the computer has drawn in the calculated path of the neutrino. Now we see the decay of the W particle in full.*

TOP

'WHERE IS top?' The question began to haunt particle physicists after the discovery in 1977 of the bottom quark hidden within the upsilon particle. Theorists knew that if the sixth quark—top or truth—exists, then it must be heavier than the bottom quark. But no one really knew how much heavier. As machines successively reached for higher energies, experimenters diligently searched for signs of top quarks. For nearly seven years they were unsuccessful. Then, in the summer of 1984, the year after its triumphant work on the W and Z particles, the UA1 collaboration at CERN released the first tentative evidence for top quarks.

Both the previous quarks—charm and bottom—had been discovered in their 'hidden' states, charmonium and bottomonium, in which they are bound to their antiquarks. Only later were charmed and bottom particles found in which charm and bottom combine with other quark 'flavours'. The opposite occurred in the case of the elusive sixth quark. The evidence garnered by UA1 was for the decay of a top particle, and it utilized a characteristic of the newly discovered Ws—illustrating once again how rapidly new particles become tools in the search for new phenomena.

Theory suggested that because the weak force acts democratically on leptons and quarks, the heavy W particles are at least as likely to decay to a top quark and a bottom quark as to an electron and neutrino, for example. The former type of decay should produce a distinctive pattern when the top and bottom quarks decay in their turn. It was this signature that UA1's computer was programmed to hunt for among the millions of proton–antiproton annihilations. Several examples were found, including the event illustrated in Fig. 9.21. The W particle, and the top and bottom quarks it decays into, all live much too briefly to be seen. To understand the distinctive

Fig. 9.21 *Tracks from a proton–antiproton annihilation in the UA1 detector contain the signature expected for a particle containing a top quark. The important features are two high-energy 'jets', a high-energy electron, and some missing energy indicative of a neutrino. These features are just about discernible in this display which shows all the low-momentum tracks (red, yellow, and green) as well as the high-momentum ones (blue and purple). The single blue track moving to the right links up with a cell in the Electromagnetic Calorimeter that has been hit (blue bars), and therefore belongs to a high-momentum electron. Two jets of particles appear to be travelling towards the top and bottom left of the picture.*

Fig. 9.22 *The diagram illustrates one sequence of events that should reveal the decay of a top particle. A proton and an antiproton annihilate, producing a W particle. The W immediately decays into a top quark and a bottom antiquark. The bottom antiquark spawns a bottom particle which decays into the jet of particles that shoots off downwards. The top quark, however, materializes as a top particle (t) which quickly decays into an electron (e^-), an invisible antineutrino ($\bar{\nu}$), and a bottom particle (b) containing a bottom quark. The bottom particle then decays to produce a second jet.*

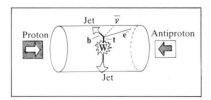

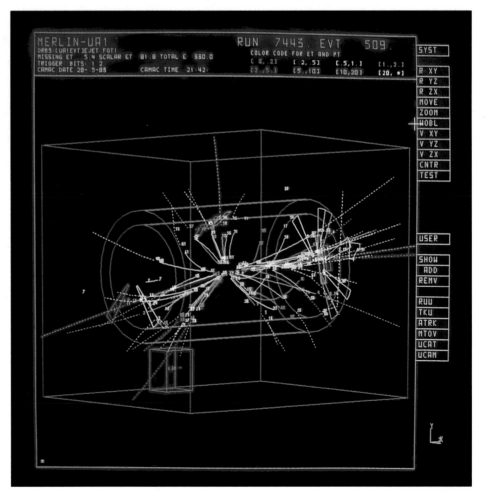

signature of the top quark, we must reconstruct what has happened, as in the diagram (Fig. 9.22).

The proton–antiproton annihilation produces a W which then decays. The resulting top quark (or antiquark) moves off relatively sluggishly on one side while the lighter bottom antiquark (or quark) travels more rapidly in the opposite direction. As the quarks separate, they materialize additional quarks and antiquarks to form a top particle and a bottom antiparticle. The top particle swiftly decays into a bottom particle, emitting an energetic electron and a neutrino. This is an exotic example of beta radioactivity, akin to the neutron's decay into a proton by the emission of an electron and an antineutrino. Now there are two bottom particles—strictly, one is antibottom—and they in turn decay into two jets of particles, one on each side of the path travelled by the original proton and antiproton. We now have all the elements of the top quark's signature: a fast-moving electron and a neutrino from the top particle's beta decay into a bottom particle, and two jets on opposite sides from the decay of the two bottom particles.

The signature can just be made out in Fig. 9.21, where the tracks have been colour-coded according to the momentum (energy) of the particles: red and yellow for the lowest values, blue and white for the highest. Amid the welter of low-energy particles (red and yellow tracks) produced in the original annihilation, two energetic jets (blue and purple) can be distinguished as they stream vertically upwards and towards the bottom left. A high-energy electron (blue track) moves off to the right. The pattern immediately becomes clear when the computer is instructed to remove all the low-energy tracks, as in Fig. 9.23.

The neutrino from the top particle's decay remains unseen, but the energy it carries can be calculated. Together, the energy in the two jets, plus that of the electron and the neutrino, add up to the mass of the W particle. The W decayed to a top quark and a bottom antiquark. So by deducting the energy of the lower jet, which resulted from the bottom antiquark, one derives the energy, and hence the mass, of the top quark. The figure that results is just over 40 GeV—more than 40 times the mass of the proton.

The bound system, 'toponium', in which a top quark and antiquark are tied together, will occur at about 80 GeV, which is comparable to the mass of the Ws and the Z. CERN's Super Proton Synchrotron has sufficient energy to produce this in its proton–antiproton collisions, but it is difficult to pick out the decays of toponium from the multitude of other particles produced in the annihilations. Electron–positron colliders would give a much cleaner picture of toponium, but it is beyond the reach of PETRA at Hamburg, the most energetic such collider currently available. Only when the new electron–positron machines now being built at Stanford and CERN commence operations towards the end of the 1980s will physicists be able to explore the world of toponium.

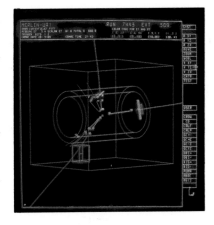

Fig. 9.23 *A display of the same event as in Fig. 9.21, but with the low-momentum tracks removed. Now the signature of the top quark is clearly visible.*

The top quark apparently completes the pattern of six leptons and six quarks, making all neat once again in the garden of fundamental particles. Some physicists are speculating that a possible further pair of even heavier quarks might show up at the new electron–positron colliders, but most are content with the symmetry achieved to date. Moreover, some astrophysicists argue that the speed of development of hydrogen and helium in the early Universe arises naturally if there are only three generations of quarks and leptons. In this way, the new electron–positron machines may directly influence our understanding of the beginnings of the Universe.

Six quarks and six leptons enable us to study the workings of three possible universes, only one of which has apparently survived. The survivor contains the lightest particles—the 'first generation'. Its nuclei consist of up and down quarks forming protons and neutrons; they are orbited by electrons; and they can decay radioactively by emitting electron-neutrinos. This is the Universe we inhabit; the stuff that we are made of. But we can now see how the alteration of a few parameters could have produced a heavier 'second-generation' universe, with strange and charmed quarks, muons and muon-neutrinos, or an even heavier 'third-generation' universe populated by bottom and top quarks, taus and tau-neutrinos.

CHAPTER 10
TO THE LIMITS

Fig. 10.1 *Snowflakes have their own exquisite hexagonal symmetry. They appear unchanged only when rotated through multiples of 60°, whereas the raindrops from which they crystallized are completely symmetrical—they look the same from every direction. Theorists believe that in a similar way, symmetries that existed in the Universe when it was younger and hotter have become hidden or 'broken' in our present cool Universe. One proposal is that there is a symmetry between the four fundamental forces of nature, which existed at the time of the Big Bang, but which is hidden today. By probing at ultrahigh energies, particle physicists hope to glimpse that symmetry.*

TECHNOLOGY LIMITS our view of nature. This is particularly frustrating to theorists whose calculations can probe extreme conditions impossible here on Earth. How can their calculations ever be tested? Theories of particle physics now deal with levels of energy that existed only in the first instants of the Big Bang with which the Universe began. Yet even the largest accelerators of the final decade of the 20th century—the Large Electron Positron collider at CERN and the Superconducting Super Collider (SSC) in the US—will reach only a million-millionth of such energies. The difference between the energy of the SSC and the energies of the early Universe is as enormous as the span from chemistry and molecular physics to the SSC. This is why physicists have started exploring other ways to test their theories.

The interest in ultra-high energies stems from a belief that the laws of nature contain an elegant symmetry, which is hidden below a certain temperature. Experiments at extremely high energies—and therefore temperatures—will reveal this symmetry; at low energies it remains concealed.

An analogy is the complete symmetry we find in a spherical drop of rainwater. It looks the same from every direction because there is a symmetry in the laws that control the behaviour of water molecules. If the raindrop freezes and forms ice crystals to become a snowflake, the complete symmetry is broken. Snowflakes have their own vivid beauty and symmetry, but this symmetry is not complete: the snowflake does not appear the same from every direction. However, if you heat the snowflake so that it melts, the complete symmetry—hidden in the frozen form—is once again revealed.

Physicists have noticed signs of such 'melting' in the ways that elementary particles behave when they are heated up. 'Heating' particles means colliding them at high energies, and it appears that the nature of their interactions is changing subtly as energy is increased. It is these observations, together with the mathematical elegance of the theories of the electromagnetic, weak, and strong nuclear forces, that have led to the idea that these three forces are intrinsically of the same strength: it is only in our 'frozen' Universe that their symmetry is hidden.

The mathematical formulations that predict and explain this unification of the electromagnetic, weak, and strong nuclear forces are resoundingly known as 'grand unified theories', or more basically as GUTs. There are quite a number of competing GUTs around, and they are undergoing constant criticism and modification. Devising the correct GUT is one of the greatest prizes in theoretical physics today. This is why it has become so important to physicists to find ways of testing the various predictions made by different GUTs. The problem is that the GUTs all agree that grand unification occurs only at temperatures above 10^{28} degrees.

Even the centres of stars, where elements are cooked in an inferno of some billion (10^9) degrees, are cool by comparison. Nature's laws and forces always appear asymmetrical to us. This is a fortunate occurrence as far as we are concerned, because our existence depends on the low-temperature disparity between the forces. The strong force grips together a compact nucleus; the less powerful electromagnetic force holds electrons remote in the periphery of atoms. Even feebler is the weak force responsible for radioactivity and the burning of stars; it is feeble enough for life on Earth to have had time to start before the Sun burns out, but not so feeble that life never started at all. The fourth and weakest of the fundamental forces, gravity, remains something of a mystery, beyond the purview of the GUTs. Perhaps one day it will be incorporated into a 'theory of everything', or TOE!

Can we ever hope to heat matter up enough to see the underlying symmetry?

Smashing particles together at high energies creates for an instant a high temperature in a microscopic volume. Physicists at CERN have done this in collisions of protons and antiprotons. When individual particles collide at energies of 600 GeV, the effective temperature is some 10^{15} degrees, and this is warm enough for the weak force to 'melt'. Weak radiation is then as 'liquid' as electromagnetic radiation, escaping in the form of the W and Z particles discovered in 1983.

The successful prediction of electroweak symmetry has given theorists confidence in the idea that the strong and electroweak forces become equivalent, in their turn, at a temperature of 10^{28} degrees. But to check this prediction with experiments at accelerators would require particle collisions at an incredible 10^{15} GeV. With present technology, you would need an accelerator whose total length approaches one light year—a hundred million circumnavigations of the globe!

Testing the GUTs requires alternative approaches conducted in parallel with 'low'-energy accelerator experiments. For instance, we can observe cosmic rays, which have been raised to extremely high energies by nature's own accelerators in the Sun, stars, and galaxies. Another approach is to look for relics of the Big Bang, when the Universe was so hot that particles collided at the extreme energies meaningful to GUTs—a kind of subatomic archaeology.

A successful GUT must explain another mysterious characteristic of the birth of our Universe. The energy of the Big Bang created protons slightly more readily than antiprotons, and this asymmetry led to a Universe that consists of matter rather than antimatter. Yet in their 'low'-energy experiments at accelerators, physicists always create matter and antimatter in equal amounts. The GUTs suggest that subtle manifestations of the original matter–antimatter asymmetry should still be around. In particular, Andrei Sakharov, who won the Lenin Prize for the Soviet hydrogen bomb and the Nobel Prize for Peace, realized several years ago that the price we pay for the excess of protons is that they are ultimately unstable. Matter as we know it should be eroding slowly away—a reversal of the process that created protons slightly more readily at the start of the Universe.

As GUTs have inspired particle physicists to take an interest in the early Universe, so have astrophysicists become aware of particles in the Universe at large. A growing symbiosis has developed between the two branches of physics. Cosmologists developing theories of how galaxies form and how the Universe evolved look to particle physics for ideas on how matter behaves. Measurements made by astronomers can in turn impose important constraints on theories such as the GUTs.

These are some of the ways that particle physicists are attempting to look beyond the limits imposed by accelerators. By hunting for rare relics of hotter times, by keeping watch for the decay of matter, by turning their detectors to the stars, they hope to reveal the symmetry between nature's forces. This chapter surveys in more detail these and other experiments that do not require accelerators.

THE DEATH OF THE PROTON

THERE IS something about bulk matter so obvious that we rarely wonder about it: it is electrically neutral. But think about it a while and it heralds a great mystery. A proton and an electron have the same *amount* of charge, the proton carrying a positive value and the electron a negative value. Any difference in the amounts of their charges is too small ever to have been measured; they are the same to better than one part in a thousand billion billion. It is most likely that they are identical. But why should this be? How do protons and electrons 'know' about each other?

This precise balancing of the electron's and proton's charges is the reason why atoms have no net charge, why matter overall is neutral. If matter were electrically charged, the Sun and the Earth and the stars would be powerfully attracted (or repelled) by electrical forces rather than by gravity. So the electron–proton conspiracy is rather crucial for the behaviour of the Universe as a whole.

Yet electrons and protons are quite dissimilar. We now know that protons are complex objects built from quarks. The electron, on the other hand, appears to be a truly elementary particle, the lightest of the electrically charged leptons. The proton and electron are fundamentally very different, so why should they work together so well to form the Universe we inhabit?

Another property of protons that is necessary to our existence is that they are exceedingly stable. If they decay at all, they do so extremely slowly. Indeed if their life were less than 10^{17} years we would be destroyed by radiation from the decaying protons in our own bodies. We would literally 'feel it in our bones'.

That our bodies, which rarely last for a century, can reveal that protons live as long as 10^{17} years may seem paradoxical. However, statistical probability provides the key. An actuary is certain that at least half of us will be dead by the age of 80, though some will survive until 100 and others will die young. So it is with protons; 10^{17} years is an *average* life. (Technically it is the 'half-life': in a large collection of protons, half will have decayed by that time). There are so many protons in our bodies—10^{27} or so—that, unless they are exceedingly stable, many protons will die young and at the same time kill us. That we are here at all shows that on average protons must be stable for at least 10^{17} years.

The GUTs cast light on both these characteristics of matter—its electrical neutrality and its basic stability. These theories not only unite the electroweak and strong forces, they also unite matter by relating the quarks and the leptons. This implies that quarks, which tend to be heavier, can become leptons; consequently, protons can decay.

The prediction of proton decay is one of the few ways that we can test GUTs with present technology. The theories generally predict that the proton's average lifetime will be some 10^{32} years. This is much more than the age of the Universe, which is about 10^{10} years. But there is hope of seeing a few protons die young. Over one year about 100 protons should die in about a million kilograms (1000 tonnes) of matter. So if we take several thousand tonnes of material, instead of the few kilograms of our bodies, and surround it with sensitive detectors, then this giant might feel proton decay in *its* bones. This is the strategy behind experiments now underway, which are seeking the first glimpse of the Universe eroding away.

To see a proton decay you have to watch very carefully. And when it happens you must be sure that what you saw was not something else faking the real thing. To catch such a faint whisper you must first blanket out as much background noise as possible. Cosmic rays hitting the apparatus can cause signals that mimic proton decays. A shield over the experiment, hundreds of metres thick, can eliminate all but the most energetic cosmic rays. Solid rock provides just such a shield, so physicists have gone down mines and under mountains in their search for the signs of a dying Universe.

Near Bangalore in India physicists have occupied a 2300 m deep mine in the Kolar gold fields. A Japanese group has apparatus in the Kamioka metal mine. In Europe, one team is using a cavern off the Fréjus tunnel in the French Alps, while another has taken over a garage off the road tunnel 3000 m below the summit of Mont Blanc. In the US, detectors are ensconced in the Soudan iron mine in Minnesota and in a salt mine 600 m below Lake Erie, near Cleveland, Ohio.

In these caverns sit huge 'swimming pools' of water or monoliths of concrete and steel. They contain billions upon billions of protons, one of which may die today, next week, next year. When it does, if it does, the circling bats and insects who live in the caverns will not notice. But an electronic detector will be triggered and record the occurrence on magnetic tape. Later, in the comfort of their offices, the physicists will be able to see what becomes of a dead proton.

The Morton Thiokol salt mine in Ohio is home to the largest of the 'swimming pool' detectors. It was begun by physicists from the universities of California (Irvine) and Michigan, and the Brookhaven National Laboratory, and is therefore known as the IMB detector. It occupies a huge cavern hewn out of the rock salt by a special mining machine. Two thousand rock bolts prevented the walls from collapsing during the excavation.

The cave is a cube with sides 20 m long—as big as a seven-storey apartment block. Two strong layers of polyethylene line the walls, in effect forming a huge sack to hold the water. It takes several minutes to fill a household bath, even with the taps on

Fig. 10.2 *The Irvine–Michigan–Brookhaven (IMB) proton decay detector occupies a huge cave, 600 m below Lake Erie. The cave is 22.5 m long, 17 m wide, and 23 m high, dwarfing the bulldozer at the bottom of this photograph taken soon after the excavation was completed in 1981. Once dug out, the cave had to be lined with plastic and filled with water—no easy task. Then came the job of installing phototubes at 1 m intervals along the walls of the cave.*

Fig. 10.3 *Strings of 13 cm diameter phototubes line the walls and floor in this picture taken by a diver inside the IMB cavern. They pick up Čerenkov light radiated by charged particles that pass through the water faster than light does.*

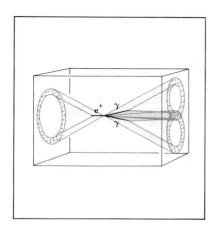

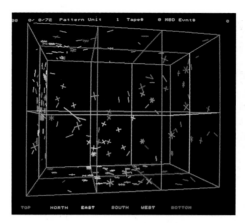

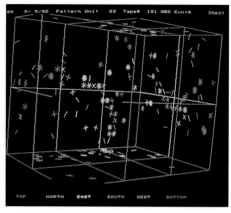

Fig. 10.4 *(left) The diagram shows the anticipated effect of a proton's decay in the IMB detector. The proton decays to a positron (e^+) and a neutral pion which, because the proton was at rest, move off back-to-back. The pion decays almost immediately to two photons (γ), and they in turn convert in the water to electron–positron pairs. Being charged particles, the original positron from the proton decay, and the two electron–positron pairs from the photons, produce cones of Čerenkov light which fall on the walls of the detector and fire rings of phototubes.*

Fig. 10.5 *(centre) This display is a simulation of a proton decay in the IMB detector. The hypothetical proton has decayed at the left. Short yellow lines mark the paths calculated for the positron and the two photons. Rings of tubes that have fired are just about visible to the top left, bottom centre, and right. The colours indicate the order in which the tubes fired, with red being the earliest, purple the latest. The number of bars in each 'star' marking the phototubes indicates the number of photons the tube detects.*

Fig. 10.6 *(right) Real data in the IMB detector displays a couple of rings of tubes that have fired—to the top right and lower back of the detector. But careful examination of the data indicates that this pattern is not due to proton decay. Instead it has come from the interaction of a cosmic ray neutrino with a nucleus in the water.*

full. Filling your local swimming pool takes several hours. But to fill the IMB tank carefully with 8000 tonnes of specially purified water took two months. And that was only at the second attempt; the first try burst the polyethylene lining. To prevent this happening again, engineers poured concrete behind the lining at the same time as the water level rose inside the bag. This helped resist the pressure from the growing wall of water and prevented it from overstressing the plastic. Once the tank was full, in July 1982, physicists and technicians became divers as they installed 2048 phototubes around its faces.

If and when a proton decays, it will produce charged particles that travel faster than light does in water. As they shoot through the water, these particles emit light at an angle to their path, similar to the shock wave of sound when a plane breaks the sound barrier. This is Čerenkov radiation (see p. 113).

The phototubes surrounding the IMB tank detect the sudden burst of Čerenkov light from a passing charged particle. A computer records how much light hits each tube and also notes the order in which the various tubes are struck. This information provides a detailed impression of the directional flow of the Čerenkov light, which the computer uses to reconstruct the trails of the original culprits.

If a single proton decays into two particles, they will move off back to back. Typically the proton might convert into a positron and a pi-zero, the latter decaying in turn to two gamma ray photons (Fig. 10.4). So the positron will give a cone of light in one direction while the photons produce two cones on the other side. Figure 10.5 shows a computer simulation of the effects of a such a proton decay in the IMB tank. Yellow lines mark the short innocuous trails of the positron and the two photons. The shock waves from these particles spread out and hit the sides of the tank. Red indicates the phototubes that fired first, blue the last to fire. The approximately back-to-back particles form rough circles at the top left and bottom of the tank.

The task of detecting the rare proton decays is made still more difficult by the fact that even 600 m of rock cannot shield the apparatus from *all* cosmic rays. Neutrinos can penetrate the Earth and interact with protons in the tank, giving signals that mimic spontaneous proton decay. These events look uncannily like genuine proton decays (Fig. 10.6). Only after the computer has analysed all the information does it become clear that an event like this does not fit with proton decay, but has been caused by a neutrino entering the tank.

According to one of the simplest GUTs, the decay of a proton to a pi-zero and a positron should be the most common decay. But so far neither the IMB tank nor any other detector has observed a convincing example. However, other detectors have produced tantalizing glimpses of events that could be due to other kinds of decay. The Japanese version of IMB is a tank of 3000 tonnes of water buried in the Kamioka metal mine. It has picked up tracks that could be due to the decay of a proton to a positive muon and a heavier relative of the pi-zero called the 'eta'.

Other possible decays have come from an example of the second type of proton decay detector—an iron monolith that operates in the bowels of Mont Blanc on the French–Italian border. The detector, known as NUSEX for 'Nucleon Stability Experiment', is housed in a garage off the side of the road tunnel through the

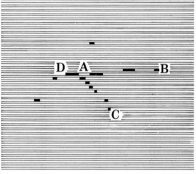

Fig. 10.7 *The NUSEX proton decay detector occupies a 'garage' off the side of the road tunnel under Mont Blanc. It is a sandwich of centimetre-thick iron plates forming a cube 3.5 m along each side. Between the plates are rows of 'streamer tubes', which give out a pulse each time a charged particle passes through. In this view, the electronics attached to the ends of the tubes can be seen bristling around the faces of the detector. Two members of the NUSEX team, Gianfranco Bologna (with the blue shirt) and Oscar Saavedra, are checking some of the circuitry. The orange structure is a crane used for assembling and disassembing the detector.*

Fig. 10.8 *Tracks of three charged particles recorded in the NUSEX detector form a pattern consistent with the decay of a proton to a neutral kaon and a positive muon. This computer display depicts a side view of part of the detector showing the sheets of iron with the layers of streamer tubes between. The tubes are seen end on, and are marked only where they have fired. A proton could have decayed at the junction of the tracks. The muon has then moved off to the right (A–B), while the kaon decayed to two pions (A–D and A–C). Physicists will be confident that they have observed proton decay only when they have accumulated several more events such as this.*

mountain. There the rumble of juggernauts gives way to the soft hum of electronics monitoring a 150 tonne assembly of centimetre-thick iron plates arranged in a cube 3.5 m per side. Sandwiched between the plates are raft-like layers of plastic 'streamer' tubes.

Each tube is filled with gas (a mixture of argon, carbon dioxide, and *n*-pentane) and has a wire held at a positive voltage running along its axis. When a charged particle passes through the tube it ionizes the gas and sets off a discharge in the strong electric field around the wire. This discharge is sensed by metal strips outside the tubes, which run both parallel and perpendicular to the tubes. The signals from these strips pass to a computer which records them, and which can use the information to build up a three-dimensional picture of tracks through the detector. Figure 10.8 shows the reconstruction of tracks that may signal the decay of a proton into a positive muon and a neutral kaon.

The lack of firm evidence for proton decay by no means sounds the death knell for the GUTs. What it does indicate is that some of the simplest GUTs are wrong, and it suggests that as of late 1985 the proton's lifetime must be in excess of 10^{32} years. The possibility of proton decay remains, and it will continue to test the ingenuity of experimental physicists for some years to come.

THE MONOPOLE MYSTERY

THE SEARCHES for proton decay test high-energy theories indirectly. However, we do have one source of high-energy particles that does not depend on our ingenuity with technology—the cosmic radiation. At Fermilab, protons are accelerated to reach an energy of 1000 GeV. But this feat pales into insignificance when compared with nature's accelerators, which can fire cosmic rays from outer space with energies as high as 100 billion (10^{11}) GeV. At such extreme energies only a few particles arrive per square kilometre in a century. These rare events provide the only direct means of studying ultra-high energies. They also allow us to search for possible super-heavy particles, with masses of up to 10^{11} GeV—billions of times heavier than the Ws and Z.

But GUTs predict that particles even more massive than this were produced during the Big Bang and that a few may still haunt the Universe today, like fossil relics of the earlier epoch. Very occasionally one of these heavy relics might fall to Earth among the cosmic rays. The 'monopole' is probably the least controversial of these hypothetical beasts. It is a particle that carries a single unit of magnetic 'charge'—in other words, it is an isolated magnetic pole.

A magnet has a north pole and a south pole. Cut it in two in an attempt to isolate a single pole and you will fail; you will simply create two smaller magnets, each with a north and south pole—two 'dipoles'. No one has ever seen an isolated north or south 'monopole', but this has not prevented people from speculating that monopoles exist. Indeed there are good reasons to believe that they do. In 1931, Paul Dirac, the theorist who had earlier anticipated the discovery of the positron, pointed out that the existence of a unit electric charge, such as the electron or proton carries, could be understood if magnetic charges—monopoles—exist. Since that day physicists have continually searched for these objects.

There are two reasons why interest in monopoles has greatly increased recently. First, GUTs predict that magnetic monopoles exist with a mass near to the energy scale of grand unification. This is in the region of 10^{16} GeV, and it corresponds to a mass of around one hundred-millionth of a gram—the weight of a small bacterium.

This massive monopole would interact with other particles. In particular it could catalyse proton decays, making them occur more readily in matter the monopole traversed. Calculations suggest that if it *could* convert all the proton's mass to energy in this way, a single monopole could release an energy of as much as 10^{14} joules per gram of matter—a factor of a thousand more than from nuclear fusion!

It would however be difficult to stop a monopole, which could in principle pass through the Earth without losing much energy. It could even penetrate into the dense matter of neutron stars and eat up the neutrons there. So, much as monopole-hunters would like to find their quarry, they do not want to discover it too easily or else the future of matter could be precarious. But there is no need for panic! If monopoles exist, they are not very abundant. We know this because they would have neutralized the magnetic fields in the galaxy and this has not yet happened.

Yet in 1982, at 1.53 p.m. on 14 February, St Valentine's Day, a tantalizing signal came from a small detector built by Blas Cabrera at Stanford University—the second reason for the present interest in monopoles. Cabrera had a coil of superconducting niobium wire 5 cm across, surrounded by a superconducting magnetic shield. If a magnetic charge passed through, it would make a current flow in the coil where there was no current before; and because the coil is superconducting the current would persist. This is precisely what happened on 14 February 1982—a current suddenly appeared in the coil. Moreover, the amount of current was exactly what Cabrera expected from a monopole.

Was this the real thing or some experimental effect that no one has yet been able to explain? Only time will tell. Physicists around the world, including Cabrera, have since set up improved versions of the experiment, with larger coils able to watch a larger area. So far, as of March 1986, only one other possible monopole signal has been seen, in an experiment at Imperial College, London; Cabrera himself has seen nothing more.

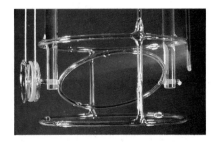

Fig. 10.9 *Blas Cabrera's detector, which picked up a signal consistent with a monopole on 14 February, 1982. The essential part of the detector is the central, tilted ring, which contains four turns of superconducting niobium wire. Normally this 5 cm diameter ring lies horizontally, like the two slightly larger calibration rings above and below it. The whole assembly is surrounded by a magnetic shield, which ensures that the residual magnetic field within is less than a millionth of the Earth's field. The passage of a monopole through the superconducting ring induces a change in current—which is just what Cabrera observed.*

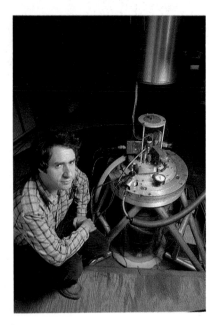

Fig. 10.10 *Blas Cabrera (b.1946), in his basement laboratory at Stanford University in 1984, with his improved three-ring monopole detector.*

LOOKING FOR FREE QUARKS

MONOPOLES ARE not the only relics from the Big Bang that may still be with us in our present-day Universe. Over the years, particle physicists have discovered hundreds of particles made from quarks. Yet no one has seen a single isolated free quark. This is not for want of looking. At every new accelerator, physicists search for free quarks. So far they have found none. One possibility is that although quarks act as if lightweight when inside the proton, they are extremely heavy when free. (This apparent paradox is a consequence of the nature of the strong nuclear force.) Thus accelerators do not have enough energy to release quarks in the laboratory. However, there was enough energy in the Big Bang, and perhaps even in some stellar systems today, to produce massive free quarks which might arrive in the cosmic rays.

The unique electrical charges of quarks should make them stand out clearly. The up, charm, and top quarks have 2/3 of a proton's charge; the down, strange, and bottom quarks each carry $-1/3$ the proton's charge. These fractional charges are the quarks' most distinctive signature; if a fractionally charged particle passed through your detector you would notice at once. This is because the ionizing power of a charged particle is related to its charge squared—double the charge and you see four times as much ionization, in other words, a track four times as dense. So a quark with 2/3 the charge of a proton should produce only 4/9 the ionization and leave a trail about half as dense.

In 1969, Brian McCusker of the University of Sydney found just such a track in a cloud chamber picture. He had decided to look for free quarks in the central 'cores' of cosmic ray air showers, where the particles are most densely packed. He installed four cloud chambers where they could be triggered by three sets of Geiger counters a few metres apart. The Geiger counters were set up so as to produce a trigger signal only when thousands of particles arrived in an area of a few square metres, in other words when the core of a shower was close by.

Figure 10.11 shows the cloud chamber picture that convinced McCusker that he may well have seen a quark. In one of the tracks, the density of drops is markedly less than in the other tracks. It has some 16 drops per centimetre as opposed to about 40 drops for the other tracks—just what you might expect for a particle with 2/3 the proton's charge. There are other possible explanations for the faintness of the track; for example, the densities of tracks always vary about an average, and the faint track could be simply a statistical fluctuation, further from the average than usual. Indeed, with the passing years the event has failed the test of time, in that no other similar experiment has come up with any equally good-looking evidence for a quark, and other physicists have remained sceptical.

Rather different hints of free quarks—or some other unusual heavy particle—have come from another cosmic ray experiment, this time high on Mt Chacaltaya in Bolivia. It is there that a team of Bolivian and Japanese physicists has set up a series of large emulsion detectors, each covering an area of some 40 square metres, and containing layers of lead as well as the emulsion. During the 1970s, the team found five examples of very unusual air showers.

The layers of lead in the detector allow the physicists to distinguish between the effects of electrons and gamma rays, which do not penetrate far, and protons and charged pions, which travel further. The five unusual showers contained very few gamma rays, but tens of charged pions and protons. This is surprising because a cosmic ray collision should produce neutral pions as well as charged pions, and the neutral pions should swiftly decay to gamma rays. In *all* five events there was probably only one pi-zero; at the energies of over 100 000 GeV that appear to have produced the showers, there should have been 20 or so pi-zeros in *each* event.

Nobody is certain yet whether these 'Centauro events' (so called because of a mismatch between information from the top and bottom halves of the detector, as in the mythical centaur) are a real phenomenon, although it is difficult to explain them away as experimental artefacts. It has been suggested that they could be due to free quarks, but without further evidence this explanation and others that invoke still more exotic particles remain unconvincing.

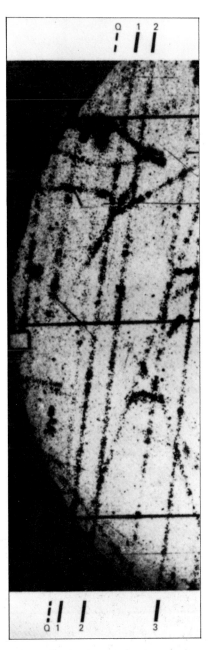

Fig. 10.11 *Brian McCusker's cloud chamber photograph, which has been claimed as evidence of a free quark. Three relatively strong tracks through the chamber are marked by solid lines top and bottom. To the left of these lies a fainter track, marked by the dotted lines. A quark with a charge of 2/3, and therefore 4/9 the ionizing power of a particle with unit charge, could have made the track. Or is it, as seems more likely, a statistical fluctuation?*

Fig. 10.12 *William Fairbank (b.1917) at Stanford University in 1984. He is holding one of the plates between which the tiny spheres of niobium are levitated in his experiment that has found evidence for fractional charge.*

If quarks do arrive in the cosmic rays, they must be extremely rare or they would have been detected in some convincing way. However, the possibility remains that cosmic quarks could have accumulated at the Earth's surface during its 4500 million year existence. Over the millennia, a small concentration of 1 quark per 10^{20} protons and neutrons could have formed in terrestrial materials—a quantity that is consistent with the limits set by the failure of modern experiments to find quarks in cosmic rays. Or, if the atmosphere is opaque to quarks, then they might be found in moonrock if not on Earth.

Astronauts brought back rocks from the Moon, but no quarks were found in them. Large amounts of various other substances have been examined on Earth. The problem is that no one knows which substances might be the most likely to contain quarks. So the search begun at Stanford University in California by William Fairbank, together with his research students Art Hebard and George LaRue, stood as good a chance as any other. They have been working with tiny balls of niobium, only 0.25 mm in diameter, and they have found some intriguing evidence for fractional charges.

Fairbank and his colleagues begin by cooling a ball until the niobium becomes superconducting, when it will conduct an electric current with no resistance at all. The researchers then switch on a superconducting coil below the plate carrying the ball. The current in the coil creates a magnetic field, which in turn induces a current and a magnetic field in the superconducting niobium ball. The two magnetic fields oppose each other, and the ball is levitated as the magnetic forces on it counterbalance the effect of gravity.

Once the ball is suspended in this way, the researchers tickle it with an alternating electric field. An electrically charged ball will move in the electric field, but an uncharged one will not. The technique then is to add single electrons or positrons, measuring the ball's movement all the time. If the ball begins with a whole number of charges, then by adding electrons or positrons—which each have a whole unit of charge—you should eventually be able to neutralize the ball so that it stops moving. But if the ball has *fractional* charge to begin with, you will never be able to neutralize it completely, because adding whole units of charge will never give zero.

This is what Fairbank's team found with several balls; they apparently had a charge of $+1/3$ or $-1/3$, as would be the case if a single free quark were sitting on them. However, no one else has managed to reproduce these effects in similar experiments. It is extremely hard to eliminate all sources of background and many physicists believe that some unknown source of error is responsible. But what?

Fig. 10.13 *The fractional charge experiment at the Rutherford Appleton Laboratory in the UK. The levitation apparatus is housed within a vacuum vessel, one end-plate of which is visible here. The magnet for levitating the spheres of niobium, steel, and other materials is the dark horizontal structure; its pole pieces point inwards, through the walls of the vacuum vessel. Lasers (the black tubes towards the front) provide beams to monitor the position of the spheres, and to measure the amount of oscillation.*

Fig. 10.14 *One of the steel balls, 0.25 mm in diameter, rests on the head of a pin.*

A team in the UK, from the Rutherford Appleton Laboratory and Imperial College, London, has recently built an improved levitation apparatus, which has tested many minerals, not only niobium. Laser beams monitor the positions of the balls and reveal the oscillation of a ball when a low-frequency electric field is applied. The experiment does not have the background problems of Fairbank's pioneering attempt and should be able to detect charges to an absolute accuracy of better than 1/20 of a proton's charge. As of Summer 1986, they have no evidence at all for fractional charge, although they have tested a wide variety of materials.

The nature of the colour forces between quarks seems to prevent the existence of individual free quarks, but no one has yet proved from first principles that nature absolutely forbids single quarks. The discovery of a free quark would not prove a major embarrassment for theorists in general and grand unified theories in particular; rather it would stimulate more questions: how much energy is needed to liberate a quark, and what law of nature dictates this energy scale?

COSMIC MESSENGERS

WE SAW in Chapters 4 and 5 how the study of cosmic rays gave birth to particle physics in the 1930s and 40s, when the rays revealed new particles such as the positron, muon, and kaon. But with the development of high-energy accelerators in the 1950s, particle physicists and cosmic ray physicists tended to go their separate ways. The particle physicists concentrated on studying the products of man-made collisions, while cosmic ray experts addressed the questions of the composition and origins of the rays. Now, however, particle physicists looking to energies beyond the reach of their accelerators, are once again taking an interest in the cosmic radiation.

Cosmic rays can have awesome power. In some regions of the cosmos, nature somehow manages to give single atomic nuclei energies as high as 150 billion (1.5×10^{11}) GeV. Clues to the origin of these ultra-high-energy rays are only now becoming apparent, from studies of what happens when they impinge upon the Earth's atmosphere.

At energies below about a billion (10^9) GeV, magnetic fields in our galaxy deflect the electrically charged cosmic ray particles, so they arrive at the Earth from all directions at random. However, the magnetic fields have less influence on more energetic rays, which tend to fly more directly to the Earth. By studying these rarer high-energy rays, physicists hope to identify the sources of the cosmic rays.

When a high-energy cosmic ray shoots down through the upper atmosphere, it generates an avalanche of subatomic particles; a primary cosmic ray with an energy of 10 billion (10^{10}) GeV creates a shower containing about 10 billion particles by the time it reaches sea level. These particles spread out sideways while preserving the direction of the main thrust. A snapshot of the shower would reveal it as a thin disc of particles moving towards the ground at nearly the speed of light. The disc can be several kilometres across, and its leading edge reaches the ground before the trailing edge. By measuring the relative arrival time of particles at several widely-separated spots, physicists can determine the direction of the shower to within two or three degrees.

Two arrays of detectors set up to study 'extensive air showers' in this way are located at Kiel in West Germany and Haverah Park in the UK. These two arrays have recently provided evidence that cosmic rays of over a million (10^6) GeV are being emitted from a region in the constellation of Cygnus. At this relatively low energy, the galactic magnetic field should scramble the directions of charged cosmic rays. So it seems likely that the cosmic rays that appear to emanate from Cygnus are in fact gamma ray photons. Being uncharged, they are not deflected by magnetic fields and so can fly straight from source to detector. These cosmic gamma rays appear to come from a particularly interesting object in Cygnus, known as Cygnus X-3.

Cygnus X-3 is extremely powerful: in X-rays alone it radiates 10 000 times more energy than the Sun's *total* energy output. Astrophysicists believe that gas is being heated as it falls from a normal star in towards a small dense companion star. The

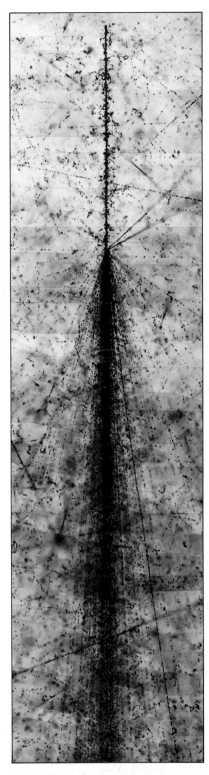

Fig. 10.15 *A very high-energy cosmic ray iron nucleus shoots into some photographic emulsion and collides with a silver or a bromine nucleus to produce a tremendous 'jet' of about 850 mesons. From the divergence of the jet, it is possible to estimate the total energy of the incoming iron nucleus as more than 15 000 GeV. But this is puny in comparison with the rarer ultrahigh energy cosmic rays. The central dark core of the jet is about 0.04 mm across.*

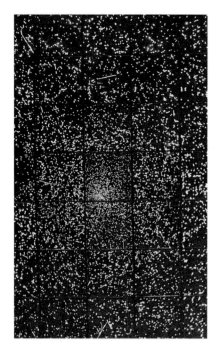

Fig. 10.16 *A 'snapshot' of the arrival of a cosmic ray air shower at ground level is given by this display of 'hits' in a 35 sq. m array of spark chambers. The shower, containing some 200 000 particles altogether, has come in more or less vertically, from in front of the page. The denser core to the shower is clearly visible. (The grid marks intervals of 1 m.) The array is operated by physicists from Leeds University.*

Fig. 10.17 *Air showers covering an area about 12 km square are picked up by Čerenkov detectors arrayed in an approximately hexagonal formation around moorland at Haverah Park, outside Leeds. The Čerenkov detectors—sealed tanks of pure water viewed by phototubes—are housed in huts marked by the red dots on this aerial photograph.*

gas becomes so hot that it emits X-rays. The compact companion is probably a 'neutron star' composed entirely of neutrons packed together as tightly as in an atomic nucleus. The intensity of X-rays fluctuates over a 4.8 hour time period as the two stars rotate about each other. The air showers from the cosmic gamma rays also show a 4.8 hour variation, so Cygnus X-3 indeed seems to be the source.

The opinion is that protons are somehow accelerated to very high energies in the vicinity of Cygnus X-3, collide with atomic nuclei in the gas, and create pions. The pions then decay; the neutral pions produce gamma ray photons, the charged pions emit neutrinos. The gamma rays produce the cosmic ray showers; the neutrino bursts remain to be detected.

Cygnus X-3 has also attracted the attention of particle physicists, particularly those operating proton decay detectors, which although underground can still pick up energetic cosmic rays. In general, penetrating cosmic rays are the scourge of teams searching for the faint signal of proton decay. But there are indications that these detectors can pin-point particles coming from Cygnus X-3.

One such experiment lies 655 m below ground in an iron mine in Minnesota. The only cosmic radiation that reaches the detector consists of neutrinos and high-energy muons; the rest is absorbed by the Earth. The physicists operating the detector, known as Soudan I after the name of the mine, have tracked over 800 000 muons. They can even plot the topography of the Earth's surface. Hills and valleys change the amount of rock that muons from various directions must penetrate to reach the detector and so affect the number recorded.

The Soudan I 'telescope' consists of a concrete block, embedded with nearly 3500 steel detector tubes. About two muons pass through the detector each minute, as expected. What is intriguing, however, is that when the team analysed muons coming from the direction of Cygnus X-3, they observed a 4.8 hour variation, just as in the X-rays and cosmic ray showers. NUSEX, the proton decay experiment in the Mont Blanc tunnel (see pp. 202–3), has also found similar effects.

The muons themselves cannot have travelled all the way from Cygnus X-3, as the galactic magnetic fields would have scrambled their directions. Instead, they must originate in the Earth's atmosphere in showers initiated by *neutral* particles from Cygnus X-3. Are these neutral particles gamma ray photons? At present there are curious discrepancies between the numbers of muons detected and the number expected from photon interactions in the atmosphere. Moreover, both teams rule out the possibility that the muons originate from neutrino interactions.

The possibility remains that the collisions in Cygnus X-3's accelerator produce new

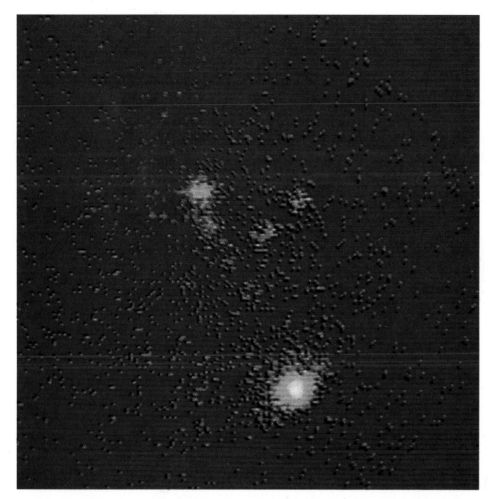

Fig. 10.18 *A false-colour X-ray image of Cygnus X-3 (large bright spot), obtained from data collected by the orbiting Einstein Observatory. One of the most powerful emitters of X-rays in our galaxy, Cygnus X-3 may be an unusual source of cosmic rays if preliminary interpretations of data from underground detectors prove correct.*

Fig. 10.19 *The Soudan I detector, located 655 m underground in an iron mine in Minnesota, was designed primarily to search for proton decays, but it has also detected muons which seem to be related to the activity of Cygnus X-3. The detector consists of a concrete block, 2.9 m square and 1.9 m high, in which nearly 3500 steel tubes are embedded. Each tube contains a gas mixture of argon and carbon dioxide, with a tungsten wire running down the centre. The wire is kept at a high positive voltage, and picks up avalanches of electrons released when a charged particle passes through the tube. In this view of the detector we see the electronics attached to the ends of the tubes.*

kinds of particle, unknown at the 'low' energies we can create in laboratories on Earth. Cygnus X-3 is for the time being a curiosity to astrophysicists and particle physicists alike.

The Soudan I and NUSEX detectors were designed to pin-point the deaths of protons, and it is in many ways a happy coincidence that they are providing useful information about cosmic rays. However, as in the 1930s, cosmic ray physicists continue to use techniques that are similar to those employed in particle physics. Cosmic ray detectors come in many guises, including Geiger counters, wire chambers, scintillators, and emulsions—all familiar tools described in earlier chapters of this book. And, as in particle physics, new possibilities have opened with advances in electronics and computing.

One detector to harness modern technology is called the Fly's Eye. Each night it watches the clear skies above the Dugway Desert in Utah in search of cosmic ray air showers. The Fly's Eye in fact consists of two 'eyes' four kilometres apart. The main 'eye' contains 67 mirrors mounted in an array of large 'cans'. Phototubes at the focus of each 1.5 m diameter mirror pick up reflected flashes of light—scintillations—produced as the cosmic rays pass through the atmosphere.

This is nothing more than an old technique put to new use. More than 70 years ago, Rutherford detected alpha particles by the faint flashes, or scintillations, that were emitted when the particles collided with atoms in a zinc sulphide screen. Nitrogen in the air also scintillates when electrically charged particles pass through, but it does so only weakly: it gives off five photons for every metre along the track of a very high-energy electron. The flashes are too faint to see, but they can be detected by modern high-quality phototubes coupled to sensitive electronics. The Fly's Eye is a child of the 1980s.

The Fly's Eye is so called because its view of the whole sky, built from the overlapping segments seen by each mirror, is like the scene from a fly's compound eye. On clear, moonless nights the detectors can watch the cosmic ray showers streaking across the sky, more than 20 km away. A computer records how much light triggers the various phototubes and in what sequence. From this information it reconstructs the flight of the shower and the direction of the primary cosmic ray.

The total energy of the shower can be determined once its proximity and intensity are known. The Fly's Eye has observed a number of showers with an energy above 10 billion (10^{10}) GeV, confirming earlier results from Haverah Park. One fundamental question the detector is attempting to answer is whether there is an upper limit to the energies in cosmic rays.

We know that the primary radiation contains protons and nuclei, but these electrically charged particles should be cut off at energies above about 100 billion (10^{11}) GeV for a rather exotic reason. The Universe is bathed in a 'background' of microwave radiation, which is a relic of the high temperatures of the Big Bang. But a proton rushing through this background at high enough velocities will be confronted with gamma rays rather than the cool microwaves that we detect on Earth.

This is because the proton sees the frequency of the microwaves greatly increased, just as the pitch of a blaring horn increases as a car rushes towards you. The microwave radiation becomes gamma rays as far as the fast proton is concerned. And when a gamma ray hits a proton, it is absorbed and pions are emitted as Fig. 3.26 shows (see p. 65). As a result, the original proton slows down. Any proton with an energy above 10^{11} GeV will not last more than about 100 million years without being slowed. This sounds a long time, but it is small compared with the age of the Universe, which is about 20 billion years. It is surprising that any cosmic rays of such high energies reach Earth at all.

However, there have been glimpses of showers with this energy and more. These have managed to reach Earth while avoiding the microwave background radiation, which implies that the particles are relatively young—less than 100 million years old. Moreover, measurements of the directions of these ultra-high-energy rays suggest that these cosmic rays come from beyond our galaxy, perhaps from the direction of the Virgo Cluster.

Determining the origin of cosmic rays has occupied many a physicist from Robert Millikan onwards, and will no doubt continue to do so for some years to come. This section has given only a taste of the many varieties of experiment being tackled. Particle physicists will follow the results of these experiments with great interest, because nature's own particle accelerators may yield valuable evidence about particle interactions at energies higher than we can hope to achieve in the laboratory, at least in the foreseeable future. Indeed, particle physicists are already looking beyond our own galaxy to the Universe at large in their quest for the ultimate theory that unifies all of nature's forces.

THE UNIVERSE AS A LABORATORY

IN 1929, the American astronomer Edwin Hubble discovered that the galaxies are rushing away from one another. This observation, which implies that the Universe is expanding, is embodied in one of the foundations of modern cosmology—the theory of the hot Big Bang. According to the theory, some 20 billion years ago all the material of today's Universe must have been crammed into a volume smaller than a clenched fist. This highly compressed Universe would have been incredibly hot—a fireball of radiation and matter. Under these circumstances the particles of matter would have been colliding at high energies, similar to the conditions that we can create today in a small region of space within a particle accelerator. In this way experiments in particle physics have made contact with our interests in the large-scale Universe. Earthbound high-energy experiments are mimicking the condition of the young Universe.

During its first 10^{-33} s the Universe was hotter than 10^{32} degrees. At these temperatures, far beyond anything achieved in an accelerator on Earth, matter and

should be about 100 neutrinos in every cubic centimetre of the Universe. This is some 100 million times the density of protons. So the visible galaxies are but islands in a sea of neutrinos. If a neutrino weighed even as little as 30 electron volts (one 30 billionth of a proton's mass), then neutrinos would dominate the Universe's gravity and the dark matter would be explained. The electron-neutrino's mass is the best measured of all, and experiments indicate that it is probably less than 30 eV. But do neutrinos have any mass at all?

There are hints from the behaviour of our Sun that they might do. Astrophysicists can calculate how many neutrinos should arrive at Earth from the Sun. But a celebrated experiment in the Homestake gold mine in South Dakota, run by Ray Davis from the Brookhaven National Laboratory, detects only one third of the neutrinos the calculations suggest. One way to reconcile the theory of the Sun's interior with Davis's results is to endow the neutrinos with a small mass. However, this is not the end of the story because neutrinos with a mass even as small as 30 eV pose problems for those astrophysicists who are attempting to understand the formation of galaxies.

If neutrinos have mass, however little, then in the first ten thousand years after the Big Bang, they would have been moving at almost the speed of light, rushing outwards with the expanding Universe. But when the Universe cooled and the neutrinos slowed, they would have begun to form clusters under the influence of gravity. These neutrino structures would have covered the whole Universe. Local instabilities within them would have formed the core of galactic superclusters and clusters, out of which individual galaxies would then have condensed. This scenario is called 'top

Fig. 10.23 *Fifteen hundred metres below ground in the Homestake gold mine in South Dakota, a huge tank of dry-cleaning fluid—perchlorethylene—lies in wait to capture neutrinos from the Sun. The neutrinos occasionally interact with a chlorine-37 nucleus in the fluid, and thereby transform it to an argon-37 nucleus. The subsequent decays of the argon nuclei reveal how many such interactions have occurred. But the detector, which has been operating for nearly 20 years, finds only about one third of the number of neutrinos predicted by the accepted theory of nuclear interactions within the Sun.*

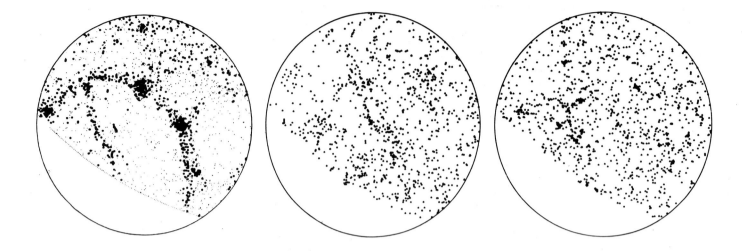

Fig. 10.24 *The nature of the dark matter in the Universe would have affected the formation and distribution of visible clusters of galaxies. The image at left is a computer simulation based on the assumption that the dark matter is hot, consisting of neutrinos with a mass of 30 eV. The picture shows much denser clustering than exists in the real Universe (centre), which was mapped by the Center For Astrophysics in Cambridge, Massachusetts. A simulation (right) which assumes that the dark matter consists of cold, slow-moving particles, on the other hand, is much closer to what astronomers actually observe.*

down' in the jargon of astrophysics, because smaller structures such as galaxies condense out of larger ones such as galactic clusters and superclusters.

Computer simulations show that if the mysterious dark matter is fast-moving and therefore hot—as would be the case if it consisted of neutrinos with a small mass—then the galaxies should form regions of dense clusters, with large volumes of void in between (Fig. 10.24). This is not quite what astronomers observe. The clustering of galaxies *is* patchy, but not nearly as patchy as the computer simulations require.

The evolution of galaxies would have been very different if the dark matter consists of cold, slow-moving particles. Such particles could be slow either because they happen to be produced that way or because they are very massive. Either way, the slow-moving particles would have been 'left behind' as the Universe expanded, and they would have been concentrated by gravity to form clusters on all scales, from clusters of stars to individual galaxies or clusters of galaxies. The small clusters would have developed first, and then merged together to form the larger clusters. This scenario is known as 'bottom up' and seems to fit the real Universe better than the 'top down' version.

So the motion of stars within galaxies strongly suggests the presence of invisible dark matter, while the distribution of the visible galaxies through the Universe seems to imply that this dark matter consists of particles not yet seen in the high-energy laboratories. This is but one example of the way that cosmology and astrophysics can make demands of particle physics and vice versa.

THEORIES OF EVERYTHING

THE GUTs describe only the strong and electroweak forces, but a valid 'theory of everything' (or TOE) must also incorporate gravity. Many theorists are therefore putting considerable effort into working out just what kind of theory can accommodate gravity. This is the cutting edge of theoretical research, and it thrives on a number of new ideas with exotic names such as supergravity and superstrings. These ideas are still very much unproven; it is too soon to know if any of them mirror the natural law of the Universe, or whether they will all end up on the large heap of abandoned scientific theories.

Some of the first exciting hints came with the suggestion of a new kind of symmetry—supersymmetry, or SUSY. The GUTs imply that there are basically two families of particle—particles of matter (quarks and leptons) and force-carrying particles (the gauge bosons). Supersymmetry, on the other hand, links all these particles within one 'superfamily'. But it does so at the expense of predicting many new particles, in the following way.

One feature that distinguishes the matter particles from the force-carrying particles is the property known as *spin*. Many particles behave like spinning tops, but quantum theory dictates that they cannot spin at just any rate. Instead, they are constrained to spin at 'allowed' rates, specific to each kind of particle, rather as electrons within an atom can have only certain 'allowed' energies. This spin can be measured experimentally, and the Table of Particles at the end of the book (see pp. 228–31) gives the value for each particle expressed in units of Planck's constant, $h/2\pi = 1.055 \times 10^{-34}$ joule seconds. The electron and the proton, for example, have spin 1/2 in these units, whereas the Ws and the Z have spin 1. And herein lies an apparent difference between matter particles and force-carriers. While the quarks and leptons all have spin 1/2, the gauge particles have spins of 1.

In linking these particles of different spin, supersymmetry requires a host of matter particles and force-carriers that we have not yet perceived. It predicts 'supermatter' built from particles with integer (0, 1, 2, . . .) rather than half-integer (1/2, 3/2, . . .) spin; and 'superforces' transmitted by agents with half-integer rather than integer spin. So far, however, there is no experimental evidence for such particles.

What has this to do with gravity? The idea of supersymmetry grew out of detailed studies into the structure of space-time. Gravity is also intimately related to this structure, and supersymmetry implies that general relativity—Einstein's theory of gravity—is but one part of a richer theory known as 'supergravity'. One consequence of this theory is that particles called gravitinos should exist, which would be related to the graviton, the hypothetical carrier of the gravitational force. One exciting possibility is that these particles might be found at the new generation of particle accelerators, which will begin to operate in the late 1980s.

The ideas of supersymmetry and supergravity may also lead theorists to understand why, for example, space has three dimensions. Einstein's theory of relativity follows from treating time as a fourth dimension; could there be further dimensions subtly intertwined with the familiar ones so that our senses do not perceive them? Some theories suggest that we may already be aware of the effects of additional dimensions. Over 40 years ago, Theodore Kaluza and Oscar Klein noticed that electromagnetism may be the effect of gravity 'spilling over' from a fifth dimension. They worked out a theory of gravity in five dimensions and then allowed one of the dimensions to 'curl up' and become imperceptibly small. The result of this curling up was something that Kaluza and Klein recognized as electromagnetism. In a similar way, the weak and strong forces may be gravity in higher dimensions.

Most recently, there has emerged the possibility of constructing a theory that contains all these bizarre ideas and more. There is the promise of a unique theory in which the Universe began with 10 dimensions, of which only four expanded to form what we now call space and time. In this theory, particles arise naturally out of the basic mathematical structure not as point-like objects, but as entities that are extended in space—albeit with dimensions of a mere 10^{-36} m. These extended particles are referred to as 'strings'. Supersymmetry is one of the vital ingredients of the theory, which has therefore become known as the theory of 'superstrings'. Physicists are excited about the theory's potential because it describes all four fundamental forces, including gravity, in a natural way, without forced juggling of the mathematics. It thus promises to produce the long-sought marriage of gravity and quantum theory—essential in any TOE.

The superstring theory comes close to being impossible. It may have only one solution; for the theory to be viable, this solution must produce the Universe we inhabit. The theory actually produces two universes in parallel. One of these may be the familiar one, where elements fuse, stars shine, and we exist. The other universe may have its own analogues of the electromagnetic and other forces, which are quite distinct from our own, so that we cannot perceive its radiations.

How can we know if such a 'shadow universe' does exist? The answer lies with gravity, the one thing our Universe and its shadow should have in common. The matter of the shadow universe should persistently tug at the stars and galaxies of our own, disturbing their motions. Science fact or science fiction? Only time and space will tell.

Chapter 11
Particles at Work

THIS BOOK has followed the development of a science that probes the very extremes of the physical world. We have encountered the minute dimensions of the electron and the vast energies of cosmic rays; huge machines that consume as much power as a small city, yet keep billions of tiny particles on a path that is accurate to fractions of a millimetre; detectors as big as a house that can register the presence of particles that leave no tracks at all. From the first blurred images of X-rays and radioactivity, we have come a long way to the coloured trails that depict the decay of a Z particle, and thereby confirm the theory that unites the different phenomena of X-rays and radioactivity within a single elegant framework.

Particle physics is an exhilarating adventure in scientific exploration. It represents one of the peaks of human intellectual activity. And as in any branch of pure science, curiosity plays a large part in motivating the subject's continuing development. But the quest for the ultimate nature of the matter and forces of the Universe offers such a heady prospect that it perhaps obscures the more practical spin-offs that have flowed from studies of atoms and the world within them. Some of these have become so pervasive in the developed world today that we take them for granted and forget their origins in the work of the 19th-century researchers.

The humble electron underlies our lives at almost every step. From digital alarm-clocks, through computer-controlled transport and communication systems, to the television, modern electronics moulds our lives every moment of the day. J. J. Thomson once spoke of how visitors to his laboratory in Cambridge advised him to put his bizarre-looking apparatus to one side and spend his time on something useful. He did not heed their advice, for he was curious about the nature of electricity, and in 1897 he was rewarded with the discovery that electricity is carried by tiny particles—electrons—that are constituents of every kind of atom.

Today, descendants of Thomson's apparatus sit in almost every living room—the ubiquitous television. More important, and more far-reaching, our understanding of the behaviour of materials in terms of the electrons they contain has led to major developments in many areas of science. Chemists have learned how to synthesize new materials and new drugs; biochemists have begun to unravel the intricate workings of the human body and brain; and in solid-state physics, the discovery of the electron has led to the invention of the transistor and the 'microchip', and the consequent revolutions in computing and information-processing.

We live in an electronic age; yet it is also a nuclear age. Turn on your television, pick up your electronically typeset and printed news magazine and you will almost certainly encounter ramifications of another discovery made in the search for the nature of matter—the atomic nucleus. Most often it is the negative aspects of the word 'nuclear' that are brought home to us. The threat of an annihilating nuclear war; the problems associated with safety at nuclear power plants; the difficulties of dealing with radioactive waste—all are crucial areas of debate in which nuclear physicists have a duty to participate, both as scientists and as concerned citizens. Yet 'nuclear' need not mean 'nasty'. Many people are possibly unaware of the real benefits that have come with the discovery of the atomic nucleus and the subsequent 'particle explosion'.

In this final chapter, we look at some of the ways that particles discovered in fundamental research have already become established as 'high-tech' tools in areas as wide-ranging as medicine and archaeology, and we glimpse other exciting possiblities that may lie ahead as a result of today's 'curiosity'.

Fig. 11.1 We live in an electronic age, which began nearly 100 years ago with J. J. Thomson's discovery of the electron. This wafer of over 150 uncut silicon chips, seen about life-size (top) and in close-up, is just one of the many consequences of that discovery. Who can say what technological changes might result in another 100 years from the pursuit of 'pure' science today?

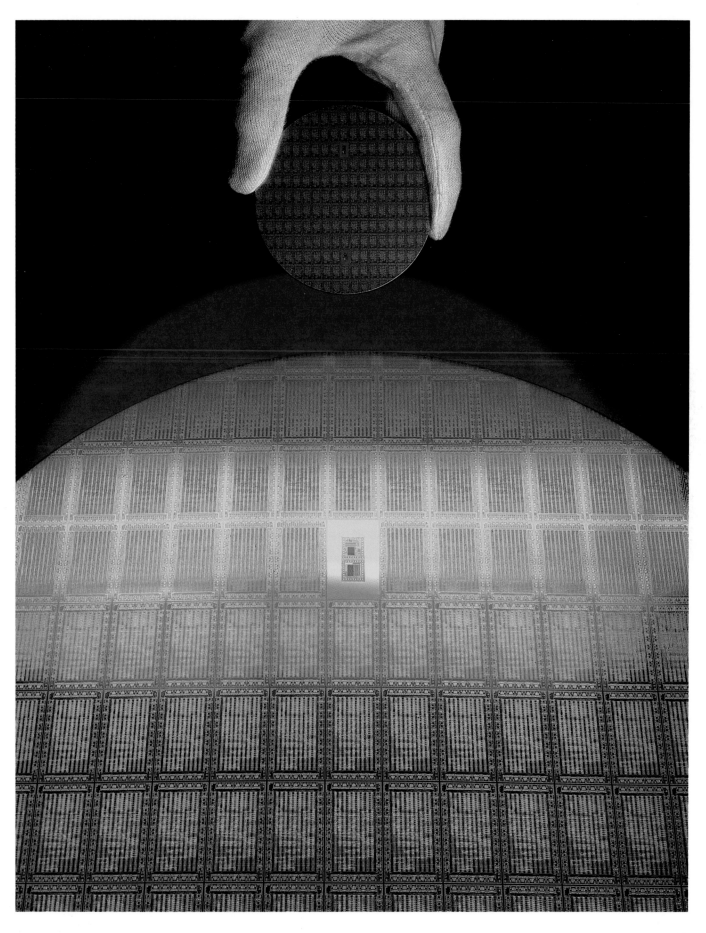

THE NUCLEUS AS A MAGNET

THE NUCLEI of many elements, such as radium, are well known for being radioactive, a property that Chapter 2 described. But some nuclei are distinguished by a less notorious characteristic—they are magnetic. Over the past 30 years or so, scientists in a variety of disciplines have been putting the magnetic properties of nuclei to use in a technique known as *nuclear magnetic resonance*, or NMR.

How can a nucleus be magnetic? The answer lies in its net positive electric charge. When an electric charge rotates, it generates a magnetic field. This is how an electromagnet works; the electric current flowing through the coil of wire is nothing more than a host of circulating charges, carried by the electrons. A nucleus can also behave as if it is rotating, spinning like a miniature top. In the jargon of nuclear physics, we say the nucleus 'has spin'.

The overall spin of a nucleus depends on the way that the motions of the individual protons and neutrons within the nucleus combine together; thus not all nuclei have spin. However, a nucleus with spin is in effect a rotating electric charge and it produces a magnetic field, just like a tiny electromagnet. This is the property that comes into play in NMR.

Put a sample of a material containing magnetic nuclei in a magnetic field, and the tiny nuclear magnets will try to line up with the field. The fact that the nuclei are spinning prevents them from lining up with the field exactly, and instead they wobble about the field direction in the way that a spinning top wobbles about the vertical. The frequency associated with this wobble—more formally known as the precession frequency—depends on the strength of the field and the type of nucleus. Herein lies the key to the practical value of NMR. If you measure the frequency at which the magnetic nuclei in a small amount of matter wobble, and if you know the strength of the field, you can identify the type of nuclei. Conversely, if you know what nuclei are in the sample, the measured frequency tells you the strength of a magnetic field

To measure the frequency, you briefly stimulate the nuclei to wobble more violently by applying an additional oscillating magnetic field—in practice, radio waves—vibrating at the precession frequency of the nuclei. (This is a resonance condition [see p. 139], so now we have all the components in the name 'nuclear magnetic resonance'.) After stimulation, the nuclei return to their normal state by radiating their newly-gained energy as radio waves at the precession frequency.

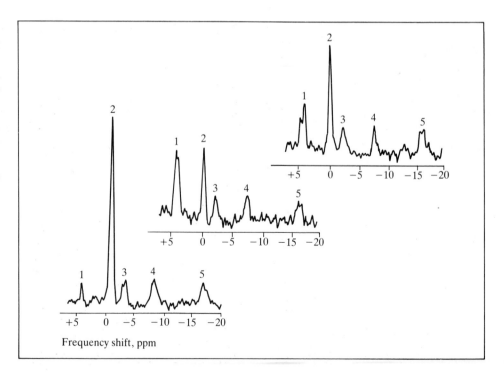

Fig. 11.2 *NMR spectroscopy on humans can reveal important metabolic changes. Here phosphorus spectra of the muscle in a person's arm show changes due to exercise. The first spectrum is for the muscle at rest, and exhibits a large peak due to phosphocreatine (peak 2), a compound characteristic of muscle tissue in vertebrates. The central spectrum was taken while the person was exercising the arm, and this shows that the amount of phosphocreatine has fallen, while the level of ionorganic phosphate (peak 1) has risen; the phosphocreatine is used to maintain the levels of adenosine triphosphate (peaks 3, 4, and 5)—the energy source in the muscle cells. The third spectrum shows that after exercising, the amount of phosphocreatine begins to rise again.*

The first work on NMR in bulk matter began in the 1940s, and physicists soon realized that they had a sensitive new tool for measuring magnetic fields. Geologists, civil engineers, archaeologists, and space scientists now use NMR probes routinely to measure magnetic fields. Chemists have found NMR useful in helping them to analyse chemicals. In the same magnetic field, magnetic nuclei of different elements wobble at different frequencies. Moreover, nuclei of the same element but in different chemical groups wobble at slightly different frequencies; this is because the magnetic environment provided by the surrounding atoms varies from one grouping to another. Thus the signal from hydrogen nuclei—protons—in a CH_2 group in a hydrocarbon is slightly offset from the signal from protons in a CH_3 group in the same molecule. In this way an NMR spectrum is an important chemical 'fingerprint' that can help in identifying chemicals and in elucidating the structures of complex molecules. NMR spectroscopy is also becoming a useful tool in biochemistry and medicine. *In vivo* spectroscopy of human limbs and organs can reveal important metabolic changes that result from exercise, for example, or the administration of drugs.

NMR has another, more widely publicized role in medicine—imaging the body's internal structure. By far the most common magnetic nucleus in the body is the nucleus of hydrogen—the proton. In NMR imaging, the technique is to make the protons in a patient's body wobble in a magnetic field that varies in strength across the body. The protons then wobble at different frequencies depending on their locations. Thus the NMR spectrum contains information about the number of protons—the amount of hydrogen—at different places. A computer can analyse this information and translate it into an image of a 'slice' through the body. The technique is known as 'tomography', from the Greek 'tome,' which means 'a cutting'.

NMR imaging is a complex business, but it does seem to have certain advantages over other techniques. In particular, it is possible to differentiate between signals from a variety of soft tissues, all of which are relatively transparent to X-rays. Moreover, NMR signals contain additional information in the time the protons take to revert to their normal state after stimulation—the 'relaxation' time. It turns out, for reasons not yet properly understood, that protons in tumours, for example, have longer relaxation times than protons in normal tissue.

Another advantage of NMR is that it does not use a potentially damaging radiation. The amounts of radio-wave radiation absorbed by a person undergoing an NMR scan are too small to induce chemical changes in the body. And the deleterious effects of

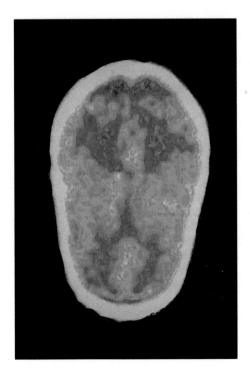

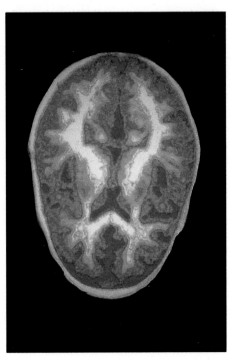

Fig. 11.3 (*left*) *Imaging of the human body with proton NMR differentiates soft tissues that are all but invisible to X-rays. This colour-coded image of the brain of a newborn human infant shows relatively little structural development.*

Fig. 11.4 (*right*) *A similar image of the brain of a 5 year old child shows significant development in contrast to the newborn child. The white areas indicate the presence of myelin, a complex material of protein and phospholipid that forms an insulating layer around the nerve fibres of certain nerve cells in the brain. Such nerves conduct impulses more rapidly than those without myelinated fibres.*

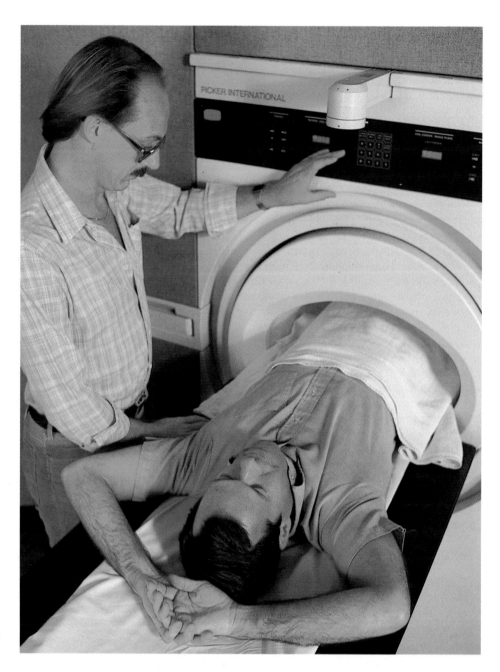

Fig. 11.5 *A patient undergoing an NMR scan lies with part of his body in the bore of a large superconducting magnet.*

magnetic fields—at least at the levels used in NMR—seem to be negligible. This is in contrast to X-rays, which in sufficient qantities can damage cells in the body. However, the two techniques have a complementary role to play in medicine, because they can highlight different conditions.

The scanning of whole bodies for NMR imaging—or indeed for *in vivo* spectroscopy—is a high-technology process. The patient must lie in a region of carefully controlled magnetic field—in practice, within the coil of an electromagnet. In medical imaging and also in NMR spectroscopy, the magnets used are often superconducting. Spectroscopy requires high fields usually over small volumes of a few cubic centimetres; for NMR imaging, the fields can be lower, but to cover a whole human body the magnets must be much larger. Superconducting magnets have proved to be the answer in both cases, being capable of producing both high fields and large regions of uniform fields. Here, we see how a technology encouraged by the needs of particle physics—for more powerful magnets in higher-energy accelerators—has become central to another area of science, which itself grew from the discovery of the atomic nucleus.

PARTICLES AGAINST CANCER

NMR IS only one example of the way in which discoveries in the search for the nature of matter have helped in medicine. The medical uses of X-rays, with their ability to produce 'internal' images, became apparent almost as soon as Röntgen had discovered the penetrating rays; the possibility of using well-defined beams of X-rays to destroy malignant tissues arose soon after. Today, radiotherapy using specially designed X-ray machines is one of the main treatments for cancer. However, X-rays are by no means always successful in destroying cancerous tissue. Cancer experts are gradually discovering that certain types of tumour respond better to other radiations—to some of the subatomic particles discovered in the decades following the discovery of X-rays.

The most well-developed alternative form of radiotherapy for cancer uses neutrons. Neutrons, like X-rays, are an electrically neutral form of radiation, and their action is therefore indirect. In both cases it is not the primary radiation that does the damage, but the electrically charged particles that the neutral radiation liberates. In the case of X-rays, the damage is caused by electrons. Neutrons, on the other hand, collide with atomic nuclei and knock out protons and alpha particles. These bulky charged particles can be much more devastating than electrons, as we can appreciate by comparing the thick tracks that nuclei leave in emulsions with the wispy tracks of electrons (see Fig. 3.12, p. 57).

The way to kill a cancer cell is to stop it reproducing *beyond repair*. The words in italics are crucial and the way to do that is to damage its DNA, because cells can repair the damage that X-rays inflict, but not the destruction that neutrons cause. Just how effective the repair mechanisms are depends on the kind of cancer cell. Some cancers are completely resistant to X-rays, and it is these that are now being treated with neutrons. Compared with X-ray therapy, neutron therapy is still in its infancy, and it will take several more years to evaluate properly the medical benefits of neutrons.

Recently pions have also begun to play a role in cancer therapy. One important limitation in conventional radiotherapy is the dose of radiation that can be delivered to healthy tissue surrounding a tumour. By carefully tailoring the beam from the X-ray machine, as few X-rays as possible are allowed into the region around the tumour. But with tumours beneath the skin, the problem still remains of damaging healthy cells between the skin and the tumour. Pion therapy provides a solution by concentrating the destructive power of the radiation at a region *within* the body. The

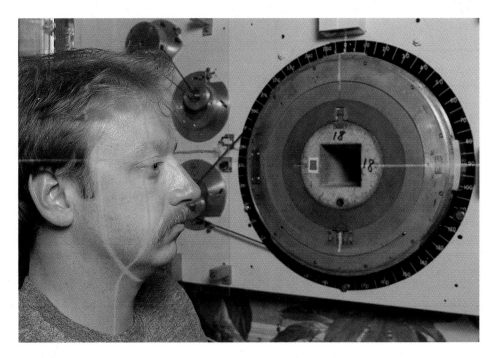

Fig. 11.6 *While the main accelerator at Fermilab shoots protons up towards 1 TeV, relatively low energy protons from the linac are used to create neutrons for use in cancer therapy. This demonstration shows how laser beams (red) are used to line up a person's head in readiness for treatment. The neutron beam emerges from the square aperture towards the right of the picture.*

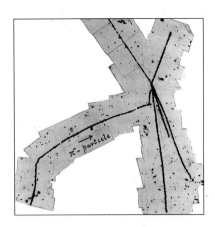

Fig. 11.7 *This 'pion star' recorded in emulsion is typical of the kind of interaction that occurs in pion beam cancer therapy. A pi-minus enters from the left, and after a wide-angle scatter hits what is probably an oxygen nucleus in the gelatin of the emulsion and breaks it up into many fragments. Two protons shoot off, one towards top left, the other to bottom right. An alpha particle moves up towards top right, while an alpha and a tritium nucleus head off down the picture, to the left of the lower proton. In cancer therapy with negative pion beams, oxygen is also the most common 'target' in the malignant tissue. The length of the alpha track at upper right is 40 micrometres.*

technique is to make *negative* pions enter the patient at just the right speed to penetrate as far as the tumour and no further.

In any kind of matter, the negative pion behaves like an exploding depth charge. It lives on average for 26 billionths of a second. While a pion is moving it does little damage to the material through which it passes. Only in the last instants of its existence does a negative pion have a large destructive effect, when it is captured by the positive nucleus in an atom in a cell. The nucleus becomes unstable and splits apart into smaller fragments, forming a 'pion star' like those first seen in the emulsion experiments of Cecil Powell and his colleagues in the late 1940s. Fragments from the star can also damage surrounding cancer cells within a small distance. In this way, a negative pion can sink through healthy tissue until it comes into contact with the 'target'—tumour cells—and 'blows up'.

Clinical experience with pion therapy, in particular at the Swiss Nuclear Institute near Zurich and at the Meson Physics Facility at the Los Alamos Scientific Laboratory in the US, appears promising. Tumours in the head and neck appear to respond particularly well.

Cancers of the eye seem to respond to yet another kind of particle therapy—treatment with protons or alpha particles. Tumours in the eye occur too near to other vital tissues for X-ray therapy to be useful. The X-rays release lightweight electrons, which can travel too far before they do their damage, thus destroying healthy tissue rather than cancerous cells. By contrast, protons and alpha particles knock out heavy particles from atomic nuclei, and these damage a well-defined narrow region around the therapeutic beam. A team at Berkeley has used beams of alphas from Lawrence's famous '184-inch' cyclotron to treat eye cancers as well as certain brain disorders. Researchers at Berkeley are also investigating the effectiveness of beams of heavy ions from the Bevatron against a variety of cancers.

THE DIAGNOSTIC POSITRON

RADIATION THERAPY harnesses the destructiveness of radioactivity by directing high levels of radiation at cancerous tissues. Another way to use radioactivity in medicine, in much smaller doses, is to 'label' interesting substances in the body, by attaching radioactive 'tracers'. The tracers are tiny amounts of radioactive elements, which are injected into the body. These radioactive nuclei can take part in biochemical reactions in the same way that the normal nuclei do, and by following their radioactive emissions we can build up a picture of what happens to the substance under study.

A famous example of a radioactive tracer is the isotope iodine-131, which emits gamma rays. Iodine is naturally taken up by the thyroid gland, but if the thyroid is overactive or underactive then it takes up more or less iodine than usual—and in both cases the person suffers from unpleasant effects, for the thyroid secretes hormones that control growth. By injecting iodine-131 into a patient and detecting the emitted gamma rays, a radiologist can measure what proportion of the iodine ends up in the thyroid, and so establish whether its activity is normal or not. The patients receive about the same tiny amount of radiation from these tests as they do from cosmic radiation during a single flight in a jet aircraft.

The first radioactive tracers to be used, such as iodine-131, were gamma emitters, but nuclei that emit positrons now play an increasingly important role. The crucial property of positrons is that they are a form of antimatter, and as earlier chapters have shown antimatter and matter annihilate upon meeting. Positrons emitted by an appropriate radioactive nucleus will annihilate with electrons close by, and in turn produce gamma rays. Because the electrons and positrons are more or less at rest when they meet, their annihilation can produce two gamma rays *back-to-back*, so as to conserve energy and momentum. The two gammas can be detected in coincidence, using the type of circuitry familiar in particle physics. Detecting *pairs* of gammas in this way yields the location of the emitting nucleus with greater precision than the single rays from a gamma emitter provide.

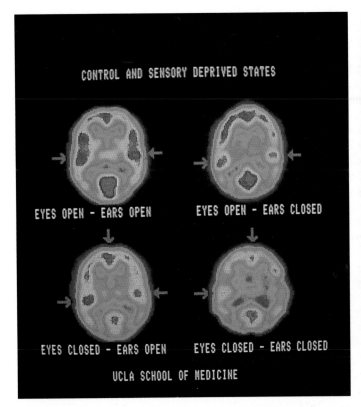

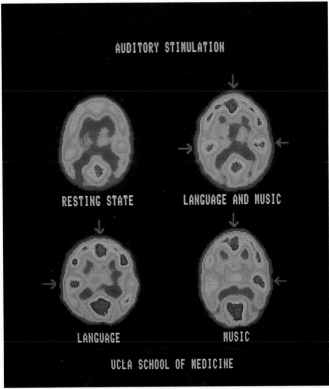

Some of the positron emitters exploited in recent years include carbon-11, nitrogen-13, and oxygen-15. These are all particularly useful because they are radioactive forms of common building blocks in the body. Radioactive oxygen nuclei can be used to label oxygen gas for the study of oxygen metabolism; carbon monoxide for the study of blood volume; or water for the study of blood flow in the brain. Similarly a radioactive form of fluorine attached to a sugar molecule can reveal the brain's sugar metabolism and hence the activity of the brain in response to different stimuli. And carbon-11, implanted in the chemical dopamine, is assisting in studies of the brain disorders that give rise to Parkinson's disease.

Detailed imaging of the brain has become possible over the past decade with the development of the technique known as positron emission tomography, or PET. The patient's head is surrounded with a 'halo' of gamma ray detectors, which feed signals to a computer. The computer uses the information to build up images of 'slices' through the brain—hence the term tomography again.

A number of centres around the world have cyclotrons specifically to produce the radioactive nuclei needed in medicine. The tracers are created when a beam of high-energy protons (or alpha particles or deuterons) collides with nuclei in a suitable element. Often the cyclotron must be located in the hospital itself, if it is to provide useful short-lived nuclei. For example, oxygen-15, which has a half-life of two minutes, can be fed directly from a production target near a cyclotron to the breathing mask of a patient undergoing a PET scan. Sometimes, the same cyclotron can be used to produce beams for neutron therapy. This was the case at the Hammersmith Hospital in London, where much of the pioneering work on neutron therapy was performed until the aged cyclotron was shut down in 1985.

Positron annihilation has also proved a valuable tool in studying industrially important materials. For example, when positrons annihilate with electrons in metals, they can reveal the onset of metal fatigue. Distortions in the atomic lattice of the metal provide 'resting sites' where the positrons survive slightly longer before annihilation. By observing this slight delay it is possible to detect fatigue before any cracks appear in the material. The ability to push turbine blades and other expensive components safely towards their ultimate breaking point, thereby narrowing large safety margins, promises to be of immense economic importance.

Figs. 11.8–11.9 *Positron emission tomography (PET) scans show the activity of the brain in response to different stimuli. The subjects have been injected with a positron-emitter that attaches to glucose in the blood. The glucose concentrates in areas where there is increased metabolic activity in the brain, and this in turn shows up as an increased amount of radioactivity.*

Fig. 11.8 *(left) The captions beneath these four scans show whether eyes are open or closed (with light-tight patches) and whether ears are exposed or covered (with rubber plugs and sound-proof headphones). Provided there is some stimulus, there is left-right symmetry in the brain's activity, but when eyes and ears are closed (bottom right), the right hemisphere decreases its activity more than the left.*

Fig. 11.9 *(right) For these four scans the subject was given different kinds of auditory stimulation. With only verbal stimuli (a Sherlock Holmes story) the left side of the brain appears more active; with non-verbal stimuli (a Brandenburg concerto) there is more activity in the right-hand side.*

DETECTIVE WORK WITH NEUTRONS

ANOTHER WAY to use radioactivity to advantage is to induce nuclei already within a sample to become radioactive and thereby reveal their identities. Nature has provided just the tool for this task with the neutron, which can easily penetrate a material. Suppose you shine neutrons onto a substance that contains several elements. Atomic nuclei capture the neutrons and are temporarily changed into radioactive forms, the nature of the new radioactive nuclei depending on the original elements. What proves valuable is that the various radioactive species decay at different rates and emit a variety of characteristic radiations. So by monitoring the spectrum of the emissions, we can identify elements even when they are present in minute quantities. The neutrons have 'activated' the original elements, provoking them into advertizing themselves.

One unlikely area for the application of neutron activation is in the art world, where the technique can detect forgeries and occasionally bring surprises with famous

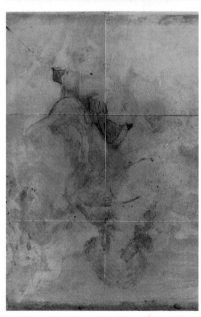

paintings. Exposing a painting to a flux of neutrons for about an hour generates a low level of radioactivity within the paints. Any electrons—beta rays—emitted will now leave an image on a specially-prepared film. Nuclei of different elements within the paints have different half-lives, so some finish emitting sooner than others. Thus photographs taken at intervals record a variety of images depending on which elements are the dominant radiators at that time.

The nuclei that give up their radiation soonest (shortest half-lives) appear only in images formed soon after activation. Longer-lived nuclei show up in later images formed after the short-lived species have all died out. Different colours of paint involve different elements and so show up at characteristic times in the post-activation images. The technique can reveal an overpainted signature or even a whole painting that has been obliterated. The would-be forger must ensure that the half-lives of his paints match those of the paints used by the master whose work is being faked!

Surprises can emerge even with well-known paintings. Figure 11.10 shows Van Dyck's *Saint Rosalie interceding for the plague-stricken of Palermo*, which is in the New York Metropolitan Museum. An X-ray image (top) shows vague hints of a face that was originally on the canvas but which has been painted over. (Turn the book upside down to find the face.) This analysis with X-rays shows up the lead-based paints as shadows. Other paints are rather transparent to X-rays, but these are the very paints that show up well after neutron activation.

The image recorded a few hours after activation (middle) reveals the presence of manganese, which occurs in umber, a dark earth-pigment. The canvas was filled with an umber base layer and the original figures drawn in, as revealed in the X-ray image. The blank areas in the neutron-activated image are where modern repairs, free of manganese, have been made. We can now see that the angel immediately behind the saint's head must have been added as an afterthought, because it does not appear in the original underdrawing.

Images taken four days after activation, when the emissions from the umber have died away, reveal a different tale (bottom). Now it is the element phosphorus that is the dominant radiator. Undersketches often used charcoal or boneblack and it is the latter that contains phosphorus and shows up in the 'four-day picture'. Now we can see the mystery face clearly. The secret is out: Van Dyck made a self-portrait and then overpainted it.

'TALKING MOONSHINE'

IN THE applications of protons, neutrons, pions, and positrons, radioactivity is often a key feature. Instability plays a part both in producing the particles (as with positrons) and in their useful effects (as in pion therapy). Another unstable particle whose death can be revealing is the muon.

The muon, like the proton, is a spinning electric charge and so behaves like a subatomic magnet. But unlike the proton, the muon quickly decays and emits an electron or a positron, depending on the muon's charge. The direction in which the electron or positron emerges depends upon the orientation of the muonic magnet, an effect that is valuable in the technique known as 'muon spin rotation'.

Once a muon has entered a material, it will survive for only a microsecond or so before decaying. However, this is long enough for the magnetic fields within the material to rotate the muonic magnet. So if you know the muon's orientation before it enters the material and measure the direction of the particle emitted a microsecond later, you can learn something about the internal magnetic fields. Here the properties of the muon help again, because a muon's magnetic orientation is fixed when it is born, in the decay of a pion, for example. So a beam of pions can create a beam of 'polarized' muons, with their magnets all pointing the same way.

Beams of positive muons are produed in this way at laboratories such as CERN and the Swiss Institute for Nuclear Physics, and used specifically for experiments utilizing muon spin rotation. Chemists, for example, use the technique to probe the structure of hydrocarbons. Positive muons behave chemically like lightweight protons and

Fig. 11.10 (*opposite*) *Neutron activation of a famous painting*, Saint Rosalie interceding for the plague-stricken of Palermo, *reveals a hidden self-portrait of the artist, Anthony Van Dyck (1599–1641).*

The large image is an ordinary photograph of the painting, which hangs in the New York Metropolitan Museum.

The X-ray of the painting (top) reveals hints of the face of a man, upside down near the bottom of the picture.

The middle image, recorded a few hours after irradiation with neutrons, shows the presence of manganese in the original umber base layer.

Four days after neutron irradiation (bottom), phosphorus is the main radiating element and we see the outlines of the charcoal sketch of Van Dyck's self-portrait. (Turn the book upside down if you have difficulty finding the face.)

Fig. 11.11 *A mystery event in Luis Alvarez's 25 cm diameter hydrogen bubble chamber gives a glimpse of muonic fusion at work. A negative muon travels up the picture (the central vertical track) and is apparently deflected first to the left and then sharply to the right, before decaying to an electron which curls to the left and down to the bottom of the picture. The explanation is that the muon becomes bound to a nucleus of deuterium—naturally present in the hydrogen at the level of 0.02 per cent—to form a small neutron-like object that can 'invade' a hydrogen atom and bind with the proton. The proton and deuterium nucleus fuse into a nucleus of helium-3, releasing the muon which is then free to catalyse another fusion. In this case the muon catalyses two fusions; each time it carries away the energy released in the fusion process, so the two short segments of track are of similar lengths.*

attach to electrons in reactive chemical groups. Measuring the effects of muon spin rotation can provide clues about the way that protons normally couple to these groups.

Some physicists entertain a more ambitious way to employ muons—in catalysing nuclear fusion and thereby releasing abundant nuclear energy free from the hazards associated with nuclear fission reactors. A typical fuel for a fusion reactor would be a mixture of two heavy versions of hydrogen—deuterium and tritium. These nuclei each contain a single proton, as hydrogen does; but deuterium also contains a neutron, while tritium has two neutrons. If these nuclei can be made to come close enough, overcoming the mutual repulsion due to their positive charges, they fuse to form a nucleus of helium-4, built from two protons and two neutrons. The third neutron from the original nuclei is 'spare'. The helium-4 requires less energy to hold it together than was needed for the separate deuterium and tritium nuclei, so the process of fusion releases energy. This is carried away by the spare neutron.

The idea with muon-catalysed fusion is to take advantage of the muon's similarity to the electron to bring the deuterium and tritium nuclei close enough to fuse. A muon can in fact replace an electron in an atom, but because the muon is much heavier than the electron, it is held more closely to the nucleus and moves in tighter orbitals. Introduce a muon into a mixture of deuterium and tritium and it will bind with a tritium nucleus to form a compact neutral object similar in some ways to a neutron (though three times as heavy). This neutral object can penetrate a deuterium molecule (consisting of two atoms) and attach itself to one of the deuterium nuclei. Then the stage is set for the fusion reaction, in which the muon is merely a bystander, and is freed to initiate more fusions before it ultimately decays after 2.2 microseconds.

Luis Alvarez and his team at Berkeley saw the first evidence for muonic fusion in 1956. They were studying the interactions of kaons in their 10 inch (25 cm) hydrogen bubble chamber, using a beam contaminated with muons. Moreover, the hydrogen naturally contained a tiny amount of the heavier deuterium nuclei, so they inadvertently had suitable conditions for muonic fusion. At first Alvarez and his colleagues thought they were observing some unusual decay sequence when they saw patterns of tracks like that in Fig. 11.11, but it soon became clear that this theory would not fit the facts. Then a fellow physicist at Berkeley, Jack Crawford, put forward another idea: the bubble chamber occasionally recorded successive fusions induced by a single muon.

What promise does muonic fusion hold? It takes energy to produce muons—you must first accelerate protons to produce pions which create muons when they decay. To be certain of obtaining more energy than you put in, each muon must yield 300-400 fusions. As of March 1986, the highest number claimed in an experiment is in the region of 150 fusions per muon—a figure that is close enough to the required number to make some physicists optimistic about building practical muonic fusion reactors.

A more futuristic proposal for using muons deals with prospecting for valuable ores such as uranium. This is in fact part of a package of schemes put forward by an idiosyncratic collection of particle physicists, comprising Georges Charpak (inventor of multiwire and drift chambers), Sheldon Glashow (who received the Nobel prize for his part in electroweak theory), Robert Wilson (ex-director of Fermilab), and Alvaro De Rújula (a theorist at CERN).

The centrepiece of their package is the GEOTRON—a proton accelerator 160 km in circumference, designed to produce intense beams of high-energy neutrinos. What would make this accelerator unlike any other is that it would float out at sea. This bizarre idea would make it possible to direct neutrinos—produced as usual from the decays of pions and kaons—at various angles and directions through the Earth's surface, even straight down if necessary.

For Geological Exploration with Muons produced in Neutrino Interactions (Project GEMINI), the aim would be to monitor muons produced by neutrinos as they travel through rock beneath the surface of the Earth. The high-energy muons created in this way would themselves be highly penetrating and could travel out through the rock to suitable detectors. Heavy material, such as a uranium ore body, would tend to produce more muons than the surrounding rock.

Another possibility would be to use the neutrino beams in prospecting for oil and natural gas. Geological Exploration by Neutrino-Induced Underground Sound, which has the pithy acronym GENIUS, would sense sound waves—pressure changes—generated as neutrinos deposited energy in interactions tens of kilometres beneath the Earth's surface. The aim would be to measure the velocity of sound in different directions and so build up a pattern of the density of the underlying rock, which could reveal pockets containing oil or gas.

Schemes such as GEMINI and GENIUS may sound like something out of science fiction. Yet we should not be too quick to dismiss them. Rutherford is often remembered for his famous remark in 1933 to the effect that anyone who believed that we could harness the energy of the nucleus was 'talking moonshine'; within a few years he was proved wrong. Perhaps Glashow and his colleagues *are* talking moonshine; but there is always the possibility that they are not. Today's high-energy particle accelerators are certainly huge and expensive. But physicists are already studying ways in which the high electric fields associated with laser beams might be tapped to accelerate charged particles. A smaller scale of particle accelerator could open up a host of possibilites, from cheaper high-technology facilities in hospitals to the GEMINI and GENIUS projects themselves.

The practical applications of discoveries in the search for the nature of matter often have a long gestation period. Neutron therapy is only now in its infancy, 50 years after the discovery of the neutron; the transistor, which spawned the electronic revolution of the second half of the 20th century, came more than 50 years after Thomson was advised to drop his work on electricity and do something more useful.

At present it is difficult to foresee what use may be made of the discovery of the W and Z particles, or of the third generation of quarks and leptons. But that is no reason to don blinkers and to stop asking questions. Particle physics may be a pure science, but it is a practical science; the beautiful symmetric theories of matter and forces now emerging are built on a solid foundation of measurement and observation. The measurements are made, the experiments performed, and technology extended to new limits, because people ask questions. We must hope that the opportunity remains to answer these questions, and that some physicists at least are spared the pressures to 'do something useful'.

TABLE OF PARTICLES

This table includes only the major particles described in this book. In a number of cases (e.g. the muon), details of a particle and its antiparticle are given in the same entry; in other cases (e.g. the positron), antiparticles have a separate entry; but in many cases details of antiparticles are not given at all. Our criteria have been to include antiparticles mentioned separately in the book, and ones whose discovery occurred separately from their matter equivalent. Note also that antiparticles such as the positron and antiproton are described as stable, although this is true only as long as they do not meet and annihilate with an electron or proton.

Laboratories where particles were discovered are referred to by their acronyms: BNL is Brookhaven National Laboratory, LBL is Lawrence Berkeley Laboratory, SLAC is Stanford Linear Accelerator Center, DESY is Deutsches Elektronen Synchrotron, and CERN is the European centre for nuclear research (originally Conseil Européen pour la Recherche Nucléaire).

LEPTONS

NAME	SYMBOL	PHYSICAL PROPERTIES				DISCOVERY				NATURE & ROLE	PAGES
		MASS	LIFETIME	CHARGE	SPIN	DATE	BY WHOM	SOURCE	DETECTOR		
ELECTRON	e^-	0.511 MeV	Stable	−1	$\frac{1}{2}$	1897	J. J. Thomson	Cathode ray tube	Fluorescent glass	Lepton of 1st generation; constituent of atoms; carrier of electricity	52–5
POSITRON	e^+	0.511 MeV	Stable	+1	$\frac{1}{2}$	1932	C. Anderson	Cosmic radiation	Cloud chamber	Lepton of 1st generation; antiparticle of electron; formed in cosmic ray showers	84–6
MUON & ANTIMUON	μ^- μ^+	105.6 MeV	2×10^{-6} s	−1 +1	$\frac{1}{2}$	1937	S. Neddermeyer & C. Anderson	Cosmic radiation	Cloud chamber	Leptons of 2nd generation; decay products of pions, kaons, etc.; components of cosmic rays	87–9
TAU & ANTITAU	τ^- τ^+	1.784 GeV	3×10^{-13} s	−1 +1	$\frac{1}{2}$	1975	M. Perl's team at SLAC	Electron–positron annihilation	Electronic	Leptons of 3rd generation	184–5
ELECTRON NEUTRINO & ANTI-NEUTRINO	ν_e $\bar{\nu}_e$	0 (?), <50 eV	Stable (?)	0	$\frac{1}{2}$	1956	E. Cowan & F. Reines	Nuclear reactor	Antineutrino capture detected by liquid scintillator	Leptons of 1st generation; produced by, & probe of, weak interaction	143–6
MUON NEUTRINO & ANTI-NEUTRINO	ν_μ $\bar{\nu}_\mu$	0 (?), <0.5 MeV	Stable (?)	0	$\frac{1}{2}$	1962	M. Schwartz & team from BNL & Columbia	Decays of pions produced at accelerator	Regenerated muon detected by spark chamber	Leptons of 2nd generation; produced by, & probe of, weak interaction	143–6
TAU NEUTRINO & ANTI-	ν_τ $\bar{\nu}$	0 (?), <70 MeV	Stable (?)	0	$\frac{1}{2}$	—	—	—	Inferred from tau decay; no direct	Leptons of 3rd generation	143–6

Quarks

NAME	SYMBOL	PHYSICAL PROPERTIES				DISCOVERY			NATURE & ROLE	PAGES
		MASS	LIFETIME	CHARGE	SPIN	DATE	BY WHOM	HOW		
UP & ANTI-UP	u $\bar{u}$	~5 MeV	Stable*	$+\frac{2}{3}$ $-\frac{2}{3}$	$\frac{1}{2}$	1964	Gell-Mann & Zweig quark model	Direct observation in 1968–72: electron scattering at SLAC, neutrino scattering at CERN	Quarks of 1st generation; up is constituent of protons, neutrons, & other particles	147–9
DOWN & ANTI-DOWN	d $\bar{d}$	~10 MeV	Variable*	$-\frac{1}{3}$ $+\frac{1}{3}$	$\frac{1}{2}$	1964	Gell-Mann & Zweig quark model	Direct observation in 1968–72: electron scattering at SLAC, neutrino scattering at CERN	Quarks of 1st generation, down is constituent of protons, neutrons, & other particles	147–9
STRANGE & ANTI-STRANGE	s $\bar{s}$	~100 MeV	Variable*	$-\frac{1}{3}$ $+\frac{1}{3}$	$\frac{1}{2}$	1964	Gell-Mann & Zweig quark model	Direct observation in 1968–72: electron scattering at SLAC, neutrino scattering at CERN	Quarks of 2nd generation; constituents of strange particles	147–9
CHARM & ANTICHARM	c $\bar{c}$	~1.5 GeV	Variable*	$+\frac{2}{3}$ $-\frac{2}{3}$	$\frac{1}{2}$	1974	B. Richter & team at SLAC, S. Ting & team at BNL	Inferred from J/psi (1974), charmed baryon (1975), charmed meson (1976), & charmonium spectroscopy	Quarks of 2nd generation; constituents of charmed particles	147–9 & 178–83
BOTTOM (or BEAUTY) & ANTI-BOTTOM	b $\bar{b}$	~4.7 GeV	Variable*	$-\frac{1}{3}$ $+\frac{1}{3}$	$\frac{1}{2}$	1977	L. Lederman & team at Fermilab	Inferred from upsilon (1977), & bottomonium spectroscopy	Quarks of 3rd generation; constituents of bottom particles	147–9 & 186–8
TOP (or TRUTH) & ANTITOP	t $\bar{t}$	>30 GeV	Variable*	$+\frac{2}{3}$ $-\frac{2}{3}$	$\frac{1}{2}$	1984 (?)	C. Rubbia & UA1 team at CERN	Inferred from decay of W into top & bottom particles	Quarks of 3rd generation; constituents of top particles	196–7

(*As quarks occur only in pairs (mesons) or triplets (baryons), their lifetimes are variable, depending on the nature of the individual meson or baryon. The up quark, being the lightest, is as stable as the proton which contains it.)

Gauge Bosons

NAME	SYMBOL	PHYSICAL PROPERTIES				DISCOVERY				NATURE & ROLE	PAGES
		MASS	LIFETIME	CHARGE	SPIN	DATE	BY WHOM	SOURCE	DETECTOR		
PHOTON	γ	0	Stable	0	1	1923	A. Compton; (implied: A. Einstein, 1905)	X-rays scattered from atomic electrons	Crystal spectrometer	Carrier of electromagnetic force; 'packet' of electromagnetic radiation	64–5
W (W-plus) (W-minus)	W^+ W^-	83 GeV	10^{-25} s	+1 -1	1	1983	UA1 & UA2 teams at CERN	Proton–antiproton annihilation	Electronic	Carriers of weak force (along with Z)	192–5
Z	Z	93 GeV	10^{-25} s	0	1	1983	UA1 & UA2 teams at CERN	Proton–antiproton annihilation	Electronic	Carrier of weak force (along with W^+ & W^-)	192–5
GLUON	g	0	Stable	0	1	1979	Teams at DESY	Electron–positron annihilation	Electronic	8 types of gluon; carriers of strong (colour) force	189–91

MESONS

NAME	SYMBOL	PHYSICAL PROPERTIES					DISCOVERY				NATURE & ROLE	PAGES
		MASS	LIFETIME	CHARGE	SPIN	QUARK CONTENT	DATE	BY WHOM	SOURCE	DETECTOR		
PION (pi-zero)	π^0	135 MeV	0.8×10^{-16} s	0	0	$u\bar{u}$ or $d\bar{d}$	1949	R. Bjorkland & team at LBL	Interactions of protons from accelerator	Tantalum converter & proportional counters	Involved in nuclear binding; decays into photons; a source of cosmic gamma rays	130–1
PION (pi-plus) (pi-minus)	π^+ π^-	140 MeV	2.6×10^{-8} s	+1 −1	0	$u\bar{d}$ $d\bar{u}$	1947	C. Powell & group at Bristol	Cosmic radiation	Emulsion	Involved in nuclear binding	90–1
KAON (K-zero)	K^0	498 MeV	Short: 10^{-10}s*, Long: 5×10^{-8}s*	0	0	$d\bar{s}$	1947	G. Rochester & C. Butler	Cosmic radiation	Cloud chamber	Strange meson	92–3
KAON (K-plus) (K-minus)	K^+ K^-	494 MeV	1.2×10^{-8}s	+1 −1	0	$u\bar{s}$ $s\bar{u}$	1947	G. Rochester & C. Butler	Cosmic radiation	Cloud chamber	Strange mesons	92–3
J/PSI	J/ψ	3.1 GeV	10^{-20}s	0	1	$c\bar{c}$	1974	B. Richter & team at SLAC, S. Ting & team at BNL	Interactions of protons from accelerator (Ting), electron–positron annihilation (Richter)	Electronic (Ting & Ritcher)	First known member of charmonium family	178–9
D (D-zero) (D-plus)	D^0 D^+	1.87 GeV	10^{-12}s 4×10^{-13}s	0 +1	0	$c\bar{u}$ $c\bar{d}$	1976	G. Goldhaber & team at LBL & SLAC	Electron–positron annihilation	Electronic	Charmed mesons	180–3
UPSILON	Y	9.46 GeV	10^{-20}s	0	1	$b\bar{b}$	1977	L. Lederman & team at Fermilab	Interactions of protons from accelerator	Electronic	First known member of bottomonium family	186–8

(*The K^0 and the $\bar{K}^0$ form a quantum system whose superposition yields two physical particles, the short-lived K^0_S and the long-lived K^0_L, which reveal matter–antimatter asymmetry (CP violation).)

BARYONS

NAME	SYMBOL	PHYSICAL PROPERTIES					DISCOVERY				NATURE & ROLE	PAGES
		MASS	LIFETIME	CHARGE	SPIN	QUARK CONTENT	DATE	BY WHOM	SOURCE	DETECTOR		
PROTON	p	938.3 MeV	Stable (?). $>10^{32}$ years	+1	$\frac{1}{2}$	uud	1911–19	E. Rutherford	Alpha scattering from atomic nuclei	Scintillator	Charged constituent of atomic nuclei	60–3

Name	Symbol	Mass	Lifetime	Charge	Spin	Quark content	Year	Discoverer	Source	Detector	Description	Page
ANTIPROTON	$\bar{p}$	938.3 MeV	Same as proton	−1	$\frac{1}{2}$	$\bar{u}\bar{u}\bar{d}$	1955	E. Segrè & team at LBL	Interactions of protons from accelerator	Scintillation & Čerenkov counters	Antiparticle of proton	111–13 & 134–7
NEUTRON	n	939.6 MeV	In nuclei: stable; Free: 15 minutes	0	$\frac{1}{2}$	ddu	1932	J. Chadwick	Beryllium bombarded with alpha particles	Ionization chamber	Neutral constituent of atomic nuclei	60–3
ANTI-NEUTRON	$\bar{n}$	939.6 MeV	Same as neutron	0	$\frac{1}{2}$	$\bar{d}\bar{d}\bar{u}$	1956	B. Cork & team at LBL	Interactions of protons from accelerator	Liquid scintillator	Antiparticle of neutron	134–7
LAMBDA	Λ	1.115 GeV	2.6×10^{-10} s	0	$\frac{1}{2}$	uds	1951	C. Butler & group at Manchester	Cosmic radiation	Cloud chamber	Strange baryon; replaces neutron in nuclei to make hypernuclei	94–7
ANTI-LAMBDA	$\bar{\Lambda}$	1.115 GeV	Same as lambda	0	$\frac{1}{2}$	$\bar{u}\bar{d}\bar{s}$	1958	D. Prowse & M. Baldo-Ceolin at LBL	Interactions of pion beam produced from accelerator	Emulsion	Antiparticle of lambda	134–7
SIGMA (sigma-plus)	Σ^+	1.189 GeV	0.8×10^{-10} s	+1	$\frac{1}{2}$	uus	1953	G. Tomasini & Milan–Genoa team	Cosmic radiation	Emulsion	Strange baryon	98–9
SIGMA (sigma-minus)	Σ^-	1.197 GeV	1.5×10^{-10} s	−1	$\frac{1}{2}$	dds	1953	W. Fowler & team at BNL	Interactions of kaon beam produced from accelerator	Diffusion cloud chamber	Strange baryon	98–9
SIGMA (sigma-zero)	Σ^0	1.192 GeV	6×10^{-20} s	0	$\frac{1}{2}$	uds	1956	R. Plano & team at BNL	Interactions of kaon beam produced from accelerator	Bubble chamber	Strange baryon	98–9
XI (xi-minus)	Ξ^-	1.321 GeV	1.6×10^{-10} s	−1	$\frac{1}{2}$	dss	1952	R. Armenteros & team at Manchester	Cosmic radiation	Cloud chamber	Strange baryon	98–9
XI (xi-zero)	Ξ^0	1.315 GeV	3×10^{-10} s	0	$\frac{1}{2}$	uss	1959	L. Alvarez & team at LBL	Interactions of kaon beam produced from accelerator	Bubble chamber	Strange baryon	132–3
OMEGA MINUS	Ω^-	1.672 GeV	0.8×10^{-10} s	−1	$\frac{3}{2}$	sss	1964	V. Barnes & team at BNL	Interactions of kaon beam produced from accelerator	Bubble chamber	Strange baryon; confirmed theory of Eightfold Way	141–2
CHARMED LAMBDA	Λ_c	2.28 GeV	2×10^{-13} s	1	$\frac{1}{2}$	udc	1975	Team at BNL	Interactions of neutrino beam produced from accelerator	Bubble chamber	Charmed baryon	180–3

Further Reading

Acknowledgements

The following is a selection of generally non-technical books that cover the same subject area as *The particle explosion*. It is not intended to be a comprehensive guide to the literature on particle physics

General Interest Books

The cosmic onion: quarks and the nature of the Universe, Frank Close (Heinemann Educational, 1983). An account of particle physics in the 20th century for the general reader.

Quantum reality: beyond the new physics, Nick Herbert (Rider, 1985). An accessible explanation of quantum theory for the general reader.

The nature of matter, ed. by J. H. Mulvey (Oxford University Press, 1981). A series of lectures by well-known experts for a general audience.

The cosmic code: quantum physics as the language of nature, Heinz R. Pagels (Michael Joseph, 1982). An account of quantum theory and particle physics in the 20th century.

Inward bound, Abraham Pais (Oxford University Press, 1986).

From X-rays to quarks: modern physicists and their discoveries, Emilio Segré (W. H. Freeman, 1980). From radioactivity to charm, a detailed account by a leading experimenter.

The physicists: a generation that changed the world, C. P. Snow (Macmillan, 1981). Snow's vivid description of leading experimental and theoretical physicists from the end of the 19th century to 1939.

The particle connection: the discovery of the missing links of nuclear physics, Christine Sutton (Hutchinson, 1984). A detailed but non-technical account of the discovery of the W and Z particles and their significance.

Building the universe, ed. by Christine Sutton (Basil Blackwell/New Scientist, 1985). A compilation of articles about the new physics from *New Scientist* magazine.

Pioneers of science, Robert Weber (Institute of Physics, 1980). Brief biographies of physics Nobel prize winners from 1901 to 1979.

Rutherford: simple genius, David Wilson (Hodder & Stoughton, 1983). An authoritative biography.

The left hand of creation, John Barrow and Joseph Silk (Basic Books, 1983). A general account of the evolution of the Universe from its beginnings to the present day, by two astronomers.

The first three minutes, Steven Weinberg (Andre Deutsch, 1977). The first three minutes after the Big Bang, described in non-technical detail by a leading theorist.

More Specialist Books

The birth of particle physics, ed. by Laurie Brown and Lillian Hoddeson (Cambridge University Press, 1983). Proceedings of a symposium on the history of particle physics in 1930–50, with contributions from many individuals active at the time.

Fundamental particles, Brian Duff (Taylor & Francis, 1986). An introduction to particle physics for students and non-specialists.

QED: the strange theory of light and matter, Richard Feynman (Princeton University Press, 1985). The theory of quantum electrodynamics, explained by one of the theorists who developed it.

An atlas of typical expansion chamber photographs, W. Gentner, H. Maier-Leibnitz and W. Bothe (Pergamon Press, 1953). Out of print, but a comprehensive collection of cloud chamber photographs.

Cambridge physics in the thirties, ed. by John Hendry (Adam Hilger, 1984). Accounts of the Cavendish Laboratory by physicists who worked there.

The study of elementary particles by the photographic method, C. F. Powell, P. H. Fowler, and D. H. Perkins (Pergamon Press, 1959). Also out of print, but the authoritative compilation of emulsion photographs.

Early history of cosmic ray studies, ed. by Yataro Sekido and Harry Elliott (Reidel, 1985). Personal reminiscences of the early days of cosmic ray research, with old photographs.

The discovery of subatomic particles, Steven Weinberg (Scientific American Books, 1983). A detailed introduction to the discoveries of the electron, proton, and neutron.

LBL News Magazine, Vol. 6, No. 3, Fall 1981. This entire issue of the Lawrence Berkeley Laboratory's news magazine is devoted to 'A historian's view of the Lawrence years'. Copies are still available from LBL.

This book would not have been possible without the generous contributions of the many individuals who have supplied photographs, given advice, checked the manuscript, and spent time helping us during our visits to DESY, Brookhaven, Fermilab, SLAC, LBL, and CERN. Not to mention the personal and professional friends who have put up with our demands. We have tried to remember everyone in the list below, but we apologize to anyone who has inadvertently been omitted:

Mike Albrow, Luis Alvarez, Herbert Anderson, Richard Ansorge, Roger Anthoine, Larry Arbeiter, Bill Ash, Lawrence Bartell, Gerard Bertin, Roy Billinge, David Blockus, Gianfranco Bologna, Chris Bowdery, Laurie Brown, Blas Cabrera, Allen Caldwell, Harry Carter, Roger Cashmore, Owen Chamberlain, Georges Charpak, Jill Cheney, Dan Coffman, Stephen Compton, Hans Courant, Hank Crawford, Albert Crewe, Don Cundy, Orin Dahl, Per Dahl, Dick Dalitz, Chris Damerell, Paul Dauncey, Raymond Davis Jr, Richard Dease, Jonathan Dorfan, Jim Dunlea, Douglas Dupen, John Emsley, Debbie Errede, Steve Errede, William Fairbank, Joe Faust, Colin Fisher, Douglas Fong, Peter Fowler, Bode Franek, Gordon Fraser, Carlos Frenk, Henry Frisch, Günter Flügge, Matt Gaines, Bill Galbraith, Angela Galtieri, Steve Geer, Murray Gell-Mann, Paolo Giromini and the INFN team working on the CDF at Fermilab, Gerson Goldhaber, Judy Goldhaber, Maurice Goldhaber, Dennis Grant, Bruce Gunderson, Jon Guy, P. L. Hain, Gail Hansen, Rick Harnden, Malcolm Harvey, Satio Hayakawa, Richard Hemingway, Michael Houlden, Nick Jackson, Gron Jones, Renee Jones, Steven Kahn, George Kalmus, Peter Kalmus, John Kinson, Roger Klaffky, M. Koshiba, Leon Lederman, Steve Leffler, Sam Lindenbaum, Owen Lock, Fred Loebinger, Patrice Loïez, Bill Love, Gwendoline Lowe, Henry Lowood, Louis Lyons, Tom Madison, John Malos, Elizabeth Marsh and the Rutherford Appleton library staff, Robin Marshall, Silvia Miozzi, Nari Mistry, Roberto dal Molin, Alex Montwill, Diane Moss, John Mulvey, Yuval Ne'eman, Aseet Nukherjee, Pierre Odone, Mitsuo Ohtsuki, Margaret Pearson, Marty Perl, Charles Peyrou, Jim Phillips, Guido Piragino, Art Poskanzer, Helen Quinn, Fred Reines, Burt Richter, George Rochester, Phila Rogers, Adib Romaya, Mona Rowe, Carlo Rubbia, Oscar Saavedra, Nick Samios, Giorgio Sartori, David Saxon, Glenn Seaborg, Emilio Segrè, Steve Sewell, Eric Shumard, Ron Sidwell, Janet Sillas, Marilyn Smith, Peter Smith, Mike Sokoloff, Godfrey Stafford, Larry Sulak, Rosemary Taylor, Sam Ting, Timothy Toohig, Art Tressler, Maria Valdata-Nappi, Louis Voyvodic, Pedro Waloschek, Lorraine Ward, Alan Watson, Steve Watts, Geoffrey Webb, Hywell White, Rolf Wideröe, Günter Wolf, Arnold Wolfendale, N. Yandagni, Richard Zdarko, Mike Zguris. THANK YOU.

PICTURE CREDITS

A number of credits have been abbreviated after the first mention, in order to reduce the length of the credits list. In particular, Science Photo Library is abbreviated to SPL, and the book *The study of elementary particles by the photographic method*, by C. Powell, P. Fowler, & D. Perkins (Pergamon Press, 1959), has been abbreviated to *Elementary particles*, Powell, Fowler, & Perkins.

Most of the photographs and other illustrations in the book, including all the pictures from the major laboratories, are available from the Science Photo Library, 2 Blenheim Crescent, London W11 1NN (Tel: 01 727 4712). Other major sources are given at the end of the credits.

We have attempted to obtain permission to reproduce all the pictures in the book, but in the case of some of the older photographs the copyright owner is unknown or could not be located.

Front cover illustration:

A computer display of an event in the UA1 detector at the CERN particle physics laboratory, Geneva, takes on a spectacular appearance as a result of special effects photography. (David Parker/SPL)

1.1 David Parker/SPL
1.2 Photo CERN
1.3 HRS experiment, SLAC/David Parker/SPL
1.4 Courtesy of the archives, California Institute of Technology
1.5 Photo CERN
1.6 Fermi National Accelerator Laboratory
1.7 Stanford Linear Accelerator Center
1.8 Brookhaven National Laboratory
1.9 Photo CERN
1.10 Brookhaven National Laboratory
1.11 Photo CERN
1.12 Lawrence Berkeley Laboratory
1.13 TASSO experiment, DESY
1.14 Mary Evans Picture Library

2.1 Neil Hyslop
2.2 David Parker/SPL
2.3 Mary Evans Picture Library
2.4 Jean-Loup Charmet
2.5 Cavendish Laboratory, University of Cambridge
2.6 AIP Niels Bohr Library, Lande Collection
2.7 SPL
2.8 Jean-Loup Charmet
2.9 The Mansell Collection
2.10 AIP Niels Bohr Library, William G. Myers Collection
2.11 Jean-Loup Charmet
2.12 *The study of elementary particles by the photographic method*, C. Powell, P. Fowler, & D. Perkins (Pergamon Press, 1959)
2.13 National Radiological Protection Board
2.14 Cavendish Laboratory, University of Cambridge
2.15 Neil Hyslop
2.16, 2.17 C. T. R. Wilson/Science Museum
2.18 I. Joliot-Curie & F. Joliot/Science Museum
2.19 *Elementary particles*, Powell, Fowler, & Perkins
2.20 Cavendish Laboratory, University of Cambridge
2.21 McGill University Archives
2.22 Manchester University/Science Museum
2.23 By permission of the syndics of Cambridge University Library

2.24 Mary Evans Picture Library
2.25 N. Feather/Science Museum
2.26 Cavendish Laboratory, University of Cambridge
2.27 Science Museum
2.28 Copyright of the trustees of the Science Museum
2.29 C. T. R. Wilson/Science Museum
2.30 Mary Evans Picture Library
2.31 P. M. S. Blackett/Science Museum
2.32 Société Francaise de Physique, Paris
2.33 Cavendish Laboratory, University of Cambridge
2.34 AIP Niels Bohr Library, Margrethe Bohr Collection
2.35 Cavendish Laboratory, University of Cambridge
2.36 P. I. Dee/Science Museum
2.37 Cavendish Laboratory, University of Cambridge
2.38 Courtesy Curie Laboratory, Radium Institute

3.1 Lawrence Berkeley Laboratory
3.2 Dr Mitsuo Ohtsuki/SPL
3.3 Science Museum
3.4 David Parker
3.5 Courtesy of Dr L. S. Bartell
3.6 David Parker
3.7 P. Auger/Science Museum; colouring by David Parker
3.8 J. G. Wilson, *Proc. Roy. Soc.*, London (A), 166, 482 (1938); colouring by David Parker
3.9 Lawrence Berkeley Laboratory
3.10 Stanford Linear Accelerator Center
3.11, 3.12, 3.13, 3.14 *Elementary particles*, Powell, Fowler, & Perkins
3.15 P. M. S. Blackett & D. S. Lees/Science Museum; colouring by David Parker
3.16 P. M. S. Blackett/Science Museum; colouring by David Parker
3.17 P. M. S. Blackett & D. S. Lees; colouring by David Parker
3.18 Lawrence Berkeley Laboratory; colouring by David Parker
3.19 K. Brueckner & W. M. Powell; colouring by David Parker
3.20 I. Joliot-Curie & F. Joliot/Science Museum
3.21 Lawrence Berkeley Laboratory; colouring by David Parker
3.22 Courtesy of Dr G. Piragino, Turin University
3.23 I.K.Bøggild
3.24 Neil Hyslop
3.25 Dept. of Physics, Imperial College, London
3.26 SLAC Hybrid Facility Photon Collaboration

4.1 David Parker
4.2 Ullstein Bilderdienst
4.3 Courtesy of the archives, California Institute of Technology
4.4 Ullstein Bilderdienst
4.5 Neil Hyslop
4.6 Courtesy of Prof. Bruno B. Rossi
4.7 Courtesy of Prof. D. Skobeltzyn
4.8 D. Skobeltzyn/Science Museum
4.9 Courtesy of the archives, California Institute of Technology
4.10, 4.11 C. D. Anderson/Science Museum
4.12 Courtesy of George Rochester
4.13 Courtesy of the archives, California Institute of Technology
4.14 Courtesy of Prof. Satio Hayakawa
4.15, 4.16, 4.17, 4.18 Courtesy of George Rochester
4.19 Mary Evans Picture Library
4.20 *Nuclear physics in photographs*, C. Powell & G. Occhialini

4.21 Courtesy of Ilford Limited
4.22 *Elementary particles*, Powell, Fowler, & Perkins
4.23, 4.24 Dept. of Physics, University of Bristol
4.25 *Elementary particles*, Powell, Fowler, & Perkins

5.1 *Elementary particles*, Powell, Fowler, & Perkins; colouring by David Parker
5.2 Lawrence Berkeley Laboratory; colouring by David Parker
5.3 O. Ritter, C. Lieseberg, H. Maier-Leibnitz, A. Papkow, K. Schmeiser, W. Bothe, Z. *Naturforsch*. 6a, 243 (1951)
5.4 JADE experiment, DESY
5.5 Irvine-Michigan-Brookhaven (IMB) collaboration
5.6 S. H. Neddermeyer & C. D. Anderson/Science Museum
5.7 Mark-J experiment, DESY/David Parker/SPL
5.8 G. Piragino, experiment PS 179, CERN
5.9 *Elementary particles*, Powell, Fowler, & Perkins
5.10, 5.11 C. Butler & G. Rochester, Manchester University
5.12 *Elementary particles*, Powell, Fowler, & Perkins
5.13, 5.14, 5.15 Lawrence Berkeley Laboratory; colouring by David Parker
5.16 *Elementary particles*, Powell, Fowler, & Perkins
5.17 Neil Hyslop
5.18, 5.19, 5.20 Lawrence Berkeley Laboratory

6.1 David Parker/SPL
6.2, 6.3, 6.4 Fermi National Accelerator Laboratory
6.5 David Parker/SPL
6.6 Lawrence Berkeley Laboratory
6.7 Neil Hyslop
6.8, 6.9, 6.10, 6.11, 6.12 Lawrence Berkeley Laboratory
6.13 Brookhaven National Laboratory
6.14 Lawrence Berkeley Laboratory
6.15 Photo CERN
6.16 Lawrence Berkeley Laboratory
6.17 David Roberts/SPL
6.18 Courtesy of Prof. Donald Glaser
6.19, 6.20, 6.21 Lawrence Berkeley Laboratory
6.22, 6.23, 6.24, 6.25, 6.26 Brookhaven National Laboratory
6.27 Novosti Press Agency
6.28, 6.29 Photo CERN
6.30, 6.31 Brookhaven National Laboratory
6.32 Courtesy of British Oxygen Company Limited
6.33 Physics Dept., Durham University
6.34 Photo CERN
6.35 AIP Niels Bohr Library
6.36 Courtesy of Prof. Y. Ne'eman
6.37 Fermi National Accelerator Laboratory
6.38, 6.39 Stanford Linear Accelerator Center
6.40 David Parker/SPL

7.1 Patrice Loïez, CERN
7.2 Lawrence Berkeley Laboratory
7.3 R. D. Leighton
7.4, 7.5, 7.6 Lawrence Berkeley Laboratory
7.7 Lawrence Berkeley Laboratory; colouring by David Parker
7.8 Lawrence Berkeley Laboratory
7.9 CDF experiment, Fermi National Accelerator Laboratory
7.10 Brookhaven National Laboratory
7.11, 7.12 Dept. of Physics, Imperial College, London

7.13 Lawrence Berkeley Laboratory
7.14 Neil Hyslop
7.15 Brookhaven National Laboratory
7.16 Neil Hyslop
7.17 Irvine–Michigan–Brookhaven (IMB) collaboration
7.18 J. Csikay & A. Szalay, *Nuovo Cimento. Suppl.*, Padova conference (1957)
7.19, 7.20, 7.21 Courtesy of Dr F. Reines
7.22, 7.23 Experiment 594, Fermi National Accelerator Laboratory/David Parker/SPL
7.24 Fermi National Accelerator Laboratory
7.25 Mark-J experiment, DESY/David Parker/SPL
7.26 Neil Hyslop

8.1 David Parker/SPL
8.2, 8.3, 8.4 Courtesy of Prof. P. I. P. Kalmus
8.5, 8.6 David Parker/SPL
8.7 Stanford Linear Accelerator Center
8.8 David Parker/SPL
8.9 Neil Hyslop
8.10 Lawrence Berkeley Laboratory
8.11 Neil Hyslop
8.12 AIP Niels Bohr Library/Orren Jack Turner
8.13 Stanford Linear Accelerator Center
8.14 Courtesy of Laboratorio Nazionali di Frascati dell' INFN
8.15 Stanford Linear Accelerator Center
8.16 David Parker/SPL
8.17 Lawrence Berkeley Laboratory
8.18 Brookhaven National Laboratory
8.19 Stanford Linear Accelerator Center
8.20, 8.21 David Parker/SPL
8.22 Cornell University
8.23 Photo CERN
8.24, 8.25 David Parker/SPL
8.26 Deutsches Elektronen Synchrotron (DESY)
8.27, 8.28, 8.29 David Parker/SPL
8.30, 8.31, 8.32, 8.33 Photo CERN
8.34 Cavendish Laboratory UA5 experiment, CERN/David Parker/SPL
8.35 UA1 experiment, CERN/David Parker/SPL
8.36 Neil Hyslop
8.37 Photo CERN

9.1 Patrice Loïez, CERN
9.2 Mark I experiment, SLAC; redrawn by Neil Hyslop
9.3 S. Ting & team's experiment, Brookhaven National Laboratory; redrawn by Neil Hyslop
9.4 Mark I experiment, SLAC
9.5, 9.6 TASSO experiment, DESY
9.7 SLAC Hybrid Facility Photon Collaboration
9.8 Mark I experiment, SLAC; redrawn by Neil Hyslop
9.9 Brookhaven National Laboratory
9.10, 9.11 TASSO experiment, DESY
9.12 Fermi National Accelerator Laboratory
9.13 CLEO experiment, Cornell University; colouring by David Parker
9.14 TASSO experiment, DESY
9.15 HRS experiment, SLAC/David Parker/SPL
9.16 Mark-J experiment, DESY/David Parker/SPL
9.17, 9.18, 9.19, 9.20, 9.21 UA1 experiment, CERN/David Parker/SPL
9.22 Neil Hyslop
9.23 UA1 experiment, CERN/David Parker/SPL

10.1 Claude Nuridsany/SPL
10.2, 10.3 Irvine–Michigan–Brookhaven (IMB) collaboration
10.4 Neil Hyslop
10.5, 10.6 Irvine–Michigan–Brookhaven (IMB) collaboration/David Parker/SPL
10.7 David Parker/SPL
10.8 NUSEX experiment; redrawn by Neil Hyslop
10.9 Courtesy of Dr Blas Cabrera
10.10 David Parker/SPL
10.11 Courtesy of Dr Brian McCusker
10.12 David Parker/SPL
10.13, 10.14 Rutherford Appleton Laboratory
10.15 Courtesy of Prof. P. Fowler
10.16 A. L. Hodson, Physics Dept., Leeds University
10.17 Courtesy of Meridian Airmaps Limited & West Yorkshire County Council
10.18 Harvard-Smithsonian Center for Astrophysics, courtesy of F. R. Harnden, Jr.
10.19 Tom Foley, University of Minnesota
10.20 G. Piragino, experiment PS 179, CERN
10.21 US Naval Observatory/SPL
10.22 Dr R. J. Allen et al/SPL
10.23 Brookhaven National Laboratory
10.24 Courtesy of Dr. C. S. Frenk, Prof. S. White, Dr G. Efstathiou & Dr M. Davis

11.1 David Parker/SPL
11.2 'Examination of the energetics of aging skeletal muscle using nuclear magnetic resonance', D. J. Taylor, M. Crowe, P. J. Bore, P. Styles, D. L. Arnold, G. K. Radda, *Gerontology* 30: 2–7; redrawn by Neil Hyslop
11.3, 11.4 Petit Format/Nestlé/Steiner/SPL
11.5 Hank Morgan/SPL
11.6 David Parker/SPL
11.7 Courtesy of Prof. P. Fowler & Prof. D. H. Perkins, *Nature*, Vol. 189, Feb. 18, 1961
11.8, 11.9 Dr John Mazziotta et al/SPL
11.10 Brookhaven National Laboratory
11.11 Lawrence Berkeley Laboratory

Major sources of pictures are:-

Science Photo Library, 2 Blenheim Crescent, London W11 1NN (Tel: 01 727 4712)

Cavendish Laboratory, Madingley Road, Cambridge CB3 0HE (Tel: 0223 66477)

Niels Bohr Library, American Institute of Physics, 335 East 45th Street, New York, NY 10017 (Tel: (212) 661 7680)

Science Museum, Photographic Section, Exhibition Road, London SW7 2DD (Tel: 01 589 3456)

CERN, CH-1211 Geneva 23, Switzerland (Tel: (022) 83 61 11)

DESY, Notkestrasse 85, 2000 Hamburg 52, Federal Republic of Germany (Tel: (040) 89 980)

Brookhaven National Laboratory, Upton, Long Island, NY 11973 (Tel: (516) 282 2345)

Fermi National Accelerator Laboratory, PO Box 500, Batavia, Illinois 60510 (Tel: (312) 840 3000)

Lawrence Berkeley Laboratory, 1 Cyclotron Road, Berkeley, California 94720 (Tel: (415) 486 4000)

Stanford Linear Accelerator Center, PO Box 4349, Stanford, California 94305 (Tel: (415) 854 3300)

INDEX

Compiled by Paul Nash

Note: references to Figures (and their captions) are indicated by *italic page numbers*